Solar Energy Capture Materials

Inorganic Materials Series

Series editors:
Duncan W. Bruce, *University of York, UK*
Dermot O'Hare, *University of Oxford, UK*
Richard I. Walton, *University of Warwick, UK*

Titles in the Series:

1: Pre-combustion Carbon Dioxide Capture Materials
2: Post-combustion Carbon Dioxide Capture Materials
3: Solar Energy Capture Materials

How to obtain future titles on publication:
A standing order plan is available for this series. A standing order will bring delivery of each new volume immediately on publication.

For further information please contact:
Book Sales Department, Royal Society of Chemistry, Thomas Graham House, Science Park, Milton Road, Cambridge, CB4 0WF, UK
Telephone: +44 (0)1223 420066, Fax: +44 (0)1223 420247,
Email: booksales@rsc.org
Visit our website at www.rsc.org/books

Solar Energy Capture Materials

Elizabeth A. Gibson
Newcastle University, UK
Email: elizabeth.gibson@ncl.ac.uk

ROYAL SOCIETY OF CHEMISTRY

Inorganic Materials Series No. 3

Print ISBN: 978-1-78801-107-5
PDF ISBN: 978-1-78801-351-2
EPUB ISBN: 978-1-78801-850-0
Print ISSN: 2472-3819
Electronic ISSN: 2472-3827

A catalogue record for this book is available from the British Library

The Royal Society of Chemistry is a charity, registered in England and Wales, Number 207890, and a company incorporated in England by Royal Charter (Registered No. RC000524), registered office: Burlington House, Piccadilly, London W1J 0BA, UK, Telephone: +44 (0) 20 7437 8656.

For further information see our web site at www.rsc.org

Printed in the United Kingdom by CPI Group (UK) Ltd, Croydon, CR0 4YY, UK

Series Preface

Materials chemistry has grown hugely in the past few decades and continues to evolve rapidly. This is especially the case since the subject now finds itself central to many of the major challenges that face global society, such as those concerning energy and the environment; but applications of materials are also found in many everyday situations, with ongoing developments in technology around electronics and the miniaturisation of devices. Although the study of materials is a highly interdisciplinary topic, the role of innovative, fundamental chemistry continues to lie at the heart of the discovery of new materials, the understanding of their atomic-scale structure and the relationship of structure to a useful property.

With the *Inorganic Materials Series*, our aim is to provide a series of themed volumes, each reflecting the diversity of the topic as it continues to develop over the coming years, with many chosen to relate closely to emerging practical applications. The use of the title *Inorganic Materials* is by no means limiting, but instead emphasises the study of the chemistry of all elements in the Periodic Table and how they impact the development of new materials. The chapters in each book will have a pedagogical flavour with a target audience from first-year postgraduate student upwards, but with wide subject coverage ranging from continuous inorganic solids, to molecular and soft matter systems. We will use expert, active researchers as editors and contributors to provide an up-to-date perspective of their field. At the same time our aim is to provide an international perspective, so to reflect the diversity and interdisciplinarity of the now very broad area of the subject.

Duncan W. Bruce, York, UK
Dermot O'Hare, Oxford, UK
Richard I. Walton, Warwick, UK

Inorganic Materials Series No. 3
Solar Energy Capture Materials
Edited by Elizabeth A. Gibson

Published by the Royal Society of Chemistry, www.rsc.org

Preface

The importance of harnessing the power from the sun to satisfy our growing energy demands without harming the planet is now widely accepted. 'Solar energy conversion' includes many categories of technology, including solar thermal, solar fuels and solar-to-electricity (photovoltaic). This book focuses on photovoltaic technology, which now contributes a sizeable percentage of power generation globally, with new markets appearing each year and with over 100 GW of solar-to-electricity capacity installed annually. The large majority of this commercial technology is based on crystalline silicon, which captures the energy from sunlight and converts it to electricity using the pholtovoltaic effect, a phenomenon first reported by Edmond Becquerel in 1839. Evolution of the technology throughout the 1950s led to applications of solar cells for space applications, and strong developments in the 1980s lead to establishment of the commercial technology that is familiar today. Research since then has focused intensively on bringing down the cost and energy required to manufacture the modules. While the efficiency of solar to electrical power conversion is the established figure of merit for solar cells, processability is becoming increasingly important as industry looks towards integrating solar power into new applications, such as sensors and portable electronic devices. New categories of photovoltaic devices have also emerged, such as thin-film solar cells and devices based on nanotechnology, which provide routes to high-throughput manufacturing methods and new applications, such as built-in photovoltaics and wearable technology.

At the heart of these devices are semiconductors, which can be made from extended structures of both inorganic and organic materials. Rapid development in recent years has focused on the discovery and engineering of

Inorganic Materials Series No. 3
Solar Energy Capture Materials
Edited by Elizabeth A. Gibson

Published by the Royal Society of Chemistry, www.rsc.org

these materials towards capturing light and converting it to electricity more efficiently. This volume describes the synthesis and properties of the inorganic materials which underpin the current state-of-the-art and possible opportunities for future research. Synthetic chemistry remains fundamental to progress in energy capture materials and targets include increasing the absorptance of materials, replacing rare elements with earth-abundant materials, avoiding the toxic components and manipulating grain boundaries or interfaces to promote charge-transport and minimise charge-recombination processes. The evolution of photovoltaic technology is discussed, including solution-processed materials for continuous manufacturing, reaching high conversion efficiency through stacked or tandem architectures, and device assembly on flexible substrates for easier systems integration. In addition to the materials, advances in characterisation and modelling of charge transfer and transport in the devices continues to enable scientists to understand the fundamental processes occurring in the devices. The knowledge generated is increasingly important as the materials and device configurations become more complex.

The five chapters of this volume cover different approaches to capturing light with inorganic materials. The first chapter introduces silicon devices, which make up most of the PV industry. These are still researched intensively to try and bring down the energy used to make the silicon wafers and to capture as much light as possible. The chapter on compound semiconductor solar cells introduces the concept of heterojunction-based devices and describes synthesis routes to the chalcogenide materials, which dominate thin-film technology. This chapter highlights how both the composition and crystalline properties of the material determine light absorption. The chapter on dye-sensitized solar cells introduces the concept of hybrid devices, which combine organic and inorganic materials. These devices differ from established technology in two ways, first by containing a semiconductor–liquid junction and second because light absorption and charge transport are performed by separate components (a dye, a transparent metal oxide and a redox electrolyte). These devices perform particularly well under diffuse light conditions and are potentially useful for indoor applications. Research in this area focuses on tandem configurations to increase the efficiency and replacing the liquid with a solid-state charge-transport layer. The chapter on solution-processable materials builds on this concept and explores the cutting-edge research into perovskite solar cells. Research in this area has exploded over the last few years as conversion efficiencies have risen steeply to match those of established PV technology. The unique properties of these materials, in terms of charge-separation and transport, are described. Finally, the application of computational methods to aid the discovery of new materials is described, using the example of copper-based absorber materials. Density functional theory is applied to investigate the structural, electronic, and optical properties of emerging copper-based chalcogenides, to identify compositions that could be applied in ultrathin light-absorbing films in very high-efficiency devices in the future.

I am delighted that very active researchers who are working at the forefront of the solar energy field agreed to prepare the chapters in this volume. I thank them for their excellent contributions across the breadth of this important research area. I hope that this book will provide a useful and up-to-date introduction to the area and the role that materials chemistry can play in improving the way we use the abundant and freely available source of energy for the welfare and security of everyone on the planet. There is still much chemical space to explore: the semiconductor industry is still dominated by only a few chemical elements!

Elizabeth A. Gibson

Contents

Inorganic Materials Series No. 3
Solar Energy Capture Materials
Edited by Elizabeth A. Gibson

Published by the Royal Society of Chemistry, www.rsc.org

CHAPTER 1

Silicon Solar Cells

P. M. USHASREE*[a] AND B. BORA[b]

[a]Energy Materials Laboratory, School of Natural and Environmental Science, Newcastle University, Newcastle upon Tyne NE1 7RU, UK; [b]National Institute of Solar Energy (NISE), Gurgaon, Haryana 122003, India
*E-mail: Palat.Meethale-Ushasree@newcastle.ac.uk

1.1 Introduction

The Sun is the major source of energy on Earth and has been producing energy for billions of years. Although the Sun is about 90 million miles away from us, it takes only about 8.5 minutes for light to travel from the Sun to Earth. Heat and light from the Sun, solar energy, is one of the cleanest and most abundant sources of renewable energy.

During most of the Sun's life, energy comes from the fusion of hydrogen nuclei. Four hydrogen nuclei are fused to form a helium nucleus and a neutron is released. From the law of conservation of energy, energy is released because the helium nucleus has a slightly lower mass than the sum of the original hydrogen nuclei. This energy eventually makes its way from the centre of the Sun to the outer regions and is emitted in the form of electromagnetic radiation. This radiation is spread in the form of electromagnetic waves and a particle of the electromagnetic radiation is known as a photon. The Sun emits photons across a broad range of wavelengths, getting smaller (higher in energy) from radio waves to gamma rays, as shown in Figure 1.1.[1] Wavelengths below 200 nm (X-rays, γ-rays and UV radiation) are absorbed by nitrogen and oxygen in the atmosphere, the region between 200 and 300

Inorganic Materials Series No. 3
Solar Energy Capture Materials
Edited by Elizabeth A. Gibson

Published by the Royal Society of Chemistry, www.rsc.org

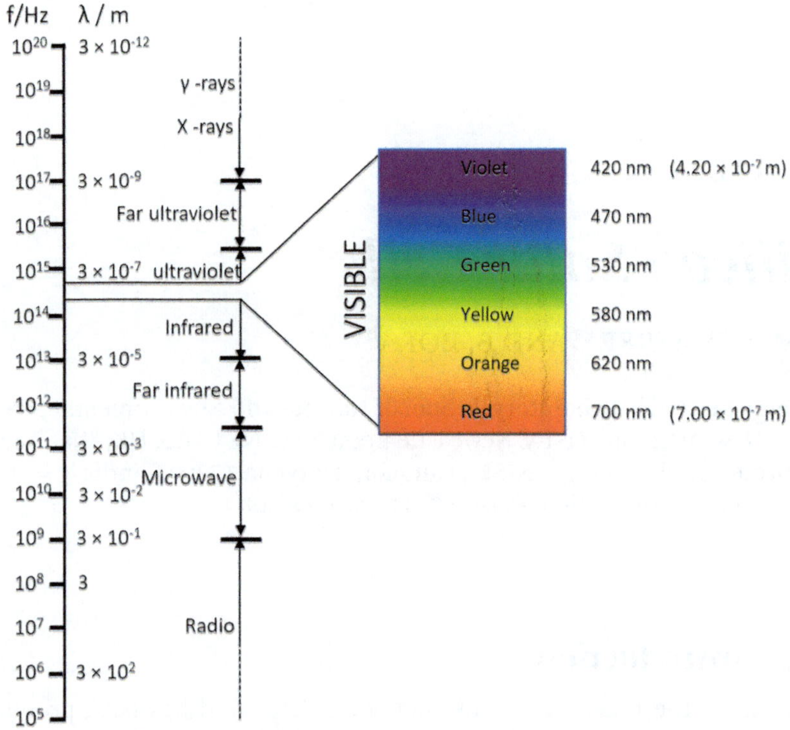

Figure 1.1 The electromagnetic spectrum adapted from the Royal Society of Chemistry Databook.[1] Reproduced from ref. 1 with permission from the Royal Society of Chemistry.

nm (UV radiation) is absorbed by O_3 in the stratosphere, wavelengths above 700 nm (IR radiation) are partially absorbed by CO_2, O_3 and water. Visible light, the region of the spectrum our eyes can detect, is composed of relatively short wavelengths in the range 400 to 700 nm. Approximately, 46% of solar radiation falls within visible light wavelengths, as shown in Figure 1.2.[2] Of this, 30% is reflected back by the atmosphere or the Earth's surface. The temperature of the Sun (5800 K at the surface) means the solar spectrum peaks at around 550 nm, in the green portion of the visible region. This is equivalent to about 2.3 eV, which can be calculated using eqn (1.1).

$$E = \frac{hc}{\lambda} \text{ or } E = \frac{1242}{\lambda} \tag{1.1}$$

where, h is Planck's constant in eV, c is the speed of light in m s^{-1} and λ is the wavelength of the light in nm.

Single-junction silicon solar cells convert light from about 300 nm to 1100 nm. A broader spectrum for harvesting the light can be achieved by stacking a number of solar cells with different operational spectra in a multi-junction configuration. For this reason, multi-junction solar cells can reach higher conversion efficiencies than traditional single-junction silicon solar cells.

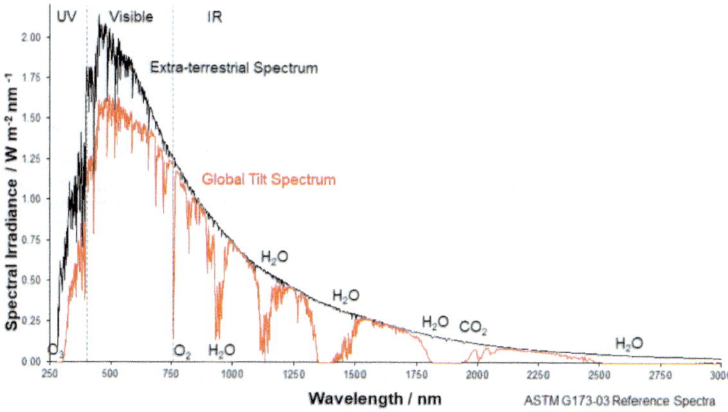

Figure 1.2 Solar irradiance spectrum above the Earth's atmosphere and at the surface. Data provided by www.nrel.gov.

The amount of solar energy incident on a given location depends on various factors, such as the time of the day, the weather and the latitude. Air mass 1.5 (AM1.5) is the most widely used solar spectrum (radiation reaching the Earth's surface) to study and compare the performances of different solar cells under standardised conditions. AM1.5 or "1 sun" is defined as equal to 1000 W m^{-2}. The number "1.5" represents that the distance the sunlight travels through the atmosphere to the Earth's surface is 1.5 times the shortest distance (when the Sun is overhead). The standard spectrum outside the Earth's atmosphere is called AM0 and is used for non-terrestrial applications.

On average, about 340 W m^{-2} of solar energy is incident on the Earth's atmosphere, of which approximately 240 W m^{-2} is absorbed by the Earth system, which includes oceans, land surfaces and the atmosphere itself.[3,4] The amount of energy the Earth receives from the Sun in a day is sufficient to cover our needs for a whole year. For example, the UK receives, on average, about 3 kWh m^{-2} of solar energy per day, or 1095 kWh m^{-2} per year. If a typical UK family uses about 30 kWh m^{-2} of electricity per year, then we get 37 times more energy per square metre from the Sun than we consume annually. In fact, enough energy arrives on a relatively small area to meet our demands for a year, if only we could harness it properly. Hence, the challenge is how to harness all of this energy efficiently and to convert it into usable energy, efficiently. This clearly implies that we do not have an *energy* problem on our planet, we have a *conversion* problem. The key to efficient conversion of sunlight into electricity is to develop technologies to collect and absorb solar energy efficiently and to convert it into electricity, efficiently. In general, the generation of electricity from sunlight by a solar cell takes place in steps as shown in Figure 1.3, wherein each step contributes to the total conversion efficiency of the cell.

Solar cells are made of semiconductor materials. In a semiconductor all valence electrons reside in the valence band and the lowest lying unoccupied

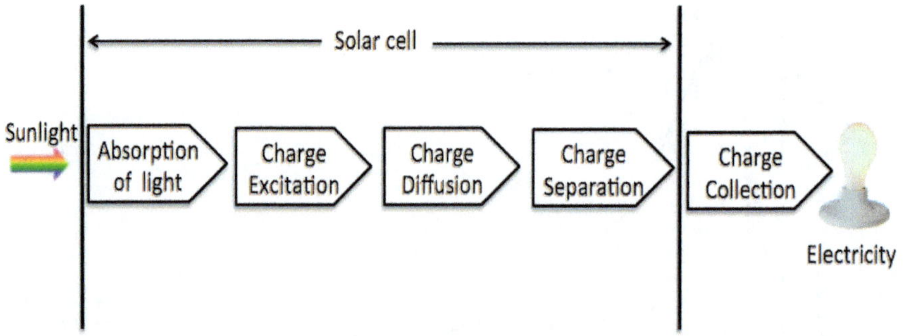

Figure 1.3 Steps involved in conversion of sunlight into electricity in a typical solar cell.

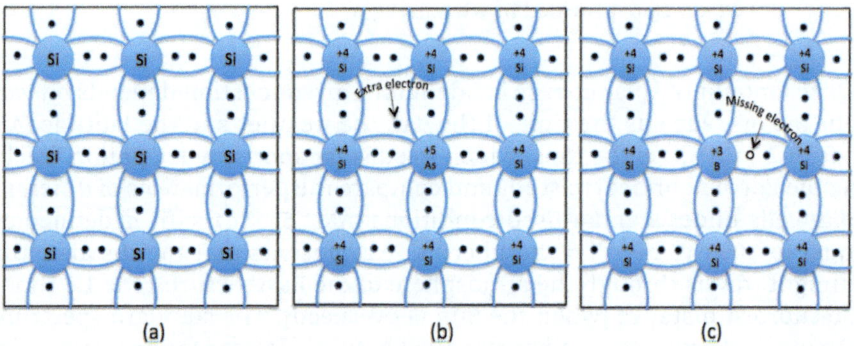

Figure 1.4 Representation of electronic structure of (a) undoped silicon (b) n-type silicon and (c) p-type silicon.

band is called the conduction band. The region in between, where there are no energy states, is called the band gap. Most of the solar cells on the market today are based on silicon crystals.[5–7] Silicon has 14 electrons (electronic configuration: $1s^2\ 2s^2\ 2p^6\ 3s^2\ 3p^2$) with 4 electrons in the outermost shell, as shown in Figure 1.4(a). If silicon is doped with pentavalent arsenic atoms (5 electrons in the valence shell) from group 15, the 1 extra electron from each arsenic atom does not chemically bond with the silicon atom and becomes a free electron. These free electrons that cannot chemically bond to the neighbouring atoms, move around freely in the conduction band, as shown in Figure 1.4(b). This type of semiconductor is called an n-type semiconductor, or in other words, a semiconductor rich in electrons. Likewise, if silicon is doped with trivalent boron atoms (3 electrons in the valence shell) from group 13, each dopant atom shares only 3 electrons with the neighbouring silicon atom as they have 1 less electron in the valence shell when compared to silicon. This will result in a vacancy (or "hole") in the valence band, as shown in Figure 1.4(c). This type of semiconductor is called p-type semiconductor, or in other words a semiconductor rich in holes.

To form a solar cell, these two types of semiconductors are placed in contact with each other. At the region of contact, the extra electrons in the n-type semiconductor combine with the extra holes in the p-type semiconductor. When this combination of electrons and holes happens, the loss of negatively charged electrons leaves positive ions in the n-type material, and the depletion of holes leaves negative ions in the p-type material. This build-up of charges near the interface (the "depletion zone") creates an electric field between the two n-type and p-type materials that stops any more combination of electrons and holes. If an electron from the n-type material tried to penetrate through this barrier, it would be repelled by the negative charges of the p-type material and the holes would be repelled by the positive charges of the n-type material. This barrier acts as a wall that keeps the electrons to the n-side and the holes to the p-side. This barrier is called a p–n junction. When sunlight strikes the p–n junction with energy greater than the band gap energy of silicon, electrons are promoted from the valence band to the conduction band, generating an electron–hole pair with different potential energy. Since an electric field has been created at the junction, the electron will move towards the n-side and the hole towards the p-side. If the solar cell is connected through wires to an external circuit, the electrons travel to the contact with the n-type material and pass through the external circuit to reach the p-side, where they recombine with the holes. This flow of electrons is the current and, by placing metal contacts on the top and bottom of the solar cell, current can be drawn off for external uses such as powering an electrical device. Each similar cell generates only a few volts of electricity, or about 1.5 W of power. A solar panel is used to combine the energy produced by many such cells to make a useful amount of electrical current and voltage. Several solar panels are combined to create a solar array, for residential and commercial use. As the current produced by a solar cell or a solar panel is a direct current (DC), solar arrays are connected to an inverter to convert the DC into alternating current. These panels are used to supply power to a wide variety of systems, from calculators, streetlights, parking meters and household appliances to space stations. They are also used in conjunction with other sources of energy, such as wind, natural gas and nuclear energies.

1.2 Silicon Substrates

Silicon has evolved as a dominant technology for the production of electricity from sunlight and, currently, accounts for more than 80% of the PV market.[6] In its pure form, silicon is a silvery grey lustrous solid. As shown in Figure 1.5, it is the second most abundant element in the Earth's crust (about 27.5% by mass) after oxygen (about 50.5% by mass), and is available in sufficient quantity to match global energy demand on the order of terawatts.[8]

In nature, silicon is not found in its elemental form and, instead, it occurs in the form of silica or silicon dioxide. Quartz, which is found abundantly in sand, is the most common form of silica. Silica is used to produce metallurgical grade silicon, which then undergoes several stages of purification

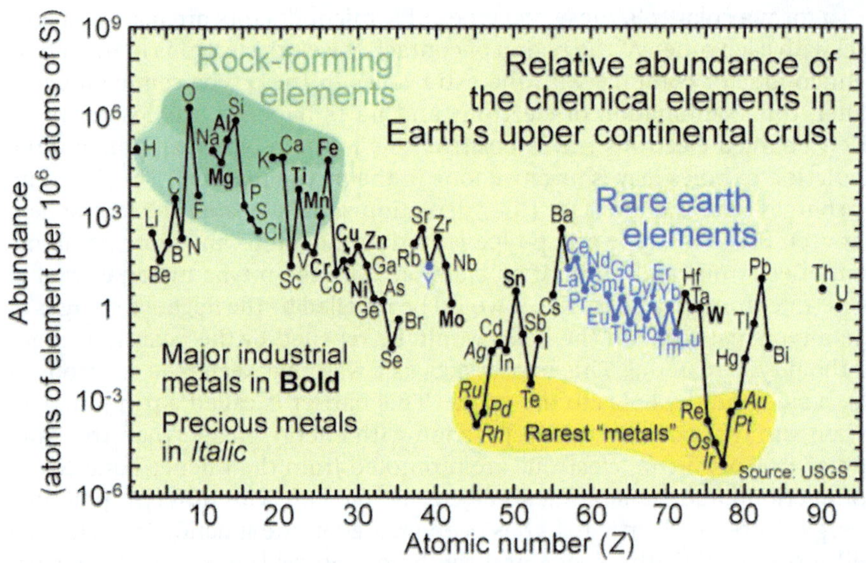

Figure 1.5 Abundance (atom fraction) of the chemical elements in Earth's upper continental crust as a function of atomic number. Reproduced from ref. 8 with permission from the Royal Society of Chemistry.

and refining steps to produce silicon of high purity for applications in the photovoltaic (PV) industry. Apart from its abundance, there are other reasons why silicon remains the material of choice for PV applications. Silicon is non-toxic in all its natural forms. Silicon is also non-corrosive and forms a very thin, chemically stable, strongly adhering protective layer of silicon oxide at the interface between silicon and the atmosphere. This naturally formed, electrically insulating oxide acts as a passivating layer and reduces the number of recombination centres at the surface, thereby improving the performance of the silicon solar cells.[9] The oxide layer also lowers the risk of degradation of a silicon wafer inside a solar module and, consequently, the long-term stability of the technology has contributed to its establishment in the PV market.

Solar cells have non-linear current–voltage (I–V) characteristics, which are very sensitive to factors like the temperature and intensity of the light incident on the device. The I–V relationship for a typical solar cell is shown in Figure 1.6. Ideally, the current–voltage behaviour follows the single exponential relationship known as the diode equation:

$$I_{\mathrm{dark}}(V) = I_0(\mathrm{e}^{qV/kT} - 1)$$

where I_0 is a constant, q is electronic charge, k is the Boltzmann constant, and T is temperature (K).

The net current (I) of the solar cell device is the difference between the dark current that flows as a result of the applied potential difference (V), and the photocurrent in the opposite (positive) direction.

$$I(V) = I_{\mathrm{SC}} - I_{\mathrm{dark}}(V) = I_{\mathrm{SC}} - I_0(\mathrm{e}^{qV/kT} - 1)$$

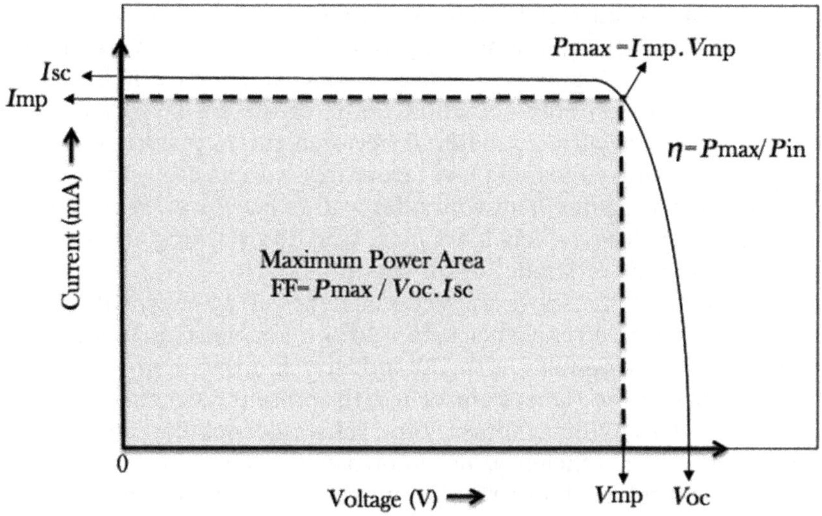

Figure 1.6 Current–voltage characteristics of a typical solar cell.

A non-ideality factor, β, is introduced to account for resistance in the device.

$$I(V) = I_{SC} - I_{dark}(V) = I_{SC} - I_0(e^{qV/\beta kT} - 1)$$

The solar parameters: the short circuit current (I_{SC}), the open circuit voltage (V_{OC}), the fill factor (FF) and the efficiency (η) of the solar cell are all determined from this characteristic. The efficiency, η, is the most commonly used parameter to compare the performance of different solar cells.

$$\eta = \frac{P_{out}}{P_{in}} = \frac{I_{SC} \times V_{OC} \times FF}{P_{in} \times A}$$

$P_{in} = 100$ mW cm^{-2} for AM 1.5, A = cell area (cm^2).

The maximum power point of the cell is the values of I and V (I_{max} and V_{max}) at which the maximum rectangle in Figure 1.6 meets the I–V curve. I_0 and γ in the diode equation limit the maximum obtainable power to approximately 75–80% of the maximum theoretical power. To account for this, a term called the "fill factor" (FF) is introduced.

$$FF = \frac{I_{max} V_{max}}{I_{SC} V_{OC}}$$

$$FF = 1 \text{ when } I_{max} = I_{SC} \text{ and } V_{max} = V_{OC}.$$

Much optimisation goes into fabricating a solar cell to achieve a high device performance or energy conversion efficiency.[10,11] This is discussed in detail in Section 1.2. Efficient energy conversion efficiency is an important issue because the efficiency influences the entire value-chain cost of the PV system, from material production to system installation.[12] The more efficient a solar cell is, the more cost effective it is to meet the current growing energy demands. Some of the earliest PV modules used circular wafers with no material between them, resulting in a low packing density and, so losing power from those areas brought down the efficiency of the modules. These circular wafers have been replaced with pseudo-square wafers, which have increased the packing density and, hence, increased the efficiency of the modules.

The chart in Figure 1.7 shows the National Renewable Energy Laboratory's (NREL) best research-cell efficiencies, as of 2018.[13] The chart tracks the growth of research and development in the PV industry and the progress in device performance across the globe, from 1976 to the present, for a range of photovoltaic technologies. William Shockley and Hans Queisser first calculated the maximum theoretical efficiency, also known as the Shockley–Queisser (SQ) limit, of an ideal solar cell in 1961. The maximum theoretical solar cell efficiency is determined by the band gap of the semiconductor materials used for the p–n junction. Semiconductor materials with lower band gaps absorb more photons, resulting in higher I_{SC} and lower V_{OC}, while the materials with higher band gap have lower I_{SC} and higher V_{OC}. For maximum output power and efficiency, a compromise between the material with low band gap and high band gap is necessary. The trade-off between higher V_{OC} with increasing band gap and decrease in I_{SC} results in an optimum band gap energy for a single p–n junction solar cell, which falls close to 1.1 eV. For silicon solar cells with a band gap of 1.1 eV, the SQ limit is calculated to be about 30%.[14] In the laboratory, the record solar cell efficiency for mono-crystalline silicon solar cells is as high as 25%, and about 20% for multi-crystalline Si solar cells.[15,16] The best commercial silicon cell efficiency is about 23% at the cell level and about 18–24% at the module level.[15,16] The existing challenge in current PV technology is to bring this gap between the research record efficiency and the efficiencies achieved in commercial production, as close as possible.

1.2.1 Refining

Silicon is obtained by thermal reduction of silica or silicon dioxide in the presence of carbon. An essential prerequisite for the growth of crystalline silicon from the raw materials is the availability of silicon of the highest purity attainable.[17] Impurities or defects in the single crystals can lower the performance of the solar cell device due to recombination of charge carriers. Depending on the impurity content, there are three grades of silicon: metallurgical grade silicon (MG-Si), solar grade silicon (lesser level of impurity than MG-Si) and semiconductor grade silicon.

MG-Si is extracted from a quartzite reduction reaction with carbon in an electrode arc, furnace heated to about 1800 °C, as shown in eqn (1.2).

$$SiO_2 + C \xrightarrow{\Delta} Si + CO_2 \quad (1.2)$$

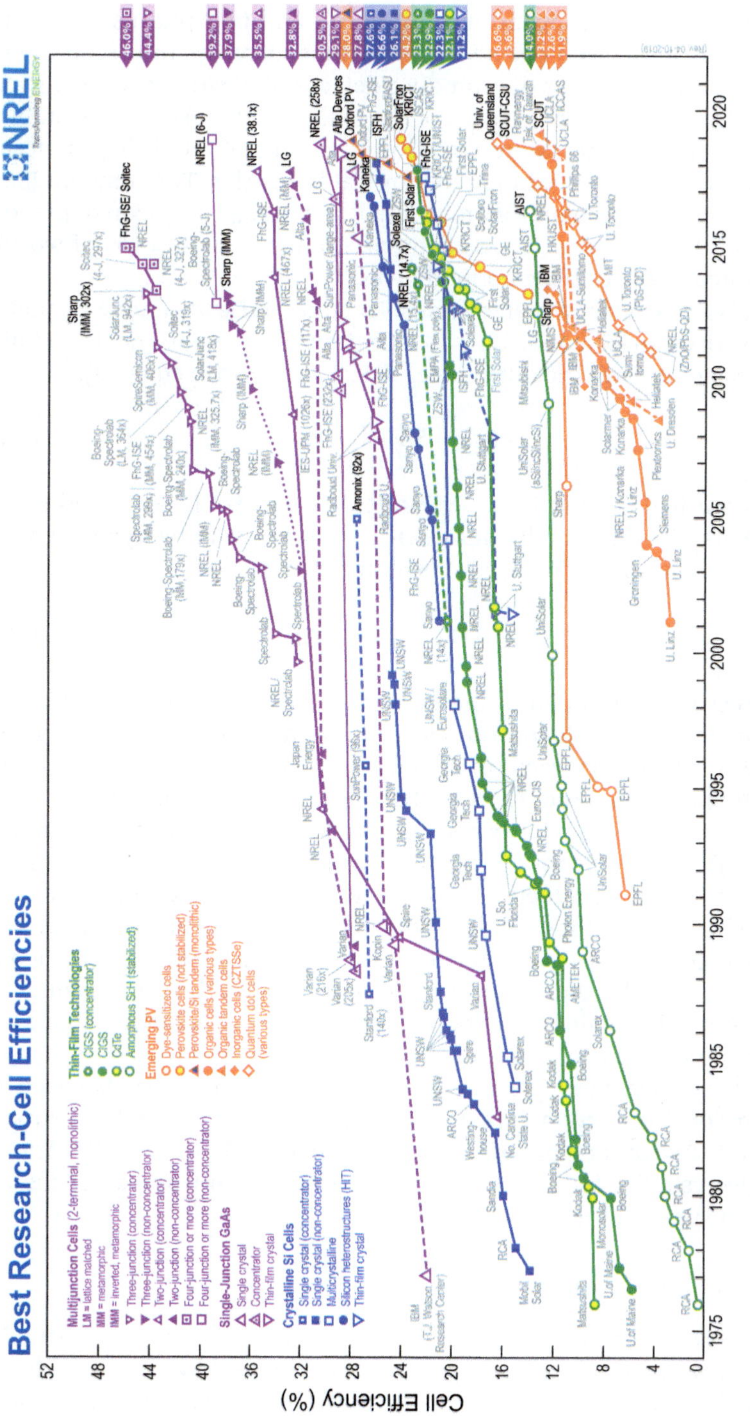

Figure 1.7 NREL's efficiency chart, 2017. This plot is courtesy of the National Renewable Energy Laboratory, Golden, CO.[13]

The purity of MG-Si is only 98%, with carbon, alkali earth metals, transition metals and a high concentration of boron and phosphorus (>50–100 ppmw) as impurities. The presence of impurities increases the recombination of the charge carriers, as they produce defect states within the band gap. This affects the electronic properties of the silicon and makes it unsuitable for electronic applications. More than 2 million tons of MG-Si is produced annually and it is largely used in metallurgical industries. The semiconductor and PV industries account for only a small percentage of silicon demand. Excluding the United States, ferrosilicon accounts for about 65% of world silicon production on a silicon-content basis.[18]

MG-Si is also used as a raw material for solar grade and semiconductor grade silicon. Pulverised MG-Si is reacted with anhydrous hydrochloric acid in a fluidised bed reactor to form trichlorosilane, $SiHCl_3$ as shown in eqn (1.3).

$$Si + 3HCl \rightarrow SiHCl_3 + H_2 \tag{1.3}$$

$SiHCl_3$ is purified from impurity chlorides, such as $FeCl_3$, $AlCl_3$ and BCl_3, by multiple distillations. Finally, pure $SiHCl_3$ is reacted with hydrogen to produce a very pure form of silicon, as shown in eqn (1.4).

$$SiHCl_3 + H_2 \rightarrow Si + 3HCl \tag{1.4}$$

The process was developed by Siemens in the 1950s and is currently the standard process for production of electronics grade silicon. The reaction takes place inside a large vacuum chamber, a Siemens deposition reactor, with two thin polysilicon rods (~5 mm diameter), and silicon is deposited on these rods to produce high purity polysilicon rods (diameters up to 300 mm) of columnar grains of silicon. The Siemens process is a very energy intensive method of polysilicon production and hence it is the most expensive when compared with other methods such as the fluidised bed reactor and upgraded metallurgical grade (UMG method). However, it produces the purest silicon of all techniques.[19] The UMG method cannot be used in the semiconductor industry, but it is becoming more popular for solar cell applications, due to the reduced production cost and time. However, to make silicon cells of reasonable performance, large-grained, multi-crystalline (grain size between 1 and 100 mm) or single crystal (grain size > 100 mm) substrates of high purity are required.[19] Two other types of crystalline silicon used in solar cell fabrication are polycrystalline (grain size between 1 μm and 1 mm) and amorphous silicon (grain size < 1 μm).[20,21]

1.2.2 Crystal Growth

Single crystals of silicon (c-Si) for the PV industry are grown by the Czochralski and float zone methods, which account for 35% of worldwide photovoltaic production.[12] Czochralski silicon (Cz-Si) is grown by gradually pulling an oriented seed crystal out of the molten silicon contained in a quartz crucible with a graphite susceptor, while simultaneously rotating it so that the inhomogeneities in the melt are not replicated in the growing crystal. By controlling the speed of rotation, pulling rate and the growth temperature, defect-free single crystals are grown. Boron, phosphorus, arsenic or antimony are added to grow p-type or n-type single crystals of silicon. Traditional silicon-based solar cells are usually p-doped with boron. This method is extensively used to obtain silicon ingots in semiconductor industries for the production of silicon wafers. The only drawback is that during the growth process a large quantity of oxygen is released from the quartz crucible into the melt. A small percentage of oxygen from the melt is incorporated into the crystal, while most of it is lost as silicon monoxide from the molten surface. This gas can interact with the hot graphite susceptor and form carbon monoxide that re-enters into the melt, introducing a tiny amount of carbon into the single crystal as an impurity. The presence of carbon and oxygen as impurities can affect the performance of the solar cell adversely by lowering the cell efficiencies.[22] Oxygen can form precipitates and act as recombination centres to reduce the lifetime of charge carriers. Crucible contamination can be avoided by using magnetic fields (magnetic field applied Czochralski, MCz) to control the convection fluid flow and create oxygen traps to slow its migration into the crystal. This technology continues to provide control of oxygen content with good radial uniformity and, hence, high quality single crystals.

Float zone silicon (FZ-Si) substrates are an alternative to Cz-Si substrates, with an extremely low concentration of impurities.[23] In this method, a polycrystalline silicon rod is passed vertically downwards through a heating zone, which starts melting the part of the rod coming under the influence of the heating field. A seed crystal is brought in contact with the melt, which initiates nucleation and growth of a single crystal. At the interface between the solid and liquid phases, impurities (low segregation coefficient ~10^{-4}) diffuse more into the liquid phase than the solid phase, resulting in the growth of a highly pure single crystal of silicon. FZ-Si crystals are doped by adding doping gases, PH_3, AsH_3 or B_2H_6, into the inert gas. Owing to the difficulty in growing crystals with large diameters compared with the Czochralski technique and a higher production cost, FZ wafers are typically only used for laboratory applications.

The process used to make multi-crystalline silicon (Mc-Si) is simpler and costs less, accounting for approximately 50% of worldwide photovoltaic production.[12] Mc-Si is grown by slow solidification of molten silicon in a fused quartz silica crucible coated with silicon nitride (to avoid sticking of silicon to the walls), and nucleation is initiated either from the bottom or walls of the ingot. Compared with single crystalline silicon, the material quality of Mc-Si is poorer due to the presence of grain boundaries. During the growth process, impurities from the walls of the crucible accumulate at the grain boundaries, locally increasing the recombination activity. This reduces the charge carrier lifetime and the performance of the solar cell.[24] High-performance Mc-Si silicon, achieved by nucleation on special seed layers at the crucible bottom, is now increasingly replacing conventional Mc-Si as it has a much-reduced dislocation density.[25]

To avoid the cutting process (Section 1.1.3), techniques have been developed to grow sheets of silicon with a defined thickness. Due to the introduction of a high level of impurities and a large number of defects during crystal growth at high temperature with higher cooling rate, these techniques are less frequently used and continue to be dominated by traditional single crystal growth from ingots.[26,27]

1.2.3 Cutting and Polishing

Depending on factors such as the required size, quality and its application, the growth of crystalline silicon can take from one week to several months. Once an ingot has been grown, it is then sliced up into smaller ingot blocks. The smaller blocks are then cut into wafers using a wire saw or diamond-edge saw to reduce the kerf loss and increase the productivity. Wire sawing may induce small cracks penetrating around 10 μm deep into the wafers, decreasing the mechanical strength of the wafers and increasing the surface recombination.[28,29] The cut wafers are then cleaned in a series of chemical baths to remove any residual slurry and refined to make them stronger and flatter. The edges are then rounded or profiled using an edge-grinding procedure, to provide strength and stability to the wafer. This reduces the probability of breakage or chipping while handling the wafer during device fabrication.

These wafers are loaded into a lapping machine that uses pressure from rotating plates with an abrasive between them to achieve a predetermined uniform thickness. Lapping the wafer allows uniform removal of saw damages, surface defects and stress accumulated in the wafer during the cutting process. Once the silicon wafers are lapped, they go through an etching and cleaning process, using another series of chemical baths to remove any residual surface damage caused by lapping.

The final crucial step is polishing and cleaning the wafer, which takes place in a clean room (Class 1 to Class 10 000). Wafers are physically and chemically cleaned using ultra-pure water and chemicals, producing a mirrored surface. This process eventually removes the surface topography, scratches, micro cracks and any other damage caused during the manufacturing process.

1.3 Cell Processing Technologies

The industrial goal for PV power is to reduce the cost of electricity generation compared to the equivalent for commercial grid electricity.[12] This includes reducing the cost at all the stages of device fabrication starting from material production, wafer development, processing techniques to system installation, while continuing to maintain the maximum conversion efficiency of the cells. Almost half of the total cost of the PV module is from the material production, crystal growth and wafer development, while the other half is from cell processing and module assembly. Improving the utilisation of silicon during the fabrication process and decreasing the kerf loss, can reduce the cost per watt of the wafers. For instance, in the case of ribbon growth, where there is no kerf loss, as there is no cutting or wire sawing involved, most of the silicon is used up. But the disadvantage of such technologies is a lower efficiency due to both the lower material quality and a larger number of defects arising from thermal stresses during the growth process, which reduces the minority carrier lifetime of the cell. The solar cell efficiency of crystalline silicon is limited by three loss mechanisms: optical losses, carrier losses and electrical losses. The back contact silicon solar cell is another high efficiency device, where all the metallisation on the front surface is removed. This reduces the optical losses such as losses due to shadowing and reflection.

Following the etching process, the surface of the silicon wafer is shiny and reflects more than 35% of incident light.[14] Interaction of visible light within a material can be described in terms of its index of refraction by the equation:

$$\overrightarrow{n_c} = \vec{n} + i\vec{k} \tag{1.5}$$

where, n is the real component of the refractive index which represents the phase velocity of the light inside the material and k, the imaginary component of the refractive index, is the extinction coefficient which contributes to the gradual loss of the light intensity as it travels through the material.

The amount of light reflected from the surface of the wafer, R is given by:

$$R = \frac{(\overrightarrow{n-1})^2 + \vec{k}^2}{(\overrightarrow{n+1})^2 + \vec{k}^2} \tag{1.6}$$

where, 1 is the refractive index of air.

At visible wavelengths, R mainly depends on the real component of the index of refraction and the above equation becomes:

$$R = \frac{(\overrightarrow{n-1})^2}{(\overrightarrow{n+1})^2} \tag{1.7}$$

Optical losses, such as reflection of light from the front surfaces, rear surfaces and from the front contacts, limit the amount of light that is incident on the solar cell and lower its efficiency. These limitations can be addressed by texturing the top surface, applying an antireflection coating, texturing the busbar (the top contacts that connect directly to external leads) and introducing a textured back surface reflector (which follows Lambertian law and allows trapping of light back into the cell by total internal reflection) followed by encapsulation. According to Yablonovitch *et al.*, for bulk silicon solar cells, there is an upper limit to the increase in optical path length for effective absorption of light, by a factor of $4n^2$, also known as the Yablonovitch limit.[30] This enhancement factor is used as the figure of merit to estimate the capability of a solar cell to absorb light. By engineering the optical properties of the front and back surface of the solar cell, we can minimise the amount of light reflected or increase the amount of light absorbed and achieve physically thin but optically thick solar cells.

The losses due to carrier recombination in any of the three regions, *i.e.*, emitter, base or space charge region, also greatly affect the short circuit current and open circuit voltage of a cell, lowering its efficiency. The average length a minority carrier moves, from the time of its generation before it recombines, is called its diffusion length (L_d)and is given by eqn (1.8)

$$L_d = \sqrt{D\tau_{bulk}} \tag{1.8}$$

where, τ_{bulk} is the carrier lifetime in the bulk and D, the diffusivity, is given by:

$$D = \frac{kT}{q}\mu \tag{1.9}$$

where, k is the Boltzmann constant, T is the temperature, q is the charge of the minority carrier and μ is the mobility of the charge carrier.

Ideally, for highly efficient solar cells, the photoexcited carriers generated must be able to move from the point of generation to the point of collection. Hence, the longer the diffusion length, the lower the recombination rate and the larger the current collection, or the larger the short circuit current density. This can be represented by eqn (1.10), where the short circuit current density, J_{SC} is directly proportional to the diffusion length, L_d

$$J_{SC} = qGL_d \tag{1.10}$$

where, G is the charge carrier generation rate.[31]

If the diffusion length of the charge carrier is smaller than the thickness of the cell, then the short circuit current will be reduced. In such cases, reducing the thickness of the absorber layer or increasing the diffusion length of the minority carriers can increase the short circuit current of the cell. However, if the diffusion length is more than the thickness of the cell, or is twice the thickness of the cell, the short circuit current tends to reach saturation.

The saturation current density, J_o, is dependent on the diffusion length and is given by eqn (1.11):

$$J_o = \frac{qDn_i^2}{L_d N} \tag{1.11}$$

where, n_i is the intrinsic carrier concentration and N is the dopant concentration.

The open circuit voltage, V_{OC} is dependent on the short circuit current density, J_{SC}, and saturation current, as shown in the eqn (1.12):

$$V_{OC} = \frac{kT}{q}\ln\left(\frac{J_{SC}}{J_o} + 1\right) \tag{1.12}$$

From eqn (1.11) and (1.12),

$$V_{OC} \propto \ln(L_d) \tag{1.13}$$

From eqn (1.10) and (1.13),

$$\eta \propto J_{SC}\,V_{OC} \propto L_d \ln(L_d) \tag{1.14}$$

Hence, the efficiency of the cell increases with the diffusion length. However, when the diffusion length is greater than the thickness of the cell, it no longer obeys the linear relationship. Defects in the bulk of a solar cell greatly influence the minority carrier lifetime. In such a case, electrons generated closer to the surface are more likely to contribute to the total current collected at the front contact than that generated in the bulk, due to recombination at the defect centres in the bulk.

External quantum efficiency (EQE) is a measure of the number of charge carriers generated for a number of photons incident on the solar cell. Figure 1.8 shows an EQE curve as a function of wavelength, for a silicon solar cell.[32]

$$\text{EQE}_\lambda = \text{electrons s}^{-1}/\text{photons s}^{-1} = 1240 \times J_\lambda/(P_\lambda \times \lambda)$$

where J_λ = short circuit current density at wavelength λ, P_λ = incident optical power density at wavelength λ.

The absorption depth is inversely proportional to the absorption coefficient, and the absorption coefficient depends on the wavelength of the

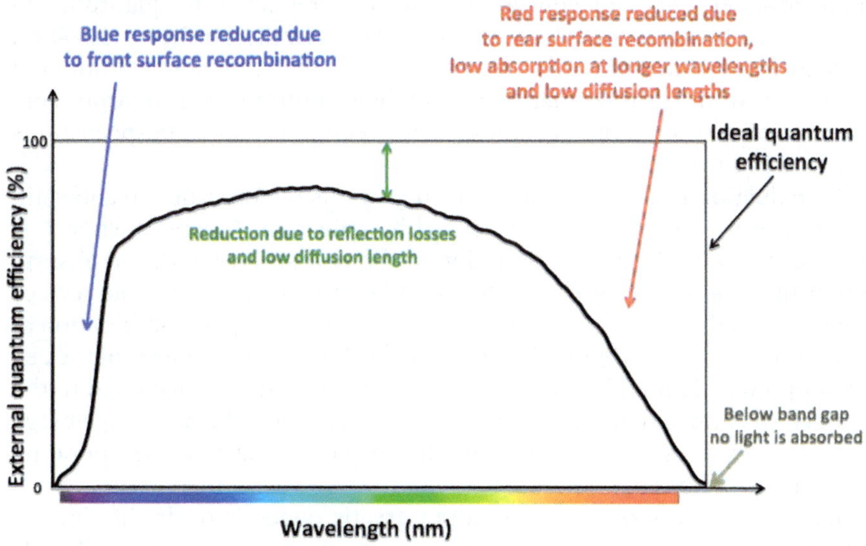

Figure 1.8 External quantum efficiency of a solar cell.[32]

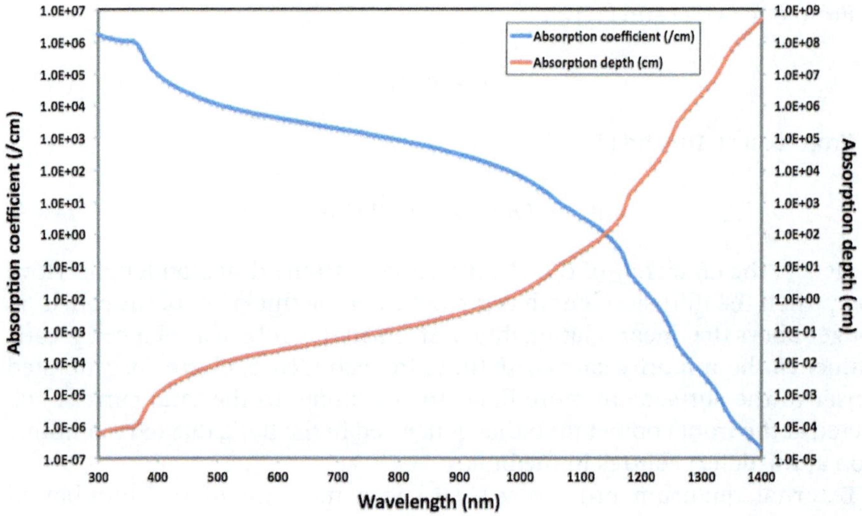

Figure 1.9 Absorption coefficient and absorption depth of silicon at 300 K as a function of wavelengths of light.[32]

light incident on the cell, as shown in Figure 1.9.[32] Consequently, light of different wavelengths penetrates different distances before being absorbed. Hence, longer wavelength light has a greater absorption depth, and shorter wavelength light has a shorter absorption depth. The blue spectrum of the incident light, which has a larger absorption coefficient, is absorbed near the front surface of the cell. Accordingly, recombination at the front surface would adversely affect the collection probability, reducing the quantum efficiency in the blue portion of the spectrum. For example, surface passivation by depositing a layer of silicon dioxide reduces the surface recombination at the front of the cell. Formation of a shallow emitter under the front contact also improves the blue response. The green portion of the spectrum is absorbed in the bulk.

The reduction in the quantum efficiency in this region is due to a low diffusion length and optical losses caused by reflection. The red response is reduced due to surface recombination at the rear surface, reduced absorption at the rear surface and low diffusion lengths. Doping the rear surface heavily keeps the minority charge carriers away from high recombination at the rear contact.[32] Photons with energy below the band gap energy of the cell are not powerful enough to excite an electron from the valence band to the conduction band and, hence, no light is absorbed below the band gap energy of the cell. As a result, the quantum efficiency is zero at the corresponding wavelengths of light.[32]

The minority carrier diffusion length greatly depends on its lifetime (τ) and mobility (μ). The minority charge carriers are metastable and will only exist for a length of time equal to the minority carrier lifetime, before they recombine. The carrier lifetime of the minority charge carriers is determined by its recombination rate, which, in turn, is dependent on its concentration. When the carrier concentration is low, the carrier lifetime is inversely related to the recombination rate, R, and is given by,

$$\tau = \frac{\Delta n}{R} \tag{1.15}$$

where, Δn is the excess minority carrier concentration.

The effective carrier lifetime, τ_{eff} is given by

$$\frac{1}{\tau_{eff}} = \frac{1}{\tau_{bulk}} + \frac{1}{\tau_{surface}} \tag{1.16}$$

The dangling bonds at the surface of the silicon introduce trap states, providing additional pathways for trap-assisted recombination at the surface.[33,34] The surface recombination rate is limited by the rate at which minority charge carriers move towards the surface; in other words, the carrier lifetime at the surface is limited by the surface recombination velocity, S, as given by eqn (1.17)

$$\tau_{surface} \approx \frac{W}{2S} + \frac{1}{D}\left(\frac{W}{\pi}\right)^2 \tag{1.17}$$

where, W is the width of the cell.

In a surface with no recombination, the net movement of carriers towards the surface is zero and hence the surface recombination velocity is zero. In a surface with high recombination, the movement of carriers towards the surface is limited by the maximum velocity they can attain.[32] Surface passivation reduces the number of dangling bonds on the surface and hence reduces the recombination at the surface of the cell. Limiting the surface recombination can lessen the rate at which minority carriers are depleted. Hence, if the rate of minority carrier depletion can be limited, the lifetime of the cell can be extended.[32]

In the bulk of a solar cell, there are three types of recombination mechanism which affect the minority carrier lifetime. They are radiative, Shockley–Read–Hall (SRH) and Auger recombination, as shown in Figure 1.10 and they are expressed as:

$$\frac{1}{\tau_{bulk}} = \frac{1}{\tau_{radiative}} + \frac{1}{\tau_{SRH}} + \frac{1}{\tau_{Auger}} \tag{1.18}$$

Radiative recombination, or band-to-band recombination, occurs when an electron drops from the conduction band into the valence band, combines with a hole and releases a photon. Radiative recombination is very slow and is rarely a dominating mechanism in silicon-based solar cells.

The presence of defects or impurity atoms in the bulk of a solar cell creates sub-band gap energy states within the energy band gap of silicon. Shockley–Read–Hall (SRH) or trap-assisted recombination occurs when an electron drops into these sub-band gap energy levels and combine with a hole. As the

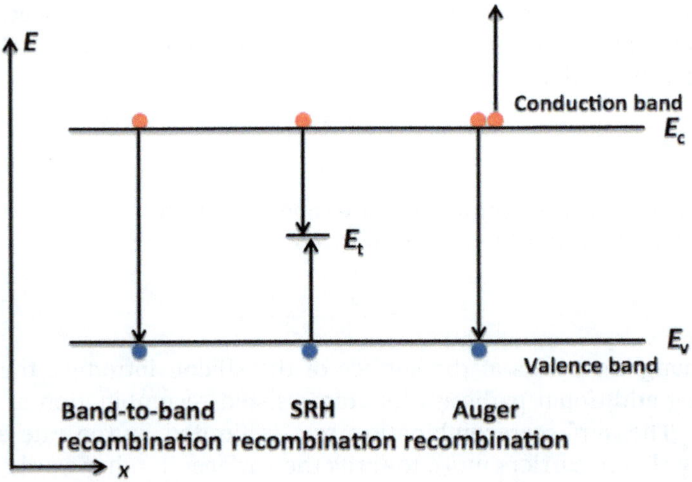

Figure 1.10 Carrier recombination mechanisms in semiconductors (E_c, E_v and E_t are the energy of the electrons in the conduction band, valence band and trap state, respectively and x is any real space parameter).

number of defects increases the SRH recombination rate also increases. Single crystals of silicon with lower concentration of impurities or defects, used for fabrication of solar cells, have a longer minority carrier diffusion length and hence lower recombination losses in the bulk.[35–37]

Auger recombination occurs when carrier concentration is high due to heavy doping or high-level injection. Since, Auger lifetime is a function of the carrier concentration, it is given by:

$$\tau_{\text{Auger}} = \frac{1}{CN_A^2} \tag{1.19}$$

where, the Auger coefficient, C, for silicon is typically: 1.66 × 10^{-30} cm^6 s^{-1}.[32,38,39]

Here, an electron and a hole recombine in a band-to-band transition, but now the resulting energy is given off to another electron in the conduction band. The involvement of another electron increases the recombination rate and decreases the carrier lifetime. Hence, the heavier the carrier concentration, the shorter is the Auger recombination lifetime and the shorter the diffusion length. Both trap-assisted (SRH) and Auger recombination limit the carrier lifetime in silicon-based solar cells.[40–42]

The final type of loss is the resistive losses, which arises due to the semiconductor bulk resistance, emitter resistance, metallisation on the top surface and metal–semiconductor contact resistance. Resistive losses due to series resistance are controlled by the design and architecture of a solar cell, while a low shunt resistance is a processing defect. Minimising these losses, increases the fill factor and, hence, the efficiency of a solar cell.[43–46] Shunt resistance between the metal–semiconductor interfaces can also affect the fill factor of the device and drop the performance of the device significantly.

Contact resistance losses occur at the interface between the metal contact and bulk silicon and depend on the doping concentration. Heavy doping under the contacts reduces this loss, while the doping is controlled by a compromise between achieving a low saturation current in the emitter and maintaining a high emitter diffusion length.[32] Increasing the number of busbars on the wafer devices minimises the series resistance, although it may cause problems with shadowing. Once again, the key design trade-off in top contact design is the balance between the increased resistive losses associated with a widely spaced grid and the increased reflection caused by a high fraction of metal coverage of the top surface.[32]

The design and optimisation of solar cells is therefore a complicated process, and small changes in one variable can have a substantial impact on the overall power produced by the system. The overall efficiency of the device is dependent on the efficiency of each individual step in the mechanism (see eqn (1.20)). Losses in efficiency (*e.g.*, charge recombination competing with charge separation, transfer or transport) at any point will limit the total efficiency of the device. In solar cell development, as one factor is made more efficient, by controlling a particular pathway for recombination or reducing the energy required for a particular step, another variable becomes limiting.

$$\eta_{\text{total}} = \eta_{\text{absorption}} \times \eta_{\text{excitation}} \times \eta_{\text{drift/diffusion}} \times \eta_{\text{separation}} \times \eta_{\text{collection}} \tag{1.20}$$

Gluntz describes this as analogous to limiting factors in ecology and has adapted Leibig's law of the minimum ("The yield potential of a crop is like a barrel with staves of unequal length and is limited by the shortest stave") for solar cells (see Figure 1.11).[10,11] In his work, Gluntz addresses the limitations in the design of a high efficiency cell structure in the PV industries for reducing the cost of solar cells. The capacity of the barrel (with water) represents the efficiency potential of the cell and each stave represents a kind of limitation which may affect the total conversion efficiency of a cell. As the capacity of water in the barrel is limited by the length of the shortest stave, the efficiency of the cell is limited by the above-mentioned losses. In this case, as shown in Figure 1.11, the shortest stave is the surface recombination velocity at the rear surface.[10,11] Current commercially produced solar cells with high efficiency have most of the features described in Box 1.1 included in their cell structures.[12]

Figure 1.11 Leibig's law of minimum illustrating the limitations on developing high efficiency solar cells. S_{back} and S_{front} are the surface recombination velocity at the rear and front surface, respectively. τ_{bulk} is the charge lifetime. R is the reflectivity. Reproduced from ref. 11, http://dx.doi.org/10.1155/2007/97370, under the terms of the CC BY 3.0 license, https://creativecommons.org/licenses/by/3.0/.

1.3.1 Texturing

Texturing the wafer surface removes the damage induced during the sawing process and is an effective light-management technique. Texturisation causes multiple reflections on the surface and enhances the absorption of light into the cell by increasing the probability of light entering the surface and by increasing the optical path length of the light. Ideally, a solar cell absorbs most of the light from the visible and ultraviolet region (the photon energy is well above the band gap energy of silicon) before reaching the rear side of the cell. Optical losses, due to high reflection in short wavelength and low absorption in long wavelength regions, limit the conversion efficiency of a solar cell. Light trapping in the cell, especially of longer wavelength photons in the infrared region, can be achieved through surface texturisation. The well-defined features selectively direct light into the cell at angles outside the escape cone of silicon increasing the absorption of light and increasing the cell efficiency.[47–50] The requirements for successful reduction in surface reflectance by surface

Box 1.1 Minimising losses in a solar cell.[12]

Optical losses

- Texturing the front surface to reduce surface reflection loss and to increase absorption of light by trapping.
- Applying single- or multi-layered anti-reflection coating to reduce reflection loss.
- Minimising the front metal contact coverage area to reduce reflection from the front contact and shadowing losses.
- Choosing back contact cell structure to eliminate shadowing losses.
- Using back surface reflector to reduce photon absorption.
- Creating a flat back surface to increase back surface reflectivity.

Electrical losses

- Deep and heavy doping under the metal contact to reduce the contact resistance losses at the interface between the surface of the silicon solar cell and the front contact.
- Fine gridline front contact and an optimum finger spacing to reduce the resistive losses.
- Minority carrier diffusion lengths longer than the base thickness for longer minority carrier lifetime.

Recombination losses

- Passivating the metal contacts to reduce the carrier recombination losses at the interface between the surface of the silicon solar cell and the metal contact.
- Surface passivation to reduce recombination at the surface.
- Passivating the back surface by using a back surface field and point contact structure to reduce carrier recombination at the back surface.
- Heavy doping under the metal contact to keep the minority carriers away from high recombination front contact.

texturisation are optimum feature size, high surface roughness and high uniform surface coverage.[50]

The texturisation in laboratory and commercial silicon solar cells can be either by physical or chemical methods.[50] Chemical methods include isotropic wet etching using either acid or alkaline solutions. Monocrystalline silicon wafers are typically textured in a weak solution of NaOH or KOH, resulting in randomly distributed pyramids, as the oriented planes ⟨100⟩ and ⟨100⟩ have different etch rates.[28,50–52] Texturing is followed by a cleaning process in acidic solution, such as HCl and HF, followed by rinsing with deionised water to remove metal impurities from the surface. The lateral uniformity and anisotropy can be improved by adding isopropyl alcohol or a surfactant to the etchant, and this can result in a reduction in the total reflection from 35% to 12%.[28,53] In Mc-Si wafers the random texturisation process is not effective as the wafers are made up of smaller silicon grains with different crystallographic orientations. Isotropic etchants, sometimes in combination with other techniques, are used in this case. Typically, acid solutions (HCl and HF) are usually used for texturisation, as they texture the wafer independently of the orientation of the grains.[50] A constant etch rate is maintained to avoid under-etching or over-etching and scrapped wafers. If the etching is too deep, the surface roughness increases thus decreasing the open circuit voltage and short circuit current of the cell. At the same time, shallow etching depth also lowers the open circuit voltage and short circuit current of the cell, due to surface recombination, as it barely removes the crystal defects on the surface.[28] The best solar cell efficiency has been found for an etching depth between 4 and 5 μm.[18–20,28]

Physical methods involve etching by dry processes such as mechanical grooving, reactive ion etching and laser ablation. Texturing by mechanical grooving is suitable for both mono-crystalline and Mc-Si silicon wafers.[54–58] The advantages of this technique are low cost and high texturisation rate. In this technique, surface texturisation is achieved by printing a structural wheel with a metal tool blade having a V-grooved surface, which is coated with abrasive diamond particles, and the cooling agent is water.[50] The spin speed is about 150–160 mm s^{-1} depending on the grain size of the diamond particles. The wafers then undergo cleaning and post-treatments with chemicals to obtain well-shaped, defect-free grooves. Reactive ion etching (RIE) is a dry etching process where the wafer surface is etched by highly reactive ion plasma with tetrafluoromethane or a combination of sulfur hexafluoride and oxygen or chlorine as a reactant gas. RIE is capable of producing highly uniform, defined features (less than 1 μm) and is suitable for both mono-crystalline and Mc-Si silicon wafers.[50] Unlike wet etching, it can be controlled easily and does not depend on crystallographic orientation. Both masked and maskless RIE texturing results in a relatively increased absorption of light compared with other methods.

Laser ablation is a texturisation technique where the wafer surface is melted and evaporated under pulsed laser-generated radiation.[50] Advantages of this technique are the possibility of obtaining features with defined patterns and accuracy. The limitation of this technique is the formation of melted silicon at the bottom or sidewalls of the groove, and amorphous silicon debris, due to the laser-generated heat, leading to the deposition of foreign materials during texturisation.[44,45] Hence, the wafers are chemically etched to remove any residue that could increase the surface recombination.[46,50]

1.3.2 Junction Formation

Doping of semiconductors can be performed during crystal growth, as mentioned in Section 1.1.2. Here, the crystal, and the wafer sliced from it, are uniformly doped with the dopants as the dopants are introduced along with the precursors. In contrast to doping during crystal growth, in the modern semiconductor industry doping is carried out on the substrate (after crystal growth and cutting), to control doping in specified regions by masking the undesired regions. In amorphous silicon devices, doping is still carried out during deposition. There are three techniques to form a p–n junction on a silicon substrate: diffusion, ion implantation and doping using alloy. Doping using alloy is rarely used in today's technology due to anomalies in the electrical properties. In this section, we will discuss the first two techniques.

Photolithography is used for patterning and defining the geometry for doping, as shown in Figure 1.12(c)–(g).[59] Prior to photolithography, an oxide layer of silicon is formed by thermal oxidation. Thermal oxidation of silicon occurs at a high temperature of about 800–1200 °C. The oxide layer can be grown in oxygen (dry oxidation) or in water vapour (wet oxidation) as shown in eqn (1.20) and (1.21), respectively.

$$\mathrm{Si} + \mathrm{O_2} \rightarrow \mathrm{SiO_2} \tag{1.21}$$

$$\mathrm{Si} + 2\mathrm{H_2O} \rightarrow \mathrm{SiO_2} + 2\mathrm{H_2} \tag{1.22}$$

In dry oxidation, oxygen diffuses through the oxide layer that is being formed and the growth rate decreases with increasing thickness. In wet

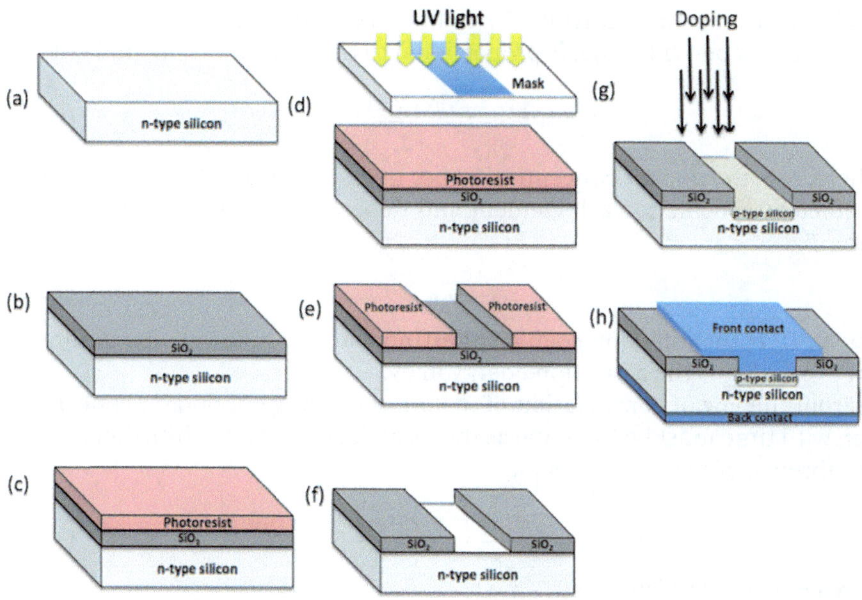

Figure 1.12 (a) An n-type silicon substrate, (b) formation of an oxide layer of silicon on the substrate by thermal oxidation, (c) photoresist deposition, (d) UV exposure through a patterned mask, (e) development, (f) SiO_2 removal, (g) formation of p–n junction by diffusion or ion implantation, (h) metallisation.[59]

oxidation, the growth rate is higher than that of dry oxidation since the hydroxide can diffuse more readily than oxygen through the oxide layer that is being formed. Hence, dry oxidation is used to form thin oxides of thickness less than 100 nm in a device structure and wet oxidation is used for thicker oxides, especially for masking, insulation or as passivation layers.

The wafer is coated with ultraviolet (UV) light-sensitive photoresist and baked for drying. The wafer is then exposed to a UV light source through a patterned mask with high sub-micrometre precision. After successful development, the resist is stripped away using chemicals. Doping is then carried out by diffusion or implantation and the p–n junction is formed as shown in Figure 1.12(g).[59] After the formation of the p–n junction, metal contacts are patterned on the wafers by a suitable metallisation process.

1.3.2.1 *Diffusion*

Doping by diffusion is a gradual mixing of atoms by random motion from a region of higher concentration to one of lower concentration. Over the years diffusion has been the most widely used technique for doping. In the diffusion method the surface of the silicon substrate, not protected by oxide layer, is exposed to a high concentration of dopant. The dopant moves into the substrate by solid-state diffusion.

Fick's law governs diffusion. Fick's first law states that the flux density, γ per unit area per unit time, is proportional to the concentration gradient:

$$\gamma = -D\frac{\partial N(x,t)}{\partial x} \tag{1.23}$$

where, N is the dopant concentration per unit volume and D is the diffusion coefficient specific to each element and depends strongly on the temperature, T:

$$D = D_0 e^{-E_a/kT} \tag{1.24}$$

where, D_0 is the diffusion coefficient extrapolated to infinite temperature in $cm^2\ s^{-1}$ and E_a is the activation energy in eV.

From the law of conservation of matter, the change of dopant concentration with time must be the same as the local decrease of the diffusion flux, in the absence of a source or a sink;

$$\frac{\partial N(x,t)}{\partial t} = -\frac{\partial J(x,t)}{\partial x} \tag{1.25}$$

From Fick's first law,

$$\frac{\partial N(x,t)}{\partial t} = \frac{\partial}{\partial x}\left(D\frac{\partial N(x,t)}{\partial x}\right) \tag{1.26}$$

When the concentration of the dopant is low, the diffusivity at a given temperature can be considered as constant, then Fick's second law is given as:

$$\frac{\partial N(x,t)}{\partial t} = D\frac{\partial^2 N(x,t)}{\partial x^2} \tag{1.27}$$

The doped regions can be created by any of the following two diffusion techniques. The constant source diffusion or pre-deposition process (dopant source is inexhaustible) which is the controlled doping of the desired dopants by solid, liquid or gas phase diffusion. The limited source diffusion or drive-in process (dopant source is exhaustible) which drives the dopants deeper into the substrate without further introduction of dopants.

When the dopant source is inexhaustible, the concentration of the dopant on the surface is kept constant. When the substrate is heated, the dopants diffuse into it to achieve equilibrium with the concentration near the surface. In other words, the surface concentration N_s remains constant and the dopants move deeper into the substrate, as the diffusion time, the diffusion temperature or both increase.[60] The concentration within the substrate is dependent on the diffusion time and diffusion temperature and results in complementary error function distribution.[60] The solution of Fick's second law that satisfies the initial and boundary conditions is given by

$$N(x,t) = N_s \mathrm{erfc}\left(\frac{x}{2\sqrt{Dt}}\right) \tag{1.28}$$

A Gaussian distribution results from limited source diffusion, *i.e.*, when there is an exhaustible dopant source on the surface, with a concentration Q cm^{-2}. As the diffusion increases with time, the dopants move deeper into the substrate and the surface concentration decreases.[60] The solution of Fick's second law that satisfies the initial and boundary conditions is given by

$$N(x,t) = \frac{Q}{\sqrt{\pi Dt}} \mathrm{e}^{-\left(\frac{x}{2\sqrt{Dt}}\right)^2} \tag{1.29}$$

For $x = 0$,

$$N(0,t) = \frac{Q}{\sqrt{\pi Dt}} \text{ or } N_0 = \frac{Q}{\sqrt{\pi Dt}} \tag{1.30}$$

Eqn (1.28) becomes,

$$N(x,t) = N_0 \mathrm{e}^{-\left(\frac{x}{2\sqrt{Dt}}\right)^2} \tag{1.31}$$

Boron (p-type dopant) is usually introduced during silicon processing and phosphorus is diffused to form an n-type junction. In a gas diffusion process for doping, phosphorus is introduced in the diffusion tube in the form of phosphine or phosphorus oxychloride gas at a high temperature of about 800 °C. Phosphorus can also be diffused using liquid phosphorus oxychloride in a heated furnace in the presence of nitrogen as a carrier gas. When oxygen is added in the diffusion tube, SiO_2 is produced at the surface of the silicon wafer. The dopant source, PH_3 or $POCl_3$, is converted to P_2O_5 and the released chlorine removes the metal impurities. This P_2O_5 reacts with SiO_2 to form a viscous liquid or a layer of phosphorus silicate glass (PSG) on the surface of the substrate.

$$2PH_3 + 4O_2 \rightarrow P_2O_5 + 3H_2O \tag{1.32}$$

$$2POCl_3 + 3O_2 \rightarrow 3P_2O_5 + 6Cl_2 \tag{1.33}$$

At the diffusion temperature, phosphorus diffuses into the wafer. A solid dopant source, triphosphorus pentaoxide, can be used in place of $POCl_3$. An alternative technique used to diffuse phosphorus is by depositing diluted phosphoric acid onto the wafer surface and moving it horizontally on a conveyer belt in a firing furnace.[61]

1.3.2.2 Ion Implantation

Recently, the ion implantation technique has gained ground in the solar industry, slowly replacing the diffusion technique that has been used for many years.[62,63] As shown in Figure 1.13, the cell efficiencies are predicted to improve with continuous development of technologies towards achieving higher efficiencies. Apart from localised and patterned doping, the main advantages of this technique include high, precise control on the amount and distribution of the dopant doses, resulting in high uniformity, reproducibility and enhanced efficiency (beyond 19%), with a much narrower cell efficiency distribution.

In the ion implantation method, the selected ions of a desired impurity are introduced into the semiconductor by accelerating the impurity ions to a high energy level and implanting the ions into the semiconductor. The

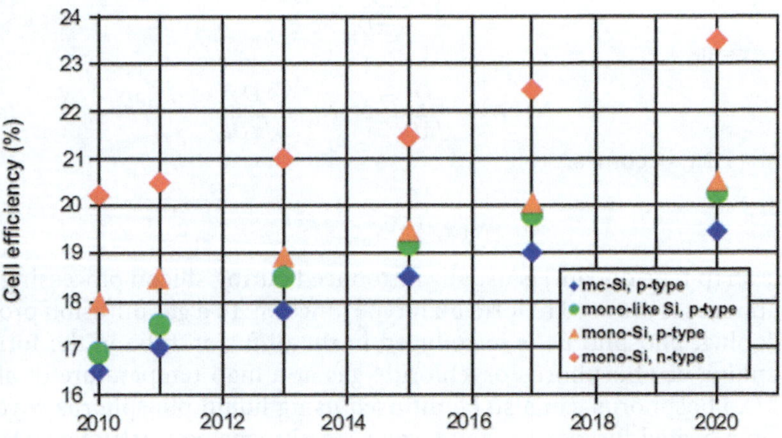

Figure 1.13 ITRPV Roadmap, March 2012 update: stabilised cell efficiency trend curves. Reproduced from ref. 62 with permission from EU PVSEC.

energy imparted to the impurity ions determines the ion implantation depth. Unlike the diffusion technique (wherein the dose of impurity ions is introduced only at the surface), in the ion implantation technique a controlled dose of impurity ions can be introduced deep inside the semiconductor. Photolithography is used for selective implantation of impurity ions, resulting in a highly reproducible and controlled doping concentration in the wafer.[64,65] For today's standard screen-printed p-type cells, ion implantation yields improved efficiencies through precision phosphorus profiles and high uniformity.[62] This technique yields higher efficiencies and lowers the cost not only by improving the technology and reducing the wafer sizes, but also by cutting the processing steps. Process simplification can be accomplished by eliminating the need for PSG and edge isolation by the use of single-sided ion implantation.

1.3.3 Edge Isolation

During phosphorus emitter diffusion from the gas phase onto the p-type substrates, the phosphorus-doped layer covers the entire surface of the cell, including the edges and the rear surface. This edge-shunt results in a reduction of the solar cell open circuit voltage and accounts for 80% of the total loss mechanism.[66] Edge isolation is performed to prevent shunting pathways between the front and the rear contacts. For screen-printed silicon solar cells, this is usually achieved by stacking the wafers on top of each other and plasma etching using the precursor gases carbon tetrafluoride and O_2. Other techniques include inserting a laser-based trench on the edges, sawing isolation trenches on the edges and grinding the wafer edge with sandpaper.[67,68] A laser edge isolation process is very commonly used in c-Si solar cell production.[69] This is done by scribing a groove at the edge of the wafer, with the groove depth extending a certain distance beyond the emitter diffusion layer, as shown in Figure 1.14.

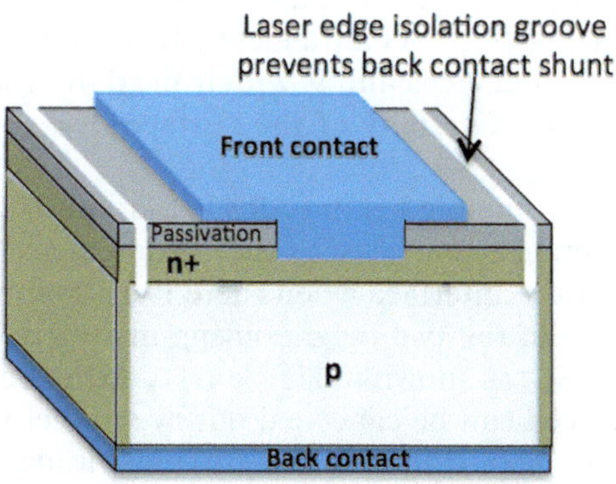

Figure 1.14 Schematic representation of edge isolation used in a solar cell structure to prevent shunting between the front and back contact.

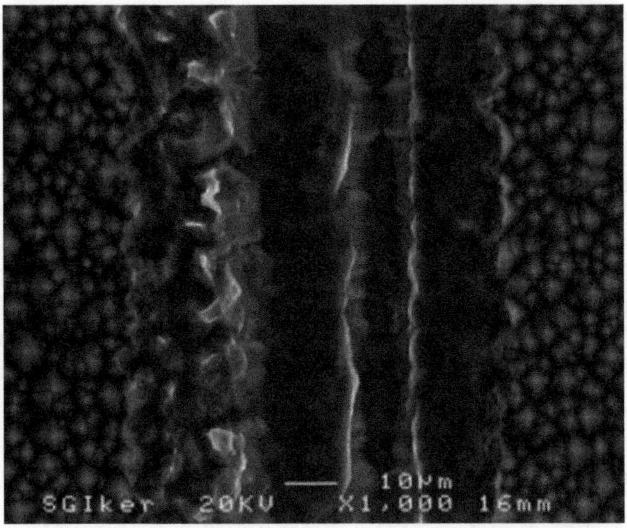

Figure 1.15 A SEM image of laser edge isolation achieved by scribing a groove with a laser of 515 nm and 10 ns pulse duration. Reproduced from ref. 69 with permission from Elsevier, Copyright 2018.

The general groove dimensions have a width of about 30 μm and a depth of about 15 μm (Figure 1.15).[69] An alternative technique used is an inline system for single-sided etching to remove the emitter from one side of the wafer with acid solution.[28,68,70–72] The advantage of this technique is the higher throughput and hence a higher fill factor.[72–74]

1.3.4 Antireflection Coating

When light falls on a bare silicon surface, over 30% of the incident light is reflected from the surface of the silicon because of the difference in the refractive indices of silicon and air, as shown in Figure 1.16.[75b–77] An anti-reflecting coating (ARC) is a dielectric thin-film coating applied to the surfaces of optical devices and is widely used in industrial solar cells to reduce the optical reflectivity of the surface in a certain wavelength range.[78,79]

The basic working principle behind ARC is that the reflected waves from different optical interfaces cancel each other by destructive interference. The reflection is large at the interface when there is a large mismatch between the refractive indices of the two neighbouring materials. On a bare silicon surface without an ARC, as shown in Figure 1.17, the reflected wave from the surface of the solar cell can be cancelled out by another wave, if that other wave can be made to be of equal wavelength, magnitude and direction, but of opposite phase (destructive interference). For this to happen, we need to create another reflected wave at a different location from the surface of the

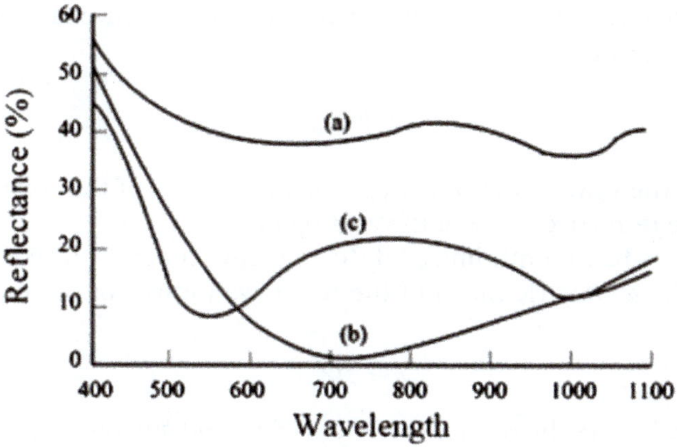

Figure 1.16 Comparison of surface reflection from (a) an uncoated silicon surface, (b) SiO as single-layer ARC and (c) MgF_2 and SiO as double-layer ARC. Reproduced from ref. 75a with permission.

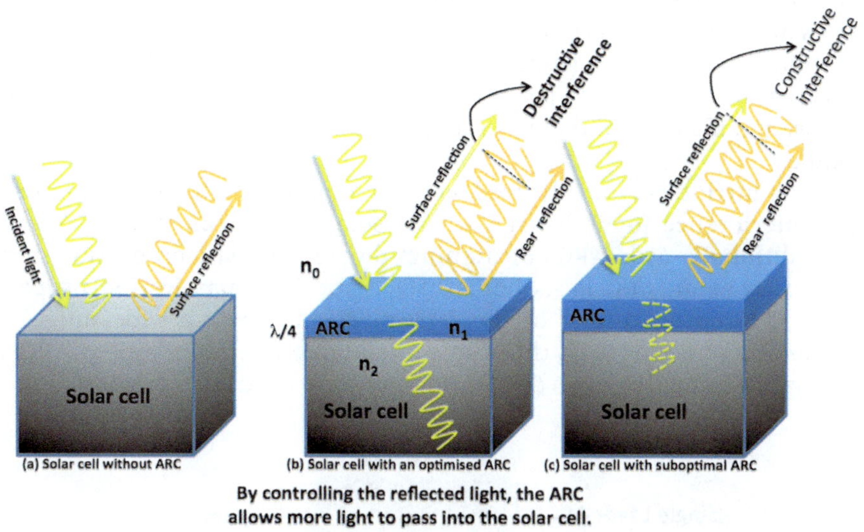

Figure 1.17 Working principle of ARC. (a) Solar Cell without ARC, (b) Solar cell with an optimised ARC and (c) Solar cell with suboptimal ARC.

silicon, such that the two reflected waves are a half-wavelength out of phase. The two reflections are exactly a half-wavelength apart, when the two reflective surfaces are a quarter-wavelength apart. Hence, anti-reflective coatings are also known as "quarter-wave" coatings. With the appropriate choice of dielectric material and its layer thickness, minimum reflection from the surface of the silicon can be achieved at a certain wavelength.

For an effective ARC, the optical thickness, *d*, which causes minimum reflection, is given by:

$$n_1 d = \lambda_o/4 \tag{1.34}$$

where, λ_o is the wavelength of the photon at the peak of the solar spectrum and n_1 is the refractive index of the coating material.

Reflection is further minimised if the refractive index of the coating material, n_1, is the geometric mean of the refractive indices of the materials on either side:

$$n_1 = \sqrt{n_0 n_2} \tag{1.35}$$

where, n_0 and n_2 are the refractive indices of air and silicon, respectively.

In a single-layer coating, minimum reflectance is obtained at a single wavelength or, in other words, reduction of reflection is possible only for a small range of wavelengths. Hence, if a reduction in reflections is required over a large range of the spectrum, several layers of ARCs of different refractive indices and thickness must be applied. The refractive index of the upper ARC layer must be lower than that of the ARC layer on the silicon substrate, as shown in Figure 1.18. Here, n_0, n_1, n_2 and n_3 are the refractive index of air, two ARC thin films and the solar cell, respectively. With a multiple layer coating, light transmission of up to just under 99% can be obtained, resulting in an increase in efficiency of the cell and, hence, reducing the total cost of the module.

Silicon nitride is typically used as the standard ARC to create quarter-wave interference for industrial silicon solar cells. This is preferred to the traditionally used ARCs of titanium dioxide and silicon dioxide thin films. TiO_2 as an ARC lacks both surface and bulk passivation properties.[80,81] Moreover, tuning of the refractive index over a wide range is not possible, making TiO_2 thin films unsuitable for multi-layered ARCs. Although, thermally grown SiO_2 thin films have good surface passivation

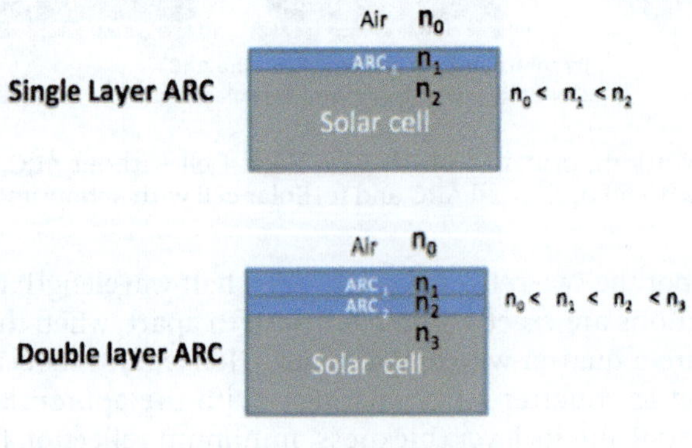

Figure 1.18 Single-layer ARC and double-layer ARC.

properties, it is not very effective for bulk passivation and its low refractive index hinders its optimal optical performance as an ARC for silicon solar cells.[82–87]

ARC thin films of silicon nitride are deposited either by chemical vapour deposition (CVD) or by sputtering. In CVD, the reaction is initiated by low pressure chemical vapour deposition, plasma-enhanced chemical vapour deposition (PECVD) or by atmospheric pressure chemical vapour deposition (APCVD), to deposit a thin film of silicon nitride coating from the precursor gases.[88,89] Here, the precursor gases are silane and ammonia and the reaction takes place as shown in eqn (1.35). Most of the hydrogen during the reaction evaporates, while some of it diffuses from the interface into the bulk of the cell. The free hydrogen that diffuses into the bulk is highly reactive and attaches itself to the defects in the bulk, passivating the defects and improving the bulk carrier lifetime.[90–93] Hence, the deposition of thin films of silicon nitride can not only reduce the optical losses due to reflection, but also simultaneously provide surface and bulk passivation.[88–94]

$$3SiH_4 + 4NH_3 \rightarrow Si_3N_4 + 12H_2 \tag{1.36}$$

The silicon nitride thin films deposited by the PECVD process can act as both a surface and bulk passivating ARC layer for crystalline silicon solar cells for two reasons. First, tuning of the refractive index is possible for the deposition of silicon nitride thin films by controlling the gas flow ratio of the precursor gases. Hence, single- or double-layered ARC for minimum reflectance can be achieved with silicon nitride thin films.[95] Second, silicon nitride thin films deposited by PECVD have larger amounts of hydrogen (about 15 to 20 atomic%) leading to a much higher passivation effect, making it also highly suitable for bulk passivation of defect-rich Mc-Si.[92–97] The APCVD method is mostly used in microelectronics. Sputtered thin films of silicon nitride have been reported to have similar surface and bulk passivation properties as films deposited by PECVD.[98,99]

1.3.5 Metallisation

Contacts are formed on the surface of the silicon solar cells for extracting charge carriers from the device and to prevent back-diffusion of the carriers into the cell. Screen printing is generally used to form front and rear contacts in commercial solar cells due to robustness and simplicity. Different printing parameters, such as paste composition, viscosity of the paste, emulsion thickness, mesh dimensions, tension of the screen, paste temperature, printing speed, snap-off distance, squeegee speed and pressure, have a high impact on the quality of the print.[100–106] The screen-printed contacts should have low contact resistance, good adhesion to silicon, high aspect ratio and no junction shunting.[28]

Almost all commercial screen printing lines have a net throughput of around 1000 and 2000 wafers h^{-1} for single and double lines, respectively, and solar cells up to 210 × 210 mm^2 in size can be processed on these lines.[28] The cells move on a conveyer belt and the silver-based paste passes through a patterned screen, leaving a uniformly spaced metal pattern on the surface. After printing the front contact, the cells are dried in the belt furnace, to avoid smudging and to remove all the organic binder that was in the paste. The cells are then loaded back to the conveyer belt for printing the rear or back contact. For the backside metallisation, aluminium is usually chosen for good ohmic contact (a low resistance junction to provide good conduction between the metal and the semiconductor). After printing, the front and the rear contact are fired simultaneously (co-firing) in a firing furnace. In addition, contacts are heavily influenced by the metal–semiconductor interface, since over firing (metal drives deep into the bulk silicon, introducing a shunting pathway in the device) or under firing (poor contacts) the contacts may result in an increase in the shunt resistance and series resistance, respectively. In a buried contact cell, on the front side, the metallisation is performed just above the low resistance or highly doped emitter (n^{++}), as shown in Figure 1.19, to prevent shunting and to reduce the contact resistance between the semiconductor and metal.[12,107] A shallow emitter, created above the p-type semiconductor improves the blue response and minimises the Auger recombination. The silver-based paste, which is used for the front contact, consists of metal oxides to facilitate silver etching through the ARC thin film. This improves the adhesion of silver to the front surface of silicon and ensures a good electrical contact. On the back surface, during the firing process, a certain amount of aluminium will diffuse into the p region to create a p^{+} region near the back surface or an aluminium back surface field. This field pushes the electrons from the back surface to the front surface preventing any back surface recombination of minority carriers. A model for the formation of the screen-printed rear contact is given by Neuhaus *et al.*[28,108] Since, it is difficult to solder onto the screen-printed aluminium contact, silver strips that can be easily soldered to the interconnects are printed on the cell.

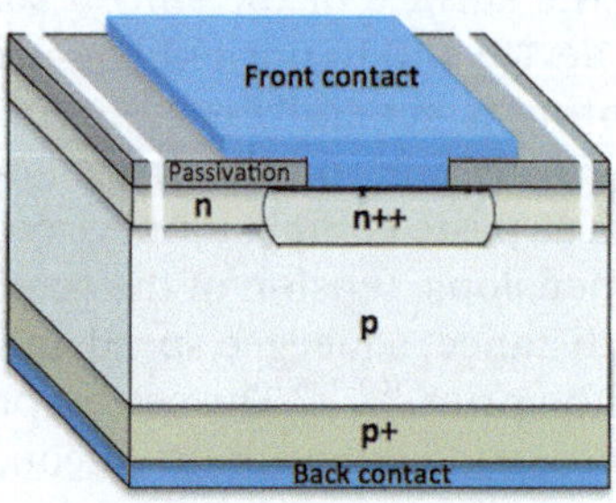

Figure 1.19 High efficiency solar cell structure showing metallisation on the front and back surface of the cell using buried contact technology.

The finer lines of metallisation connected to busbars are called fingers. The fingers collect the current generated by the solar cell and deliver it to the busbars. The critical features of the top contact design that determine the magnitude of the losses here are: the finger and busbar spacing, the metal height-to-width aspect ratio, the minimum metal line width and the resistivity of the metal. Hence, the size and grid separation of the top metal contacts have a key role in determining the performance of a solar cell. Minimising the top contact coverage of the cell surface can reduce optical losses due to reflection, but this may result in increased series resistance. A balance between the optical and electrical losses is a trade-off for the design of the top contact. Hence, the design of the top contact involves reducing the resistive losses in the emitter, the resistive losses in the metal top contact and the shadowing losses. An alternative approach to reduce both the surface recombination and the resistive and shadowing losses is to develop a back contact solar cell that is completely contact-free on the top surface of the cell.[28]

1.3.6 Testing and Sorting

Testing and sorting the solar cell devices according to their performance and appearance are the final steps in solar cell manufacture. This involves determining the solar cell parameters of an individual cell (which will eventually be connected in series to other cells to develop a module), determining its reverse breakthrough characteristics (to avoid hot-spot heating within the module) and sorting the cells (to minimise any possible mismatch losses in a module).[109–111] The modules are exposed to 1 Sun radiation under AM1.5 and the current–voltage characteristics measured at 25 °C. Most current–voltage testers consist of a halogen flash lamp that can stay constant for more than 50 ms (1000 W m^{-2}, reproducibility ± 1.5%, spectra class A and uniformity of ± 2% IEC60904-9, satisfying the standard solar simulator performance requirement International Electrotechnical Commission (IEC) standard).[107] Individual wafer flatness and surface particles undergo inspection to assure wafer quality before packaging.

1.4 Solar PV Industry

Solar power is a source of clean energy with great potential, but at present it provides only a minor portion of our global energy requirements. According to the World Energy Council report, the global primary energy consumption is about 12 billion tonnes of coal equivalent per year. This is forecast to grow to 16 billion tonnes of coal equivalent by 2020 based on the social, political and economic developments worldwide.[112] In 2016, renewable energy provided an estimated 19.3% of global final energy consumption, in which the contribution from solar PV technology was only 1.5%.[113] Figure 1.20 shows a 2 MWp PV power plant installed in India to feed a centralised grid. The power plant is installed using a multi-crystalline silicon PV module with time-based single-axis tracking.

Figure 1.20 A 2 MWp PV power plant from India, installed using a multi-crystalline silicon PV module with time-based single-axis tracking.

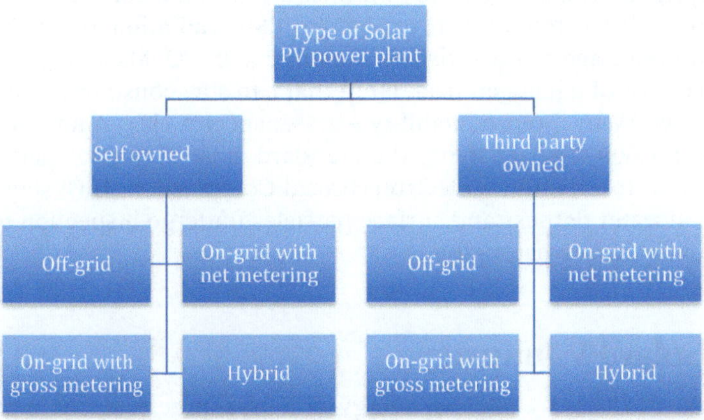

Figure 1.21 Different configurations of a PV power plant.

1.4.1 PV Systems

In general, PV systems are classified based on the loads designed to operate their connections with utility systems and the sources, as shown in Figure 1.21.

1.4.1.1 Off-grid Systems

An off-grid system is not connected to the electrical grid. The main components of an off-grid system are an inverter, battery, PV module and charge controller as shown in Figure 1.22. In residential systems, the power plant should be correctly sized so that it can generate enough power with enough battery capacity to meet all the requirements of the user. The cost of the battery and the inverter for an off-grid system is high and is more expensive than for an on-grid system of the same size. At present, the lifetime of the battery is lower than the PV module. However, the cost of batteries is reducing rapidly and the lifetime is increasing. Appliances use the solar power produced by the PV modules and then the excess power charges the battery. The charging or discharging is usually controlled by the charge controller. During the night, or when solar irradiance is low, appliances can draw power from the battery. In off-grid systems, an electric generator can be used as a back-up. The size of the generator should be adequate to supply electricity and charge the battery at the same time.

1.4.1.2 On-grid Systems

On-grid systems are not coupled to a storage system (a battery) and are directly connected to the electrical grid. The electricity generated from solar power is sold directly to the grid. The main components are an inverter, PV module, meter and electricity grid, as shown in Figure 1.23. There are two types of meter used: net meter and gross meter. Net metering schemes allow consumers to use the excess electricity that is not used in the premises when it is generated, at a later time. Net meters record the amount of energy exported by the PV system to the grid and the amount energy drawn from the grid. In gross metering schemes, consumers export all the electricity produced by their PV system to the grid in return for a generous feed-in-tariff

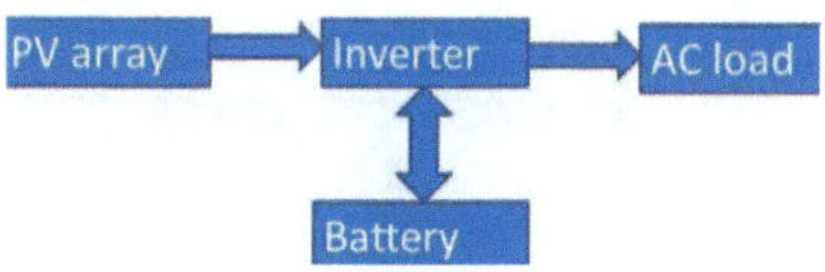

Figure 1.22 Configuration of an off-grid system.

Figure 1.23 Configuration of on-grid PV systems.

(FIT), which is normally higher than the cost of the electricity that they buy back from their electricity retailer. FITs are therefore used as a policy incentive to invest in renewable energy and, as the costs associated with PVs have decreased so the FITs offered have decreased. The on-grid inverter should have an anti-islanding function, which senses the occurrence of a grid power outage and shuts the inverter automatically.

1.4.1.3 *Hybrid Systems*

A hybrid system is connected to both the electrical grid and a battery. The system can store energy for use by the consumer and it can also export all or some of the generated electricity to the grid. The main components are an inverter, battery, PV module, meter and electricity grid, as shown in Figure 1.24.

All these types of system can be installed to produce electricity from solar power. However, for decentralised applications of solar energy, storage is very much a requirement. Details of the energy storage used in solar PVs are discussed in the next section.

1.4.2 Solar PV Energy Storage

Storage of energy in different forms has existed for centuries. Energy storage plays an important role in making the best use of electricity from sunlight. Energy can be stored when demand is low and generation is high thus ensuring its availability when demand is high. There are different types of energy storage used for solar PVs, including batteries, pumped water, hydrogen, heat storage, compressed air and flywheels. Typical types of batteries used in solar PV systems include lead acid, lithium ion, nickel cadmium and nickel iron. The characteristics needed for battery or storage systems to be used with solar PVs are: low cost, high efficiency, low maintenance, long lifetime, high reliability, wide range of operating temperatures and low

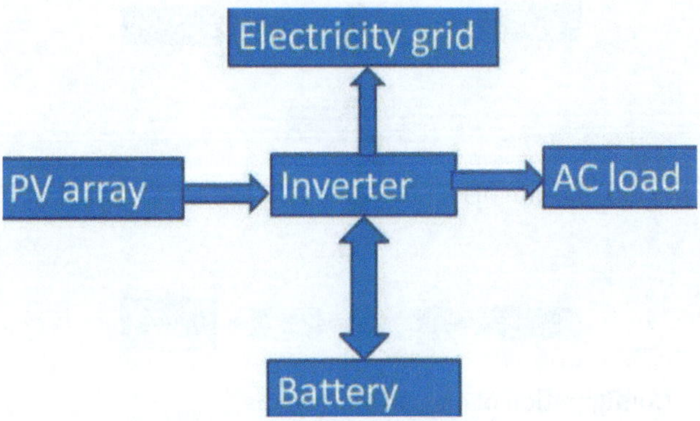

Figure 1.24 Configuration of hybrid PV systems.

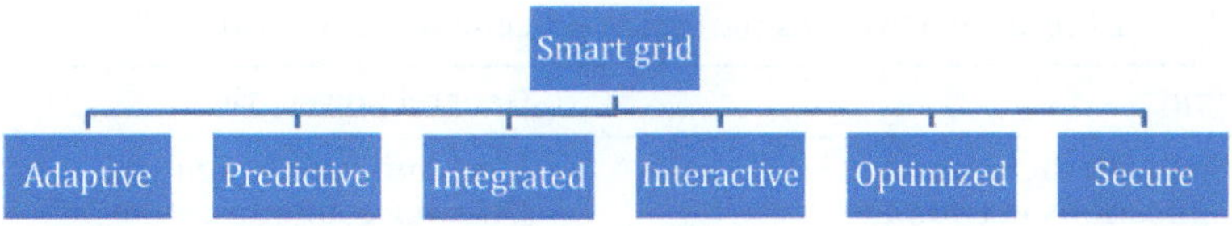

Figure 1.25 Characteristics of smart grids.

self-discharge. Hydroelectric storage can also be used for solar PVs, in which water is pumped to a reservoir at high elevation and electricity regenerated when required by releasing the water through turbines. Coupling electricity storage with solar PVs is critical for the technology to be a viable alternative to fossil fuels. However, the cost of storage can be, relatively, very high.

1.4.3 Smart Grids

For smooth generation, transmission and consumption of solar electricity, the smart grid concept is generally used. A smart grid can be defined based on its capabilities and operational characteristics. It allows the operator seated in a control room to operate the grid more effectively with distributed intelligence, sophisticated information and communication technologies. The balance between supply, consumption and demand is maintained to improve reliability and availability. Smart grids should have the characteristics shown in Figure 1.25.

The smart grid should be automatically responsive to changing conditions in the grid in order to improve reliability. The system should have the ability to detect faults before their occurrence in the power plant, so that speedy recovery of the system can be achieved. To keep the grid stable, there should be integration between real-time communications from the field to the control room and control function. Instant feedback is possible *via* interaction in the system between the customer and the controller. Thus, one of the main characteristics of smart grids is optimisation of the generation, transmission, distribution and consumption of solar electricity. Table 1.1 shows the difference between a smart power grid and a general power grid.

1.4.4 Global Status of Solar PVs

The energy from solar PVs is a distributed form of energy. Different countries have taken policy initiatives that encourage the transition towards "green" energy. Table 1.2 shows the status of solar PVs for some countries. Most of these are emerging markets for solar PVs. Details for some of the biggest global markets for solar photovoltaics are as follows.

India: India is an energy negative country. According to the Central Electricity Authority (CEA), the total electricity generation (utilities and non-utilities) in the country is 1,433.4 TWh.[116] Under the National Solar Mission,

Table 1.1 Difference between smart grid and general power grid.

Smart grid	General power grid
Operation with digitisation	Mechanisation operation
Communications in two way	One-way communication
Distributed power generation	Centralised power generation
Network topology	Radial topology
Lots of sensors and monitors	Number of sensors is low
Automatic monitoring	No automatic monitoring
Semi-automatic and automatic recovery of fault	Manual recovery
Preventive protection measures	Pay attention to failures and disruptions
Remote supervisory controlling equipment	Manual checking equipment
Decision support system and reliable prediction	Handling emergencies through staff and telephone
Control system with lots of options	Finite control
User option is high	Few user options

the target for Grid Connected Solar Power Projects is 100 000 MW by the year 2021–22. The target is distributed as 60 GW of large- and medium-scale grid connected and 40 GW of solar roof-top projects. According to a study by the National Institute of Solar Energy, India, the total potential of the country is 768 GW. As of 31 December 2018, solar installed capacity in India is more than 26 GW. India's roof-top solar market has expanded significantly in recent years. The Ministry of New and Renewable Energy, Government of India, has given individual targets to each state to set power plants. Tamil Nadu, Andhra Pradesh and Telangana have emerged as the fastest growing states in terms of installation of PVs.[117] One of the major problems in India's solar sector is congestion in the grid and curtailment. To resolve this problem, India is trying to develop a green energy corridor to carry power from solar-rich areas to the centralised grid. The Government of India launched a scheme named Ujwal DISCOM Assurance Yojana for financial and operational reform of the distribution company (DISCOM). Under this scheme, state governments will take over 75% of their debt and pay back lenders by selling bonds. The Government of India has sanctioned financial support of $1.2 billion for the development of a 40 000 MW solar park by the year 2020. Emerging markets continue to drive growth and the Indian PV market is growing rapidly.[115] According to a report by Bridge to India, a consulting and knowledge service provider in the Indian renewable market, the module price has declined by 29% year on year from 2015 to 2017.[115] Currently, the price of a module is $0.33 per Wp, and it continues to decline. However, it is essential to check the reliability of PV modules with decreasing cost. The price of inverters has also declined by 21% from 2015 to 2017. Currently, the typical price of an inverter is around $0.029 per Wp in the Indian market. The typical cost of the battery per ampere hour (Ah) rises from $0.17 to $0.23 based on the size of the Ah. Indian domestic modules tend to be 10–20% more expensive than Chinese modules. According to a MERCOM report, in

Table 1.2 Status of solar PV installation and policy for some countries.[113,114]

Country	Installed capacity (year)	Target announced (year)	Remarks about solar policy
Austria	1.08 GW (2016)	30 million euro additional budget for 2018–19	FIT, subsidy programme
Belgium	3.4 GW (2016)	3,544 GWh by 2020	Regional regulations, net metering scheme
Denmark	900 MW (2017)	3.4 GW by 2030	Subsidy, suspended net metering scheme
France	7.1 GW (2016)	20.2 GW by 2023	Smart meters, FIT, tender for system above 100 kWp
Greece	2445 MW (2017)	40% of total energy consumption from renewable energy	FIT for system above 100 kWp, net metering
Hungary	288 MW (2016)	14.65% of total energy consumption from renewable energy	FIT
Italy	19.3 GW (2016)	30 GW for the 2020–2030 period.	Self-consumption Scheme, tax break
Netherlands	2040 MW (2016)	20 GW by 2035	Net metering scheme, SDE+ programme
Spain	7.13 GW (2017)	8367 MW by 2020.	Auction
Switzerland	1.65 GW (2016)		Investment subsidy, elf-consumption, FIT, tax incentives
UK	11.7 GW (2016)	20 GW by 2020	Generation tariff, export tariff, renewable obligation certificate scheme
Australia	5.9 GW (2017)	33 GW renewable energy by 2020	FIT, subsidy
Israel	900 MW (2016)	1.6 GW of rooftop PV capacity by 2020	FIT, tendering
Malaysia	389 MW (2017)	1.25 GW being installed by 2020	FIT, net metering
Pakistan	1 GW (2017)	9.7 GW renewable energy by 2030	FIT, net metering
Philippines	0.9 GW (2017)	15.4 GW renewable energy by 2030	FIT, net metering
South Korea	4.5 GW (2017)	58.5 GW by 2030	Renewable portfolio standard, Korea voluntary emission reduction programme
Taiwan	1 GW (2016)	20 GW PV power by 2025	FIT, subsidy
Thailand	2.5 GW (2017)	20% of renewable energy by 2022	FIT, power development plan, New regulatory framework

(*continued*)

Table 1.2 (*continued*)

Country	Installed capacity (year)	Target announced (year)	Remarks about solar policy
Bangladesh	190 MW (2016)	1 676 MW by 2021	FIT
Indonesia	90 MW (2016)	6.4 GW by 2025	FIT, negotiation can be done between the country's state-owned power utility and independent power producers
Kazakhstan	60 MW (2016)	665 MW by 2020	FIT
Myanmar	16 MW (2016)	100% rural electrification by 2030	
Singapore	126 MW (2016)	2 GW of PV by 2025	
UAE	38 MW (2016)	21 GW of solar PV by 2030	Co-investment project, tendering
Vietnam	7 MW (2016)	12 GW by 2030	FIT, minimum efficiency requirement of PV module
Brazil	400 MW (2017)	13.1 GW by 2026	PPA projects, auctions
Canada	2.7 GW (2016)	10.7 GW of non-hydro renewable energy system by 2021	FIT, micro FIT, net metering
Mexico	420 MW (2016)	7 GW by 2020	PPA, auction
Algeria	380 MW (2016)	13.5 GW by 2030	Tender, regulatory framework
Cape Verde	12 MW (2015)	340 MW by 2020	PPA,
Egypt	76 MW (2016)		FIT, tender
Ethiopia	30 MW (2016)		Tender
South Africa	1.45 GW (2016)		Tender
Nigeria	20 MW (2016)	500 MW (2025)	PPA, tendering

2017–18 solar imports grew by 36% and exports declined by 60%, compared with 2016–17. India has also imposed an anti-dumping duty on tempered solar glass.[118]

Roof-top solar power plants are installed in India under capital expenditure route accounts and the operating expenditure model. It has been projected that 11.9 GW of new roof-top solar capacity will be added by 2021.[115] Currently, India has adopted a net metering system. Under this scheme residential, institutional and government consumers can install power plants under a subsidy scheme and sell the electricity to the state electricity board. The Ministry of New and Renewable Energy has sanctioned up to a 90% subsidy for solar roof-top power plants. Figure 1.26 shows a ground-mounted PV system of 1 kWp size. It is an off-grid system with a battery to meet the energy requirement of a rural household in India.

An IRENA report on storage estimates that India has a stationary power storage capacity equal to 7 GW. Of this installed capacity 110 MW is in the form of electrochemical batteries.[119] The solar water pump is a major technology in India, which is used for drinking water and irrigation purposes. Solar-based lanterns are a common decentralised application of solar PVs. In India more than 30% of people do not have access to a supply of electricity. To address this, the Government of India launched the Million Soul Urja Lamp Program. This is an effort to take solar lamps to rural India to address the primary energy needs of the residents, through installation and skill development of local people.

In India, solar PV power plants are monitored using the supervisory control and data acquisition (SCADA) system.[120] Smart grids are a new concept for controlling a centralised grid, which have yet to be widely adopted. For supply and demand management of electricity, the smart

Figure 1.26 An off-grid, ground-mounted PV system of 1 kWp size, with a battery to meet the energy requirement of a rural household in India.

grid concept is currently only applied on a small scale and further development is required before smart grids can be used to control the centralised grid. For example, so far there no defined standards and guidelines exist for the regulation of smart grid initiatives in India, although work to standardise continues.[117] The current policy and regulatory frameworks are typically designed to deal with the existing networks and utilities. India is developing a green corridor to distribute electricity from regions with high solar potential and, here, the smart grid concept will be a working agent.

China: China is the biggest PV market, globally. The Global Status Report 2017 estimates the total solar PV capacity to be 77.4 GW, far more than any other country.[113] In July 2017, China announced a target of 240 GW of solar installations all over the country by 2020. The guide published by the National Energy Administration predicted an installation of 86.5 GW, consisting of 54.5 GW of ground-mounted systems and 32 GW of "top runner programme" installations, and 45 GW of PV capacities foreseen under the Poverty Alleviation Programme of the 13th Five Year Plan during 2017 to 2020.[121] As per the data of 2016 in China, Xinjiang province (3.3 GW) was the top market in China, followed by Shandong (3.2 GW) and Henan (2.4 GW) provinces. However, the rapid increase in solar PV capacity has created grid congestion problems. To address these problems, China has set minimum guaranteed utilisation hours (purchase requirements) for solar and other renewable power plants and has built high-voltage transmission lines. The average cost of a PV module is $0.28–0.40 per watt.[122] According to IRENA, the installation cost of a PV system in China declined by up to 71% from 2010 to 2017. Although the PV industry of China is booming, it faces three pressing issues: solar curtailment, subsidy arrears and difficulties in financing. The Chinese government is addressing these through policy to incentivise greener energy.

At present, China is the largest market for smart grid technology. The two main reasons are: first, China is committed to green development and, second, its rapid growth in the smart grid sector. In China, automated measuring systems are widely used and the industry has acquired practical knowledge in real-time monitoring; for example, through SCADA, remote monitoring and load management. There are examples of different pilot projects in China that have successfully applied smart grids; however, four major barriers to the widespread adoption of smart grids have been identified by Yanshan *et al.*[123] These barriers are as follows.

1. Unified goals and emphasis on smart grid development have not been established.
2. Old systems need to be upgraded to satisfy the demand for information, automation and interoperability in the smart grid.
3. Power operating mechanisms need to improve.
4. Standards and business models of smart grids have to be improved or established in some areas.

USA: The USA is one of the top markets for PVs in the world. In 2017, the total installed capacity was more than 47 GW, which supplies more than 1% of electricity demand in the USA. The Department of Energy (DoE), USA, launched the SunShot Initiative in 2011 to achieve a cost target of $0.06 per kWh for utility-scale PV solar power by 2020.[124] Subsequently, in 2016, the DoE announced further targets for the cost of electricity of $0.03 per kWh from utility-scale PV, $0.05 per kWh from the residential sector and $0.04 per kWh from the commercial sector, by 2030. In 2016 NREL reported the benchmark cost of a string inverter at $2.78 per W, the DC power optimiser cost at $2.94 per W and the cost of a micro inverter at $3.28 per W. The average cost of a battery is $471 per kWh.[124]

The top ten states with the highest percentage of market share in the USA are: California, Utah, Georgia, Nevada, North Carolina, Texas, Arizona, Massachusetts, Florida and Colorado. Many states apply federal programmes and policies to attract customers for PVs. The PV utility projects are mainly based on power purchase agreements. Financial incentives, such as tax credits and direct legislative mandates are used to encourage the development of markets.[124,125] To achieve the target set by Sunshot, the following approach has been applied: improvements in module price and efficiency, non-module hardware and soft costs, system reliability, and operations and maintenance costs. The DoE has invested in strategic partnerships to accelerate investments in grid modernisation. They are supporting research on synchrophasors, advanced grid modelling and energy storage. In 2016, the DoE launched the Grid Modernisation Initiative (GMI), a comprehensive effort to help shape the future of the grid with primary funding support coming from the Office of Electricity Delivery and Energy Reliability and the Office of Energy Efficiency and Renewable Energy. The DoE has invested $4.5 billion to help modernise the nation's energy infrastructure.[124,125] The technologies and targets are as follows.

1. Integration of centralised and distributed renewable resources and other new technologies.
2. Meeting customer demands for power with high reliability.
3. Participation of customers in electricity markets and to manage their consumption.
4. Infrastructure required to provide for convergence of information and communication technologies with electricity control systems.
5. Provide new and better grid capabilities and reduce costs.

Japan: The PV market in Japan was the third largest in 2016. By 2030, solar power is expected to meet 7% of the country's electricity requirements. The Japanese government set a target of 28 GW for PVs by 2020 and 53 GW by 2030. However, the Japanese Photovoltaic Energy Association has targeted 200 GW of installed PVs by 2050. The total installed capacity of PVs in Japan was estimated to be 50 GW in 2017.[112,113] To expand solar PVs in the country, the Japanese government uses subsidies and feed-in-tariff policies.

In 2017 the government launched a tender scheme for solar PVs. They have awarded 140 MW to different companies to install solar PV plants. The lowest bid is $0.16 per kWh; however, the FIT rate is $0.19 per kWh. The cost of modules manufactured by Japanese companies is around $1.34 per W. The total installation cost for Japan is $3.2 per watt for 10–50 kWp systems, $2.86 per watt for 50–500 kWp systems, $2.60 per watt for 500–2000 kWp systems (Wp = peak watt).[113]

Issues related to the solar PV sector in Japan include:

1. The high cost of modules produced by Japanese manufacturers, due to the low production scale and the high labour costs.
2. Construction and other expenses are high; demand is not sufficient in the country to make it more cost effective.

The large volume of solar PV projects and their output has begun challenging Japan's fragmented electric power grid, which requires control though a centralised grid. To solve this problem, the government has revised regulations, which has lead some utilities to refuse new interconnections and to curtail output from existing plants without compensation.[128] Globally, Japan is one of the top markets for smart grids. The Japanese government has introduced supportive policies for the development of energy efficiency, smart grid and micro grid sectors, and the renewable energy sector is rapidly growing. There are some opportunities in terms of smart grids.

1. Integration of distributed sources of energy into Japan's grid and distribution management.
2. Consumer engagement for meeting energy efficiency.
3. Grid reliability in outage systems and micro grids.

Germany: Germany is the second top PV installer after China. The total installed PV capacity was 43.41 GW in 2018. Germany plans to meet most or all of its energy demand with renewables by 2050. To achieve this, an average of 4–5 GW of PV power plants must be installed every year, reaching a total capacity of 150–200 GW by 2050. To promote electricity generation from distributed sources of energy, the German government passed a series of laws known as the Renewable Energy Sources Act and launched the first feed-in-tariff (FIT) programme in the world. The PV electricity produced in MW power plants starts at $0.046–0.057 per kWh and the module cost is $0.057 per W. Depending on the system size, the feed-in-tariff for small roof systems put into operation by April 2018 can be up to $0.14 per kWh. For medium-size systems, from 750 kW up to 10 MW, the feed-in-tariff is $0.046 per kWh.[126]

Germany wants to develop their grid infrastructure to make it smarter. It has been forecast that Germany's investment on smart grid infrastructure will grow to $23.6 billion between 2016 and 2026.[126] The German government has combined smart grid activities under the auspices of the national government initiative "E-energy – IT-based energy systems of the future". They have launched a funding programme "Smart Energy Show-cases - Digital Agenda for the Energy Transition" (SINTEG) for developing and demonstrating new approaches to safeguarding secure grid operation in model regions with high shares of intermittent power generation on the basis of renewable energy.

1.5 Amorphous Silicon and Thin-film Technology

Thin-film photovoltaics make up approximately 10% of the global PV market. The thickness of these devices ranges from a few nanometres to tens of micrometres, which is about 10 to 100 times thinner than silicon solar cells. The three most widely commercialised thin-film solar cells are based on amorphous silicon (a-Si), CdTe or CIGS. CdTe and CIGS are described in more detail in Chapter 2.

Amorphous silicon (a-Si), CdTe and CIGS behave as direct band gap semiconductors with very high optical absorption coefficients (α).

$$\alpha_{\text{direct}} = A(h\nu - E_{\text{g}})^{1/2} \tag{1.37}$$

Where, A is a constant, h is Planck's constant, ν is the frequency of absorbed light at the onset of absorption ($h\nu$ is the energy of the photon at frequency ν) and E_{g} is the band gap.

$$A = \frac{q^2 x_{\text{VC}}^2 (2m_r)^{3/2}}{\lambda_0 \varepsilon_0 \hbar^3 n} \tag{1.38}$$

$$m_r = \frac{m_h^* m_e^*}{m_h^* + m_e^*} \tag{1.39}$$

Where, m^*_e and m^*_h are the effective masses of the electron and hole, respectively, m_r is the reduced mass, q is the elementary charge, n is the refractive index, ε_0 is the vacuum permittivity and x_{vc} is a matrix element.

Silicon behaves as an indirect band gap semiconductor, where the absorption coefficient is related to the band gap as follows:

$$\alpha \propto \frac{(h\nu - E_g + E_p)^2}{\exp\left(\frac{E_p}{kT}\right) - 1} + \frac{(h\nu - E_g + E_p)^2}{1 - \exp\left(\frac{E_p}{kT}\right)} \tag{1.40}$$

Moreover, silicon has a very high refractive index (>3.4) near the band gap edge and absorbs light inefficiently.

Direct transitions have a high probability (large value of α) because the momentum of the electron in the conduction band is the same as the hole in the valence band. Indirect transitions are less probable (low value of α) because the momentum of the electron in the conduction band is not the same as the hole in the valence band and the absorption of a phonon as well as a photon is required. The absorption coefficient of an absorber layer is inversely proportional to its thickness and, since silicon has a very low absorption coefficient, it requires cells thicker than 10 mm.[129] For this reason, the same thickness of a-Si can absorb much more light than its crystalline counterpart.

There are several advantages of thin-film solar cells over the thicker solar cells. The amount of semiconductor material used for thin-film fabrication and the power consumed during this process is low, making it more feasible for fabrication of large-area devices at a faster processing rate and better cost efficiency (low material consumption, processing, handling and capital costs). Although SiO_2 is abundant, the production of silicon wafers is an energy intensive and laborious process. A large amount of energy is consumed in manufacturing silicon solar cells, so it takes a long time to get the energy back from the silicon panel that was put in to making it. For instance, crystalline silicon PV systems have an energy payback time of *ca.* 1.5–2 years for south European locations and of *ca.* 2.7–3.5 years for middle European locations.[128] In contrast, thin-film solar cells are usually grown using vacuum-based deposition techniques, including CVD, and physical vapour deposition techniques such as thermal evaporation, e-beam evaporation, sputtering, vapour transfer deposition, closed space sublimation, pulsed laser deposition, atomic layer deposition and molecular beam epitaxy. Some of these techniques are very expensive but there is a trade-off between the quality of the thin film and the time taken to deposit it. Solution-based deposition techniques, such as spin coating, electro-deposition, screen printing, layer-by-layer deposition, spray pyrolysis and colloidal synthesis are also used to deposit thin films. However, thin films deposited by the vapour deposition technique are more viable, due to uniform surface coverage, high crystallinity and high purity and, hence, they tend to improve device reproducibility and stability.

The energy payback time for CIGS-based thin-film PV modules produced in China (1700 kW h m^{-2} solar irradiation per year) is significantly lower than that of conventional silicon modules and is about 0.78 years.[130,131] Another major advantage is that the material for thin-film solar cells can be deposited on flexible substrates, resulting in the potential for roll-to-roll deposition. Some of these materials have better radiation hardness than silicon and resist degradation of device performance, making thin-film PVs applicable in space applications.

The choice of deposition technique depends on the specific material system. Thin-film compounds are typically binary, ternary, quaternary or multinary semiconductors, a few of which have received lots of attention for PV applications, but have not been economically successful for large-scale commercial production. The common issue with these types of materials, with increasing numbers of components, is the complexity of phase stability and achieving a perfect stoichiometry. For example, this is clearly observed in one of the most promising quaternary semiconductor materials, Cu_2ZnSnS_4, for which there is a very small chemical potential region in which the desired stoichiometric phase forms.[133,134]

The structural, electrical and optical properties of the thin films grown, strongly depends on various deposition conditions, such as growth temperature and other physical parameters. By monitoring and controlling the growth processes or deposition conditions, reproducible thin films, with similar characteristics can be developed. Both an improvement in stability and in band gap tuning is possible by materials engineering and optimising the processing techniques. For example, a-Si is usually grown by plasma-enhanced CVD at low temperatures (200–500 °C) and, depending on the deposition conditions, a range of a-Si materials is grown with different charge carrier mobilities between 0.001 and 0.1 $cm^2 V^{-1} s^{-1}$. As deposited, a-Si has a band gap of 1.7 eV and this can be tuned between 1.3 eV and 2.0 eV by introducing GeH_4 or CH_4 during deposition. The film can be made n- or p-type by adding B_2H_6 or PH_3.

Another advantage of thin-film solar cells, including a-Si, over c-Si solar cells is their better temperature coefficients (the loss of power at higher operating temperatures). Figure 1.27 refers to the effects of temperature on the fill factor of CIGS, CdTe, thin-film Si, α-Si, and a-Si/micro-Si. With the exception of a-Si, all the other module technologies exhibit a linear behaviour with increasing temperature.[132, 135]

The efficiency of CdTe and CIGS have exceeded 22% (see Figure 1.7), and most of the increase in cell efficiency comes from generating high current. Ideally, thinner devices with a proper light-management technique will tend to have a higher concentration of charge collected in the thin layer of the cell and, hence, have a higher charge carrier density. This results in a higher separation of the quasi-Fermi energies (a shift in Fermi energy levels towards the band edges, due to enhanced carrier densities under illumination) resulting in higher output voltage for the device. The

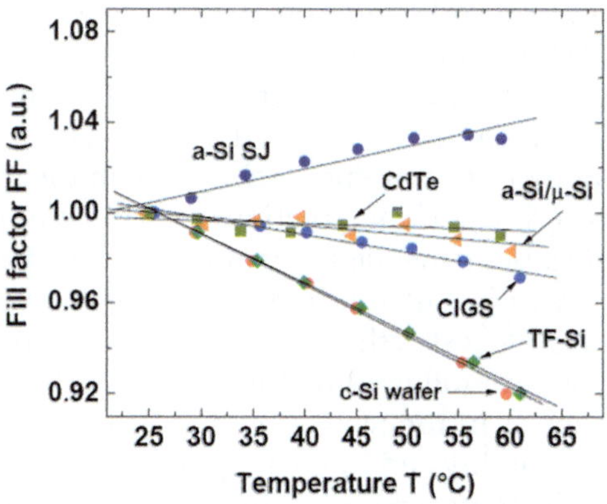

Figure 1.27 Experimentally determined comparison of the fill factor of different module technologies as a function of temperature. Reproduced from ref. 118 with permission from EU PVSEC.

present drive is to increase the output voltage of the cell to increase the device efficiency. Although the stability of industrial thin-film solar cells has improved over the years, their degradation with time due to the moisture present in the atmosphere still remains sensitive to factors such as the quality of encapsulation. Using toxic materials like cadmium in the primary steps remains less of a concern as cadmium is present in stable forms in the final device, such as CdS and CdTe, which are insoluble in water. The material constrained growth of installed capacity in the year 2020 is estimated at about 20 GWp per year for CdTe, 70 GWp per year for CIGS and 200 GWp per year for a-Si : Ge.[132]

The efficiency of a-Si thin-film solar cells has reached 14%, which is lower than c-Si because of lower carrier mobilities and internal losses due to the disordered structure. Furthermore, a-Si devices suffer from light-induced degradation. Overcoming the light-induced metastable defects (or Staebler–Wronski defects) in a-Si thin-film solar cells, and their effect on the degradation of material performance under sunlight with time, has been the focus of substantial research effort. These defects affect the device architecture by limiting the intrinsic layer thickness. The thinner the intrinsic layer, the lower the carrier recombination losses and, hence, the lower the loss in the device performance.[132,135–137] This may mean that not all of the incident light is collected, so multi-junction device architectures are being developed.

Over time, the cost of solar modules has decreased due to increasing efficiency, improving technology, economies of scale and oversupply of modules. However, the total cost of solar power is not the cost of modules alone, but also includes the balance of system (BOS) costs. BOS is all the upfront costs associated with a photovoltaic system, excluding the cost of the module. Since there is less potential for reducing the BOS, the key is to develop technologies that would further enhance the efficiency while keeping the cost of manufacturing low. In multi-junction devices, different layers are stacked preferentially to absorb light from different regions of the solar spectrum to exceed the SQ limit. By choosing materials with very similar lattice constants but having different energy gaps, defect-free interfaces could be achieved and interface recombination could be reduced.

The global development of thin-film technology in the current market is due to the expanding growth in construction, utility and building integrated PV markets and applications that demand more flexible, lighter weight solar modules that are easy to install and have low manufacturing costs. Of the three candidates, CdTe possesses the largest market share (representing ~5% of the total world market), followed by CIGS and a-Si. Progress in research and development, such as introducing more robust materials and cell architectures to improve reliability and by creating new materials to decrease the dependence on rare elements, may allow further market growth of thin-film technologies.[132] Key developments in these areas will be explored in the following chapters.

References

1. http://www.rsc.org/education/teachers/resources/databook/data/print/electromagnetic_spectrum.pdf, accessed: April 18, 2019.
2. J. A. Herron, J. Kim, A. A. Upadhye, G. W. Huber and C. T. Maravelias, *Energy Environ. Sci.*, 2015, **8**, 126.
3. K. E. Trenberth, J. T. Fasullo and J. Kiehl, Earth's global energy budget, *Bull. Am. Meteorol. Soc.*, 2009, 311.
4. R. Lindsey, Climate and Earth's Energy Budget, United States Geological Services (USGS) fact sheet, Image retrieved from, https://Earthobservatory.nasa.gov/Features/EnergyBalance/printall.php.
5. M. A. Green, *Sol. Energy Mater. Sol.*, 2008, **92**, 1305.
6. W. P. Hirshman, G. Hering and M. Schmela, Market survey on cell and module production 2006, *Photon International*, 2007, p. 136.
7. A. G. Aberle, B. B. Mathew, B. Hoex and T. Mueller, *Green*, 2012, **2**, 135.
8. R. M. Izatt, S. R. Izatt, R. L. Bruening, N. E. Izatt and B. A. Moy, *Chem. Soc. Rev.*, 2014, **43**, 2451.
9. M. A. Green, *High Efficiency Silicon Solar Cells*, Trans Tech Publications, Switzerland, 1987.
10. A. R. Whitson and H. L. Walster, *Soils and Soil Fertility*, St. Paul, MN, Webb, 1912, p. 73, OCLC 1593332.
11. S. Gluntz, *Adv. OptoElectron.*, 2007, 97370.
12. T. Saga, *NPG Asia Mater.*, 2010, **2**, 96.
13. NREL's Research Record Efficiency Chart, 2018, https://www.nrel.gov/pv/assets/images/efficiency-chart.png, accessed: June 10, 2019.
14. M. T. Castaner, *Solar Cells Materials, Manufacture and Operation*, Elsevier, Oxford, UK, 2005.
15. M. A. Green, Y. Hishikawa, W. Warta, E. D. Dunlop, D. H. Levi, J. Hohl-Ebinger and A. W. Y. Ho-Baillie, *Progr. Photovoltaics Res. Appl.*, 2017, **25**, 668.
16. M. A. Green and P. B. Stephen, *Nat. Mater.*, 2017, **16**, 23.
17. R. M. Swanson, *Progr. Photovoltaics Res. Appl.*, 2006, **14**, 443.
18. U.S. Geological Survey, Mineral Commodity Summaries, January 2017.
19. B. Gao, S. Nakano and K. Kakimoto, *J. Cryst. Growth*, 2011, **314**, 239.
20. A. F. B. Braga, S. P. Moreira, P. R. Zampieri, J. M. G. Bacchin and P. R. Mei, *Sol. Energy Mater. Sol. Cells*, 2008, **92**, 418.
21. P. A. Basore, *Progr. Photovoltaics Res. Appl.*, 1994, **2**, 177.
22. R. Falster, D. Gambaro, M. Olmo, M. Cornara and H. Korb, *Proc. Mater. Res. Soc. Symp.*, 1998, **510**, 27.
23. G. Fisher, M. R. Seacrist and R. W. Standle, *Proc. IEEE*, 2012, **100**, 1454.
24. S. Woo, M. Bertoni, K. Choi, S. Nam, S. Castellanos, D. M. Powell, T. Buonassisi and H. Choi, *Sol. Energy Mater. Sol.*, 2016, **155**, 88.

25. M. Trempa, I. Kupka, C. Kranert, C. Reimann and J. Friedrich, Fraunhofer IISB, Fraunhofer THM, *Photovoltaics Int.*, 2017, **35**.
26. H. E. LaBelle, *J. Cryst. Growth*, 1980, **50**, 8.
27. J. M. Serra, J. Maia Alves and A. M. Vallera, *J. Cryst. Growth*, 2017, **468**, 590.
28. D. H. Neuhaus and A. Münzer, *Adv. Optoelectron.*, 2007, 24521.
29. D. L. King and M. E. Buck, *Proceedings of the 22nd IEEE Photovoltaic Specialists Conference (PVSC '91)*, Las Vegas, Nev, USA, October 1991, vol. 1, p. 303.
30. E. Yablonovitch and F. D. Cody, *IEEE Trans. Electron Devices*, 1982, **29**, 300.
31. M. A. Green, *Solar Cells – Operating Principles, Technology and System Applications*, Prentice Hall Inc., 1982.
32. http://www.pveducation.org, accessed: June 08, 2018.
33. L. K. Wagner and J. C. Grossman, *Phys. Rev. Lett.*, 2008, **101**, 265501.
34. A. G. Aberle, S. W. Glunz, A. W. Stephens and M. A. Green, *Prog. Photovoltaics Res. Appl.*, 1994, **2**, 265.
35. S. K. Estreicher, M. Sanati and G. N. Szwacki, *Phys. Rev. Lett.*, 2008, **77**, 125214.
36. M. Sanati, G. N. Szwacki and S. K. Estreicher, *Phys. Rev. B*, 2007, **76**, 125204.
37. M. Sanati, G. N. Szwacki and S. K. Estreicher, *Phys. B*, 2007, **401**, 105.
38. R. A. Sinton and R. M. Swanson, *IEEE Trans. Electron Devices*, 1987, **34**, 1380.
39. P. P. Altermatt, R. A. Sinton and G. Heiser, *Sol. Energy Mater. Sol. Cells*, 2001, **65**, 149.
40. J. Schmidt, M. Kerr and P. P. Altermatt, *J. Appl. Phys.*, 2000, **88**, 1494.
41. S. Steingrube, O. Breitenstein, K. Ramspeck, S. Glunz, A. Schenk and P. Pietro, *J. Appl. Phys.*, 2011, **110**, 014515.
42. K. Yoshikawa, H. Kawasaki, W. Yoshida, T. Irie, K. Konishi, K. Nakano, T. Uto, D. Adachi, M. Kanematsu, H. Uzu and K. Yamamoto, *Nat. Energy*, 2017, **2**, 17032.
43. M. Müller, G. Fischer, B. Bitnar, S. Steckemetz, R. Schiepe, M. Mühlbauer, R. Köhler, P. Richter, C. Kusterer, A. Oehlke, E. Schneiderlöchner, H. Sträter, F. Wolny, M. Wagner, P. Palinginis and D. H. Neuhaus, *Energy Procedia*, 2017, **124**, 131.
44. G. Schubert, J. Horzel, R. Kopecek, F. Huster and P. Fath, *Proceedings of the 20th European Photovoltaic Solar Energy Conference (EU PVSEC '05)*, 2005, p. 934.
45. M. M. Hilali, A. Rohatgi and B. To, *Proceedings of the 14th Workshop on Crystalline Silicon Solar Cells and Modules*, Winter Park, Colo, USA, 2004.
46. B. Fischer, *Loss Analysis of Crystalline Silicon Solar Cells Using Photoconductance and Quantum Efficiency Measurements*, PhD thesis, Cuvillier, University of Gottingen, Germany, 2003.
47. Y. Yang, M. A. Green, A. Ho-Baillie, H. Kampwerth, S. Pillai and H. Mehrvarz, *Sol. Energy Mater. Sol. Cells*, 2013, **115**, 42.

48. K. Dasgupta, S. Ray, A. Mondal and U. Gangopadhyay, *Mater. Today: Proc.*, 2017, **4**, 12698.
49. E. Manea, E. Budianu, M. Purica, I. Cernica and F. Babarada, *Sol. Energy Mater. Sol. Cells*, 2006, **90**, 2312.
50. M. F. Abdullah, M. A. Alghoul, H. Naser, N. Asim, S. Ahmadi, B. Yatim and K. Sopian, *Renewable Sustainable Energy Rev.*, 2016, **66**, 380.
51. H. Saha, S. K. Datta, K. Mukhopadhyay and S. Banerjee, *IEEE Trans. Electron Devices*, 1992, **39**, 5.
52. U. Gangopadhyay, K. Kim, A. Kandol, J. Yi and H. Saha, *Sol. Energy Mater. Sol. Cells*, 2006, **90**, 3094.
53. M. Moynihan, C. O. Connor, B. Barr, S. Tiffany, W. Braun. G. Allardyce, J. Rentsch and K. Birnann, *35th IEEE Photovoltaic Specialists Conference (PVSC)*, 2010.
54. I. Zubel and M. Kramkowska, *Sens. Actuators, A*, 2001, **93**, 138.
55. L. Pirozzi, G. Arabito, F. Artuso, V. Barbarossa, U. Besi-Vetrella, S. Loreti, P. Mangiapane and E. Salza, *Sol. Energy Mater. Sol. Cells*, 2001, **65**, 287.
56. P. Fath, P. Borst, C. Zechner, E. Bucher, G. Willeke and S. Nayaranan, *Sol. Energy Mater. Sol. Cells*, 1997, **48**, 229.
57. C. Gerhards, C. Marckmann, R. Tolle, M. Spiegel, P. Fath, G. Willeke and E. Bucher, *IEEE Photovoltaic Spec. Conf.*, 1997, 43.
58. B. Terheiden and P. Fath, *World Conference Photovoltaic Energy Conversion*, 2003, p. 1443.
59. M. A. Green, *Sol. Energy*, 2003, **74**, 181.
60. R. C. Jaeger, *Introduction to Microelectronic Fabrication*, Addison-Wesley Publishing Company, New York, 1993, vol. V.
61. C. Voyer, D. Biro, K. Wagner, J. Benick and R. Preu, *Proceedings of the, 21st European Photovoltaic Solar Energy Conference (EU PVSEC '06)*, 2006.
62. H. Hieslmair, I. Latchford, L. Mandrell, M. Chun and B. Adibi, *Ion implantation for silicon solar cells*, Intevac, Santa Clara, California, USA, 2012, http://www.pv-tech.org.
63. M. Fischer, A. Metz and S. Raithel, *Proceedings 27th European Photovoltaic Solar Energy Conference (EU PVSEC)*, Frankfurt, Germany, 2012.
64. C. Dub, *Energy Procedia*, 2011, **8**, 706.
65. M. Jeon, *Mater. Sci. Eng., B*, 2011, **176**, 1285.
66. I. M. Hamammu and K. Ibrahim, *Proceedings of IEEE International Conference on Semiconductor Electronics*, Penang, Malaysia, 2002, pp. 230–232.
67. M. D. Abbott, T. Trupke, H. P. Hartmann, R. Gupta and O. Breitenstein Prog, *Photovoltaics Res. Appl.*, 2007, **15**, 613.
68. A. Hauser, G. Hahn, M. Spiegel, H. Feist, O. Breitenstein, J. P. Rakotoniaina, P. Fath and E. Bucher, *Proceedings of the 17th European Photovoltaic Solar Energy Conference*, 2001, p. 1739.
69. A. Otaegi, V. Fano, M. A. Rasool, J. R. Gutiérrez, J. C. Jimeno, N. Azkona and E. Cereceda, *Sol. Energy*, 2017, **155**, 847.

70. I. Melnyk, E. Wefringhaus, F. Delahaye, G. Vilsmeier, W. Mahler and P. Fath, *Proceedings of the 19th European Photovoltaic Solar Energy Conference (EU PVSEC '04)*, Paris, France, 2004, p. 416.
71. G. Emanuel, E. Schneiderlöchner, J. Stollhof, J. Gentischer, R. Preu and R. Lüdemann, *Proceedings of the 17th European Photovoltaic Solar Energy Conference (EU PVSEC '01)*, Munich, Germany, October 2001, p. 1578.
72. E. Schneiderlöchner, A. Grohe, S. W. Glunz, R. Preu and W. Willeke, *Proceedings of the 3rd World Conference on Photovoltaic Energy Conversion (WCPEC '03)*, Osaka, Japan, May 2003, vol. 2, p. 1364.
73. E. Cornagliotti, M. Ngamo, L. Tous, R. Russell, J. Horzel, D. Hendrickx, B. Douhard, V. Prajapati, T. Janssens and J. Poortman, *Energy Procedia*, 2012, **27**, 624.
74. J. Horzel, A. Lorenz, E. Cornagliotti, A. Uruena, J. John, M. Izaaryene, D. Habermann, P. Jaffrennou and J. Penaud, *26th European Photovoltaic Solar Energy Conference*, DOI: 10.4229/26thEUPVSEC2011-2CV.4.25.
75. (a) İ. G. Kavakli and K. Kantarli, *Turk. J. Phys.*, 2002, **26**, 349; (b) H. K. Raut, V. A. Ganesh, A. S. Nair and S. Ramakrishna, *Energy Environ. Sci.*, 2011, **4**, 3779.
76. M. A. Green and M. J. Keevers, *Prog. Photovoltaics*, 1995, **3**, 189.
77. G. Shao, C. Lou and D. Xiao, *J. Lumin.*, 2015, **157**, 344.
78. B. S. Richards, *Prog. Photovoltaics Res. Appl.*, 2004, **12**, 253.
79. N. S. Beattie, P. See, G. Zoppi, P. M. Ushasree, M. Duchamp, I. Farrer, D. A. Ritchie and S. Tomic, *ACS Photonics*, 2017, **4**, 2745.
80. F. Duerickx and J. Szlufcik, *Sol. Energy Mater. Sol. Cells*, 2002, **72**, 231.
81. A. Rohatgi, P. Doshi, J. Moschner, T. Lauinger, A. G. Aberle and D. S. Ruby, *IEEE Trans. Electron Devices*, 2000, **47**, 987.
82. P. E. Gruenbaum, J. Y. Gan, R. R. King and R. M. Swanson, *Proceedings of the 21st IEEE Photovoltaic Specialists Conference*, Kissimimee, 1990, p. 317.
83. A. W. Stephens, A. G. Aberle and M. A. Green, *J. Appl. Phys.*, 1993, **76**, 363.
84. S. W. Glunz, A. B. Sproul, W. Warta and W. Wettling, *J. Appl. Phys.*, 1994, **75**, 1611.
85. P. E. Gruenbaum, R. A. Sinton and R. M. Swanson, *Proceedings of the 19th IEEE Photovoltaic Specialists Conference*, Las Vegas, 1988, p. 423.
86. A. G. Aberle, *Crystalline Silicon Solar Cells: Advanced Surface Passivation and Analysis*, Centre for Photovoltaic Engineering, University of NSW, Australia, 1999.
87. A. G. Aberle, *Prog. Photovoltaics Res. Appl.*, 2000, **8**, 473.
88. G. Aberle, *Sol. Energy Mater. Sol.*, 2001, **65**, 239.
89. W. Soppe, H. Rieffe and A. Weeber, *Prog. Photovoltaics*, 2005, **13**, 551.
90. Z. Chen, A. Rohatgi, R. O. Bell and J. P. Kalejs, *Appl. Phys. Lett.*, 1994, **65**, 2078.

91. J. Hong, W. M. M. Kessels, W. J. Soppe, A. W. Weeber, W. M. Arnoldbik and M. C. M. Van De Sanden, *J. Vac. Sci. Technol., B: Microelectron. Nanometer Struct.*, 2003, **21**, 2123.
92. K. Kimura, *Proceedings of the 1st International Photovoltaic Science and Engineering Conference*, Kobe, 1987, p. 37.
93. K. V. Ravi, R. C. Gonsiorawski and A. R. Chaudhuri, *Proceedings of the 7th IEEE Photovoltaic Specialists Conference*, Las Vegas, 1985, p. 1222.
94. F. Duerickx and J. Szlufcik, *Sol. Energy Mater. Sol.*, 2002, **72**, 231.
95. H. Nagel, A. G. Aberle and R. Hezel, *Prog. Photovoltaics*, 1999, **17**, 245.
96. L. Mittelstadt, A. Metz and R. Hezel, *Sol. Energy Mater. Sol.*, 2002, **72**, 255.
97. M. P. Hughey and R. F. Cook, *Thin Solid Films*, 2004, **460**, 7.
98. W. Wolke, A. Jäckle, R. Preu, S. Wieder and M. Ruske, *Proceedings of the 19th European Photovoltaic Solar Energy Conference (EU PVSEC '04)*, Paris, France, June 2004, p. 419.
99. W. Wolke, J. Catoir, G. Emanuel, J. Liu, M. Ruske and R. Preu, *Proceedings of the 20th European Photovoltaic Solar Energy Conference (EU PVSEC '05)*, Barcelona, Spain, June 2005, p. 733.
100. P. J. Holmes and R. G. Loasby, *Handbook of Thick Film Technology*, Electrochemical Publications, Glasgow, Scotland, UK, 1976.
101. G. Schubert, F. Huster and P. Fath, *Proceedings of the 19th European Photovoltaic Solar Energy Conference (EU PVSEC '04)*, Paris, France, 2004, p. 813.
102. Dr. Alfred Hüthig, *Hybridintegration*, ed. H. Reichl, Heidelberg, Germany, 1986.
103. C. Ballif, D. M. Huljić, A. Hessler-Wysser and G. Willeke, *Proceedings of the 29th IEEE Photovoltaic Specialists Conference (PVSC '02)*, New Orleans, La, USA, 2002, p. 360.
104. C. Ballif, D. M. Huljić, G. Willeke and A. Hessler-Wysser, *Appl. Phys. Lett.*, 2003, **82**, 1878.
105. G. Schubert, B. Fischer and P. Fath, *Proceedings of Photovoltaics in Europe Conference (PV '02)*, Rome, Italy, 2002, p. 343.
106. G. Schubert, F. Huster and P. Fath, Current transport mechanism in printed Ag thick film contacts to an n-type emitter of a crystalline silicon solar cell, in *Proceedings of the 14th International Photovoltaic Science and Engineering Conference (PVSEC '04)*, Bangkok, Thailand, January 2004, p. 441.
107. A. Ebong and N. Chen, *The 9th International Conference on High Capacity Optical Networks and Enabling Technologies (HONET)*, 2012, p. 102.
108. F. Huster, *Proceedings of the 20th European Photovoltaic Solar Energy Conference*, Barcelona, Spain, 2005, p. 560.
109. D.-H. Neuhaus, R. Mehnert and G. Erfurt, *Proceedings of the 20th European Photovoltaic Solar Energy Conference*, Barcelona, Spain, 2005, p. 1947.

110. D.-H. Neuhaus, J. Kirchner and R. Mehner, *Proceedings of the 21st European Photovoltaic Solar Energy Conference*, Dresden, Germany, 2006, p. 2556.
111. D.-H. Neuhaus, F. Dreckschmidt, and R. Lüdemann, *Proceedings of the 19th European Photovoltaic Solar Energy Conference*, Paris, France, 2004, p. 817.
112. World Energy Scenarios, 2017, ISBN: 9780946121663.
113. *Renewables 2017 Global Status Report*, REN21, 2017, http://www.ren21.net/gsr-2017/.
114. *PV Status Report*, 2017, DOI: 10.2760/452611.
115. *India Solar Handbook*, Bridge to India, 2017.
116. http://cea.nic.in/, accessed: May 01, 2018.
117. www.MNRE.gov.in, accessed: May 01, 2018.
118. www.store.mercom.mercomcapital.com/product-category/solar-reports/.
119. IRENA, *Electricity Storage and Renewables: Costs and Markets to 2030*, International Renewable Energy Agency, Abu Dhabi, 2017.
120. *Research, development, demonstration and deployment of smart grids in India, June 2017*, Department of Science and Technology, Government of India.
121. http://en.ndrc.gov.cn/policyrelease/, accessed: May 01, 2018.
122. http://www.exa-watt.com, accessed: May 01, 2018.
123. Y. Yu, J. Yang and B. Chen, *Energies*, 2012, **5**, 1321–1338.
124. https://www.energy.gov/eere/solar/sunshot-initiative, accessed: May 01, 2018.
125. K. Ardani, E. O'Shaughnessy, R. Fu, C. McClurg, J. Huneycutt and R. Margolis, *Installed Cost Benchmarks and 126. Deployment Barriers for Residential Solar Photovoltaics with Energy Storage: Q1*, NREL/TP-7A40–67474, 2016, https://www.nrel.gov.
126. Recent Facts about Photovoltaics in Germany, https://www.ise.fraunhofer.de/en/publications/studies/recent-facts-about-pv-in-germany.html.
127. Renewable Power Generation Costs in 2017, http://www.irena.org/publications.
128. E. A. Alsema, M. J. De Wild-Scholten and V. M. Fthenakis, *21st European Photovoltaic Solar Energy Conference*, Dresden, Germany, 2006.
129. K. L. Chopra, P. D. Paulson and V. Dutta, *Prog. Photovoltaics*, 2004, **12**, 69.
130. M. J. De Wild-Scholten, *Sol. Energy Mater. Sol. Cells*, 2013, **119**, 296.
131. M. Powalla, S. Paetel, D. Hariskos, R. Wuerz, F. Kessler, P. Lechner, W. Wischmann and T. M. Friedlmeier, *Engineering*, 2017, **3**, 445.
132. A. Virtuani, D. Pavanello and G. Friesen, *Proceedings of the 25th European Photovoltaic Solar Energy Conference and Exhibition/5th World Conference on Photovoltaic Energy Conversion*, Valencia, Spain, 2010, p. 4248.

133. S. B. Zhang, S.-H. Wei, A. Zunger and H. K. Yoshida, *Phys. Rev. B*, 1998, **57**, 9642.
134. S. Chen, X. G. Gong, A. Walsh and S. H. Wei, *Appl. Phys. Lett.*, 2010, **96**, 021902.
135. D. T. Lee and A. U. Ebong, *Renewable Sustainable Energy Rev.*, 2017, **70**, 1286.
136. D. L. Staebler and C. R. Wronski, *Appl. Phys. Lett.*, 1977, **31**, 292.
137. M. Gostein and L. Dunn, *Proceedings of the 37th IEEE PVSC*, Seattle, WA, USA, 2011, p. 3126.

CHAPTER 2

Compound Semiconductor Solar Cells

AMITA SINGH[a] AND AJAY SINGH*[a,b]

[a]Materials Physics and Applications Division: Center for Integrated Nanotechnologies, Los Alamos National Laboratory, Los Alamos, New Mexico 87545, USA; [b]Molecular Foundry, Lawrence Berkeley National Laboratory, 1 Cyclotron Road, Berkeley, California 94720, USA
*E-mail: ajaysingh@lbl.gov, ajay18march@gmail.com

2.1 Introduction

The current photovoltaic (PV) industry is mainly based on crystalline silicon (c-Si, >90%) wafer technology (first generation solar cells). As Si is inherently an indirect band gap absorber material with a low absorption coefficient in the red and near-infrared range, thick layers (~180–200 μm) are required to absorb the majority of incident solar radiation and to provide mechanical support. Despite Si being abundant, the high manufacturing costs related to processing the absorber material on a large scale and with sufficient purity (almost 40% of the cost of solar cell fabrication)[1] remain a key challenge to marketing solar cell technology suitable for domestic applications. One way of reducing the cost of the module is to reduce the amount of the absorber material. Hence, research and development has focused on an alternative to Si-based technology, known as thin-film solar cells (second generation solar cells).[2]

Inorganic Materials Series No. 3
Solar Energy Capture Materials
Edited by Elizabeth A. Gibson

Published by the Royal Society of Chemistry, www.rsc.org

Thin-film PV technologies were originally introduced in the 1970s but significant progress has been made only in the last decade, when not only has the fabrication technology flourished, but also thin-film manufacturers have grown and expanded. This is mainly due to two factors, (1) they minimise active material usage—the demand for less material and (2) they are easier to manufacture on various kinds of substrates, such as glass, plastic, and stainless steel, which not only lowers the overall cost and makes thin-film PVs more affordable to both homeowners and commercial consumers, but also opens up new avenues for solar building integration.

Currently, thin-film PV technologies are predominantly based on direct band gap compound semiconductor materials[3] such as copper indium gallium diselenide (CIGS), copper indium diselenide (CIS), copper zinc tin sulfide (Cu_2ZnSnS_4; CZTS), gallium arsenide, and cadmium telluride (CdTe), and they have reached the commercialisation stage with a highest reported module conversion efficiency of ~19% (in the case of CIGS).[4] The compound semiconductor is a class of materials that shows great flexibility in terms of materials engineering, as the entire gamut of optoelectronic and physical properties are compositionally tuneable. This compositional and stoichiometric versatility provides significant scope to tune the material band gap over a wide range. More significantly, a number of these compounds (from binary to quaternary) maintain a direct band gap with a high absorption coefficient and good photostability under long-term irradiation over the entire composition, which, ultimately, allows them to interact strongly with light and to radiate more efficiently.[3] However, the module efficiency for second generation technology based on compound semiconductors is likely to reach the Shockley–Queisser limit in the coming decade, due to the quantum conversion process in addition to fabrication constraints. This leads to another avenue of "third generation" PVs to be pursued such as multi-junction (tandem), dye-sensitised, quantum dot, and perovskite solar cells based on compound semiconductors. These will be discussed in Chapters 3 and 4.

In this chapter, we capture and discuss progress in terms of material properties, fabrication process, PV design and the outstanding problems associated with compound semiconductors solar cells.

2.2 Materials and Growth

2.2.1 II–VI Semiconductor Compounds

The synthesis and characterisation of binary semiconductor compound materials belonging to the group II–VI family has been a point of research for decades. The large direct band gap and high absorption ($>10^4$ cm^{-1}) in thin films have made these materials prime candidates for applications in solar cells. In particular, the success of CdTe films in photovoltaic devices can be attributed to the almost ideal band gap of 1.45 eV, the availability of various film deposition techniques (vacuum evaporation, chemical vapour deposition (CVD), electrodeposition, closed-space sublimation, and physical vapour deposition (PVD)), a choice of n- or p-type dopants, and the ability to act as a nearly perfect absorber material (*ca.* 2 μm thick film will absorb nearly 100% of incident solar radiation).[5–8] The most common thin-film CdTe-based solar cell configurations involve the formation of heterojunctions between p-type CdTe and n-type CdS. The miscibility and reaction between these two layers form an interfacial layer of $CdS_{1-x}Te_x$.[9] The high deposition temperature (>550 °C) has been assumed to be the key to the formation of this interfacial layer, which lowers the defect densities at the interface and results in high-efficiency devices. The high deposition temperature and post-deposition heat treatment of a CdTe/CdS structure in the presence of $CdCl_2$ also promotes grain growth and re-crystallisation of CdTe films and improves the material's defect passivation/carrier lifetimes. These factors, together, have accounted for the high efficiency of CdTe/CdS solar cell devices.[10] CdTe-based solar cells are generally constructed in a superstrate configuration, namely, glass substrate coated with a conductive window layer of SnO_2, then a CdS buffer layer (a thin CdS film is deposited as intermediate layers to improve the interface properties and as a chemical buffer layer to protect the CdTe during the subsequent processing), and then a CdTe light-absorbing layer (Figure 2.1).

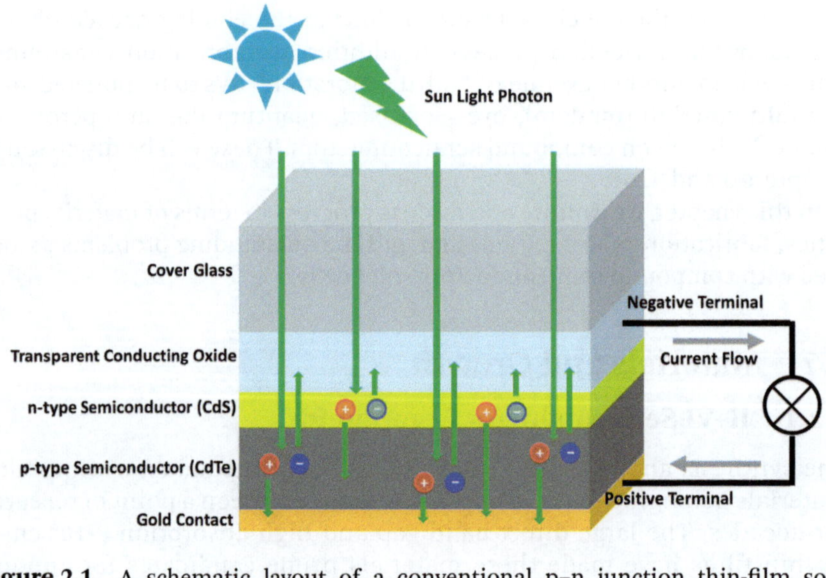

Figure 2.1 A schematic layout of a conventional p–n junction thin-film solar cell. Adapted from ref. 11 with permission from the Royal Society of Chemistry.

Even though CdTe-based solar cells using this cell structure have reached high efficiencies (16.5% for solar cells prepared by closed-space sublimation),[6,12,13] the use of CdS as a buffer/window layer is a barrier to achieving maximum performance. The band gap of CdS (2.5 eV) lies where a significant portion of the solar spectrum falls (below 515 nm). The excitons generated in this layer recombine and do not contribute to the final photocurrent. This absorption feature of CdS causes approximately 7 mA cm^{-2} of the photocurrent to be lost.[5] To solve this problem, a thin layer of CdS (less than 100 nm) has been used. This reduces the absorption and enhances the final current, but the thin layers of CdS are prone to pin-hole formation and the formation of continuous layers poses a significant challenge. These pin-holes can lead to a direct contact between the CdTe and the transparent layer (typically SnO_2) resulting in the formation of CdTe–SnO_2 microjunctions. Due to the inefficiency of these microjunctions and their ability to act as microshunts, they reduce the overall efficiency of the device. Researchers took a lateral approach to solve this problem by forming a bilayer transparent contact where a stack of transparent films (with high/low resistivity) minimised the losses occurring due to the use of thin CdS film.[6] The low-resistance contact with lateral current collection, and a thinner high-resistivity layer minimised the forward current through pin-holes and eliminated the formation of microjunctions. Different resistive oxide materials (SnO_2, In_2O_3 or ZnO) were tried for this purpose and the CdS window layer was reduced to <20 nm.[14,15]

Another point of consideration for the performance of CdTe-based solar cells is the material chosen to form the back contact. The processing of a back contact in CdTe devices has been a challenge as there are not many materials available with an appropriate work function.[16] Nickel,[15,17] HgTe, ZnTe,[18] metallic bi-layers of Cu/Au, and doped graphite have shown promising results. Another important step involved in the formation of the back contact is the cleaning and surface modification, which is typically carried out using a Br_2/methanol or nitric–phosphoric (NP) acid mixture.[19,20] The acid etching leaves the CdTe surface Te-rich, which increases the surface conductivity and aids in the formation of a tunnelling junction.[19] Copper, which is used in most of the back contacts, forms a thin layer of Cu_2Te on the surface and this interlayer appears to be critical to the formation of an effective ohmic contact.[21–23]

The II–VI semiconductors in the form of quantum dots (QDs; nanometre scale semiconductor particles with discrete and bound electronic states) have also been utilised for solar cell applications. Of particular interest with regard to CdS or CdSe is the relatively small band gap energy in these materials, which overlaps with the visible region of the spectrum.[24,25] The optical properties of the materials can be tuned by both selection of the material and quantum confinement. Control over the size and shape of the quantum dots during their synthesis also influences the end performance of the device. The properties of quantum dot solar cells have been utilised in several device configurations for capturing solar energy. In structured polymer hybrid

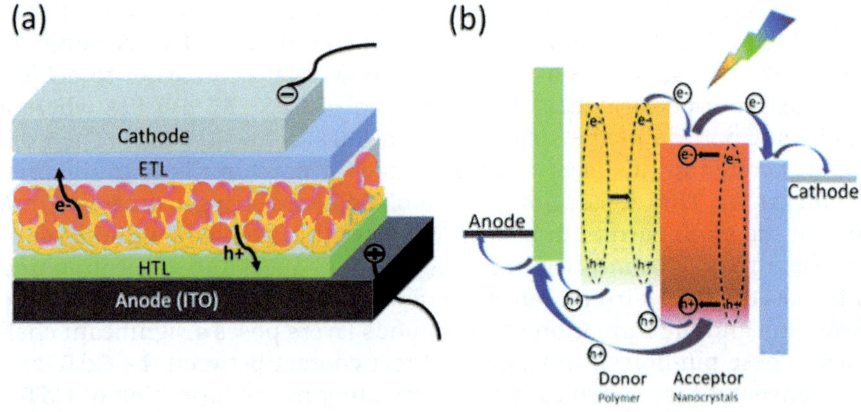

Figure 2.2 (a) Device structure of polymer–nanocrystal solar cells (HSCs) with formal structure. (b) The fundamental process in polymer–nanocrystal solar cells. Adapted from ref. 31 with permission from the Royal Society of Chemistry.

solar cells (Figure 2.2),[26–29] blends of conducting polymers (*e.g.*, poly(3-hexylthiophene)) and QDs facilitate charge separation and charge transport. The QD-based solar cells take advantage of size confinement. Additionally, the use of a highly structured substrate such as porous TiO_2 can increase the optical path for photon absorption of QDs, which will increase the optical absorption due to strong internal light scattering. This is achieved by boosting the light-trapping capabilities, which ultimately leads to a further reduction in the absorber layer thickness in nano-structured solar cells. This can be as thin as 150 nm (ultrathin) instead of several micrometres as in traditional solar cells.[30]

In general, the fundamental process in polymer–nanocrystal HSCs involves five parts, as shown in Figure 2.2b. (1) When the active layer is illuminated, excitons are generated in both polymer and nanocrystals. (2) Excitons diffuse to the polymer/nanocrystals interface. (3) Excitons dissociate or transfer charge into free carriers. An energy offset between the lowest unoccupied molecular orbital (LUMO) level of the polymer and the conduction band (CB) of the nanocrystals is required for efficient electron transfer from polymer to nanocrystals. (4) The free carriers are transported to the electrodes through bicontinuous pathways. (5) When the free carriers reach the corresponding electrodes, charge extraction will occur.

In quantum dot-sensitised solar cells (QDSSCs) (Figure 2.3),[32–37] QDs act as sensitisers instead of dyes (see Chapter 3). The photoelectrodes in QDSSCs are typically fabricated through adsorption or deposition of QDs onto wide band gap semiconductor nanomaterials (*e.g.*, TiO_2; QDs) that have been sintered to form mesoporous films. These films act as electron conductors (or acceptors) and transport layers. Under illumination with sunlight, electron–hole pairs form in the QDs, leading to the injection of electrons from the

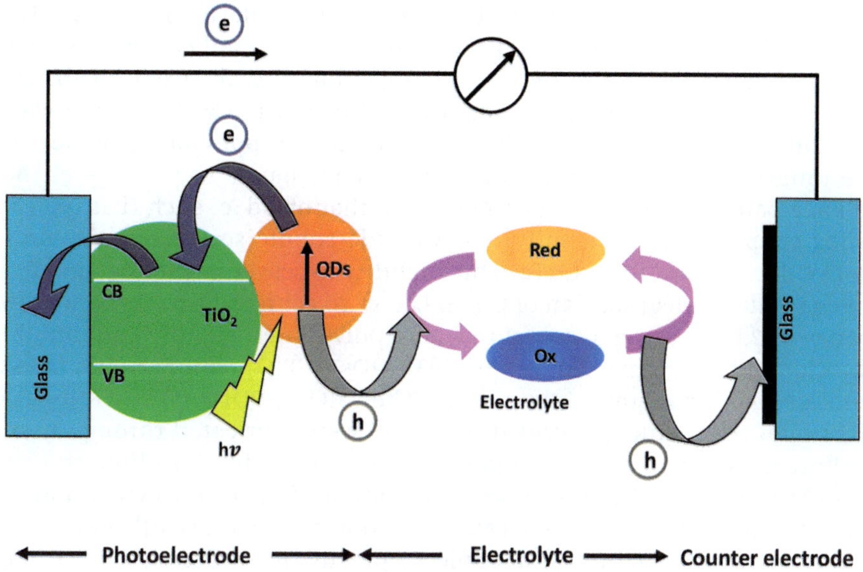

Figure 2.3 Operating principle of a typical QDSSC. Adapted from ref. 38 with permission from the Royal Society of Chemistry.

conduction band, excitonic states, or trap states of the QDs into the CB or other acceptor states of QDs (Figure 2.3). The original state of the QDs is restored through electron donation from the electrolyte, which is usually a redox system.

An advantage over dyes is that QDs may utilise the quantum confinement effect to generate multiple electron–hole pairs per photon through the impact ionisation effect (the process in a material by which one energetic charge carrier can lose energy by the creation of other charge carriers).[39–42] Over the years, studies have been performed in QD-sensitised solar cells using a range of device configurations, *e.g.*, Sun *et al.* reported an efficiency of 4.15% for an electrochemical cell using a CdS-sensitised TiO_2 nanotube array photoelectrode.[43] Toyoda and co-workers prepared a CdSe-sensitised photoelectrode based on a TiO_2 inverse opal, and an energy conversion efficiency of 2.7% was obtained.[44] By using a CdSe-sensitised nanocrystalline TiO_2 photoelectrode in a photoelectrochemical cell, Hodes and his team achieved a 2.8% conversion efficiency under 1 sun illumination.[36] Lee *et al.* reported energy conversion efficiency of 1.84% under one sun illumination for a CdS-sensitised solar cell where CdS QDs were deposited onto a mesoscopic TiO_2 film from alcohol solutions using a chemical bath deposition (CBD) process.[45] Furthermore, they showed an energy conversion efficiency of 4.22% by co-sensitising with both CdS and CdSe.[46]

CdTe QDs possess a narrower band gap and a higher conduction band edge compared with CdSe QDs of the same size, which extends the light-absorption range of CdTe to longer wavelengths and gives a faster electron injection rate from CdTe QDs to TiO_2. However, the development of CdTe QD-based solar cells is relatively unexplored. Kamat and co-workers have shown that even with a faster electron injection rate, the CdTe QD-based solar cells exhibit a very poor quantum efficiency (<3%) compared with that of CdSe QDs (>70%).[47] They attributed this poor performance to the unfavourable position of the CdTe valence band. The energy of the valence band edge of CdTe is higher than that of CdSe, such that typical redox couples such as a sulfide/polysulfide couple scavenge holes only from CdSe and not from CdTe. This inability to scavenge photogenerated holes results in degradation of CdTe. Lan *et al.* achieved a conversion efficiency of 2.02% using a structure where poly(dimethyldiallylammonium chloride) (PDDA) was used to link CdTe QDs to TiO_2 nanoparticles (NPs) through an electrostatic attraction.[48] The $(TiO_2)_m$-PDDA-$(QDCdTe)_n$-FTO electrodes (FTO, fluorine-doped tin oxide) were fabricated through thermal processing and an ionic liquid (1-ethyl-3-methylimidazolium thiocyanate) was used as the solvent for preparation of the electrolyte. Another reason why CdTe-based solar cells have not been successful previously is that it is not easy to find a stable Te precursor to directly grow CdTe QD sensitisers onto TiO_2 film electrodes *via* the chemical bath deposition or successive ionic layer adsorption and reaction route. CdTe QDs are susceptible to oxidation, thereby reducing the performance of the resultant cells. Chang *et al.*,[48] and Yu *et al.*[49] have made contributions to utilising CdTe QDs in QDSSCs, and obtained energy conversion efficiency of 2.02% (CdTe QDs on the electrode surface by conducting simple heating processes in ionic liquid solvent) and 3.8% (a one-step linker-assisted chemical bath deposition method to prepare CdTe/CdS core/shell QD-sensitised TiO_2 electrodes), respectively.

Further attempts to make solar cells using CdTe have used sintered QDs. Alivisatos and co-workers reported the first all-inorganic nanocrystal-based cell utilising a sintered CdSe/CdTe bilayer, which had a power conversion efficiency of *ca.* 3%.[50] They briefly annealed the CdTe film to remove residual solvent and paved a way to efficiently depositing a CdSe film. In this way, high-quality bilayer structures were formed with minimal intermixing at the interface. Improvements on this method were made by Olson and co-workers[51] who used a Schottky-diode (p-type CdTe heterojunctions, with CdSe or CdS serving as the n-type material) structure and reported 5% conversion efficiencies. They argued that sintering of the CdTe QD films was necessary to facilitate grain growth and enhanced the optical absorption. Mulvaney *et al.* reported conversion efficiencies of >7% by using a layer-by-layer approach to deposit the absorber and a heterojunction architecture to make a CdTe/ZnO solar cell.[52,53] Panthani *et al.* had used a similar structure to Mulvany and co-workers (ZnO nanocrystals (NCs)), in which they deposited sintered CdTe nanocrystals using a layer-by-layer method to prepare a thick (500–600 nm) sintered NC absorber film while forming a heterojunction using sol–gel ZnO.[54] They reported a conversion efficiency of 8.54%.

2.2.2 III–V Semiconductor Compounds

The III–V compound semiconductors have been used in photovoltaics since the 1970s, when the first AlGaAs/GaAs heterojunction was made.[55] Since then gallium arsenide (GaAs), aluminium arsenide (AlAs), indium phosphide (InP), and their alloys have been researched and widely used for various optoelectronic applications, largely because of their very high electron mobilities. GaAs can be doped with either a p- or n-type dopant, and its direct band gap of 1.42 eV with a strong absorption cross-section matches well with the solar spectrum. The high absorptivity, large charge carrier diffusion length, ideal band gap, and lattice-match with most other III–V semiconductor alloys make GaAs one of the ideal materials for solar cells. There is only a 0.12% lattice mismatch between GaAs (a' = 0.5653 nm) and AlAs (a' = 0.566 nm) at room temperature.[56,57] At the epitaxial growth temperatures of solid solutions (T = 600–900 °C), the lattice parameters of the substrate and absorber layers become even closer (due to the different thermal expansion coefficients of GaAs ($\alpha_T = 6.5 \times 10^{-6}$ deg^{-1}) and AlAs ($\alpha_T = 5.2 \times 10^{-6}$ deg^{-1})). This facilitates the growth of high-quality AlGaAs/GaAs solid solutions, free from misfit dislocations, on a GaAs substrate. However, as the temperature decreases to ambient, elastic strain arises because of the different α_T values. Decreasing the thickness of the AlGaAs/GaAs layer can reduce the strain in the active region of an AlGaAs/GaAs cell. In the very first AlGaAs/GaAs heterojunctions, GaAs served as base layer and substrate and AlGaAs was used as a wide gap window for illumination. Alloying with Al increased the band gap of GaAs without changing the lattice constant and this resulted in a defect-free heterojunction interface. Basically, p–n-junctions are formed in an absorbing semiconductor material, surrounded by heterostructures that are higher band gap barrier layers, to prevent photogenerated minority charge carriers from leaving the active region of the device. The researchers at Spectrolab and NREL have demonstrated an efficiency of 32.3% in a terrestrial concentrator cell (high current densities under concentrated light) using a $Ga_{0.5}In_{0.5}P$/GaAs/Ge triple-junction solar cell.[58] Later on, King *et al.* from Spectrolab reported an efficiency of >40% in an upright metamorphic triple-junction GaInP/GaInAs/Ge cell under the standard spectrum for terrestrial concentrator solar cells.[59–61]

In a triple-junction solar cell structure, the components are designed in such a way that the combination of electrical and optical properties results in the equivalent photocurrent carried by a single cell. In a typical p–n junction derived from Ge, GaInAs and GaInP, all materials are stacked on each other, with each layer having a higher band gap energy than the layer below it. The layers are assembled with low resistive tunnel junctions to connect the cells. Different epitaxial techniques can be used to grow the monolithic device in series stacks, such as liquid-phase epitaxy (LPE) or epitaxy from metal–organic compounds.[62–66] LPE was the main method of AlGaAs solar cell fabrication in the 1970s and 1980s. Since the late 1970s, AlGaAs/GaAs heterostructures have been produced by metal–organic chemical vapour deposition (MOCVD), using metal–organic compounds of group III elements and hydrides of group V elements.[67–71] The MOCVD process is carried out in reactors, with the substrate holder heated to 650–700 °C, at atmospheric or lower pressure, and with hydrogen as the carrier gas. An advantage of MOCVD is the possibility of fabricating multilayer structures in high-yield reactors, with layers of specified composition and a thickness varying from 1–10 nm to several micrometres. MOCVD is capable of producing GaAs single-crystal layers on silicon and germanium substrates, so it can potentially be used to fabricate low-cost, high-efficiency AlGaAs solar cells on these substrates and cascade cells (Figure 2.4). The epitaxy technique delivers low defect density layers and interfaces with the appropriate doping and band gaps. Due to this structure of the cell, the top p–n junction has the highest band gap and can collect light from the shortest wavelengths. The lower layers (middle and bottom) absorb the light from the subsequent longer wavelength (lower energy) region of the solar spectrum.

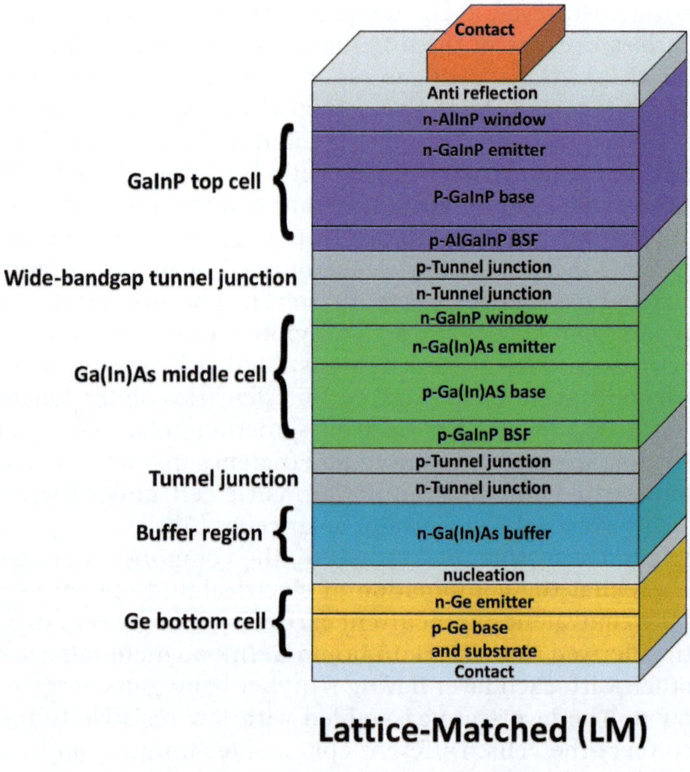

Figure 2.4 Physical schematic of a monolithic triple-junction n-on-p solar cell deposited epitaxially upon a substrate. Adapted from ref. 72 with permission from the Royal Society of Chemistry.

Bett and co-workers made theoretical predictions of high power conversion efficiencies in a triple-junction tandem cell where there is a lattice mismatch between the substrate and the semiconductor layers.[73] To summarise, the GaInP and GaInAs layers are lattice-matched to each other on a lattice-mismatched substrate (Ge).[59,74] The mismatched design of the tandem cell allows the upper two junctions to be grown with lower band gaps which results in a more efficient band gap combination for converting the solar spectrum. To grow a more efficient monolithic solar cell structure with an improved band gap combination, lattice-mismatched alloys with low defect densities are required. Jones and co-workers had reported a photovoltaic conversion efficiency of 40.8% at 326 suns concentration.[75] They demonstrated a monolithically grown, triple-junction solar cell structure in which each active junction is composed of an alloy with a different lattice constant chosen to maximise the theoretical efficiency. The authors developed a structure in an inverted configuration, where $Ga_{0.51}In_{0.49}P$ top junction lattice-matched to the GaAs substrate, a metamorphic $In_{0.04}Ga_{0.96}As$ middle junction, and a metamorphic $In_{0.37}Ga_{0.63}As$ bottom junction using an organometallic vapour-phase epitaxy technique. Takamoto *et al.* have reported a InGaP/GaAs/InGaAs inverted triple-junction solar cell for applications under concentrated sunlight.[76] They reduced the series resistance and improved the short circuit current density by reducing the shadow loss from the contact fingers to achieve an efficiency of 44%. Additional junctions can improve the performance of III–V solar cells. Essig *et al.* demonstrated that the theoretical optimum for a four-junction solar cell is reached for a set of declining band gap energies with 1.9, 1.4, 1.0, and 0.5 eV.[77] This structure was realised with a GaInP/GaAs/GaInNAs/Ge four-junction cell, but the performance of this device was not satisfactory due to the insufficient quality of the dilute nitride material grown by metal–organic vapour-phase epitaxy.

More promising results were reported for inverted metamorphic devices (reaching AM0 space solar spectrum) one sun efficiencies up to 33.6%.[78] In this configuration, two lattice-mismatched $Ga_{1-x}In_xAs$ bottom junctions are combined with a $Ga_{0.5}In_{0.5}P$/GaAs top tandem by using buffer layers to relax the crystal lattice. Dislocations cannot be completely avoided in this configuration, and this leads to constraints in device performance. Dimroth and co-workers reported later on a wafer-bonded four-junction GaInP/GaAs/GaInAsP/GaInAs concentrator solar cell with 44.7% efficiency.[79] This method requires two epitaxial structures with low surface roughness and a specific surface preparation of the materials, but offers the advantage of combining the best defect-free semiconductors in a multi-junction solar cell device.[80–82] The direct wafer bond is performed at room temperature and leads to a conductive and transparent interface. This allows the fabrication of high-efficiency monolithic tandem solar cells with active junctions. The authors reported that the wafer bonding permits the defect-free combination of two GaAs–based and InP-based tandem solar cell structures with a lattice mismatch of 3.7%. More than

90% of the substrate area with a diameter of 100 mm was bonded successfully and showed no indication of interface defects, such as voids. Nanostructures have also been used and Samuelson *et al.* reported an efficiency of 15.3% for a nanowire array solar cell of GaAs fabricated by a surface-based epitaxy technique.[83]

2.2.3 I–III–VI$_2$ Semiconductor Compounds

Among I–III–VI$_2$ compounds, Cu-based chalcopyrites have attracted attention and have been utilised successfully in solar cells. $CuInSe_2$ (ABX_2), which has a zinc blende-type crystal structure, is one of the materials that has been studied extensively.[84] Both the cations (A and B) are tetrahedrally coordinated to four neighbouring anions (X) with different bond lengths. The usefulness of this material stems from the high optical absorption, arising from a direct band gap of 1.04 eV, and the overall electronic properties that are well matched for solar cell applications.[84] This band gap can be easily tuned (*ca.* 1.14 eV) by introducing another element like gallium or sulfur to form the quaternary $Cu(In,Ga)Se_2$ or the pentenary $Cu(InGa)(SSe)_2$ alloy.[85] Rau *et al.*,[86] Zhang *et al.*[87] and Schock *et al.*[88] have shown that the $Cu(In,Ga)Se_2$ semiconducting material has tolerance to structural defects due to the existence of large off-stoichiometric compounds (a number of compounds exist with different Cu/In/Se ratios). These materials can also be doped as n- or p-type by internal defects without introducing external impurities. This can be easily achieved by making the material either Cu-poor or Cu-rich in the stoichiometry.[87,89,90] Another important aspect is the negligible electrical effect of the structural defects or grain boundaries.

Störger and co-workers first synthesised CIS material in 1953,[91] and then, almost two decades later, Wagner *et al.* reported an efficiency of ~12% in a CIS single crystal with an evaporated CdS film.[92,93] This was a significant development in the field of solar cells as only three other materials (Si, GaAs/$Ga_xAl_{1-x}As$, and InP/CdS) had shown efficiencies >10% at that time. Subsequently, Morgan and co-workers reported the first thin-film CIS/CdS heterojunction solar cell with an efficiency of *ca.* 5% by evaporating CIS powder in the presence of excess Se vapour.[94] The co-evaporation technique was further used by Devaney *et al.*, who reported a solar cell with 11.4% efficiency.[95] Another report by Kazmerski and Sanborn was published around the same time, in which CIS thin-film homojunction solar cells (a semiconductor interface that occurs between layers of similar semiconductor material, these materials have equal band gaps but typically have different doping, *e.g.*, inorganic p–n junction) with an efficiency of 3.62% were described.[96] They reported that forward dark *I vs. V* characteristics were consistent with a recombination-generation mechanism at the junction dominating the device operation. Mickelsen and Chen demonstrated that an efficiency of 9.4% can be achieved by using co-evaporation from elemental sources (*e.g.*, Se, Cu, S, In, *etc.*).[97] Subsequently, different techniques

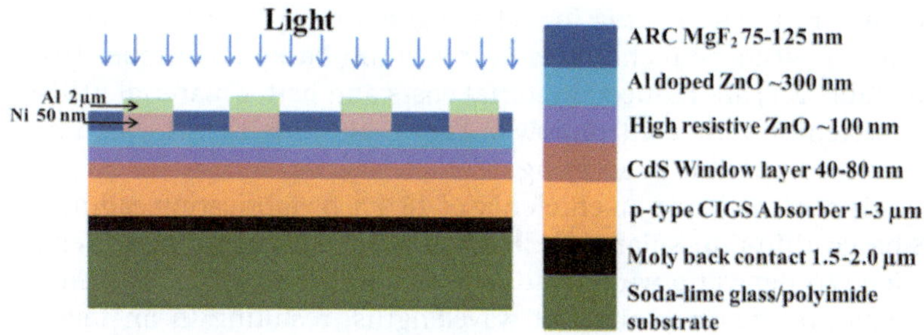

Figure 2.5 Schematic cross-section of CIGS-based thin-film solar cells. Adapted from ref. 110 with permission from the Royal Society of Chemistry.

and methods have been developed to increase the efficiency of CIS-based solar cells, for example, by alloying with Ga or In to tune the band gap, sodium (Na) doping, and reducing the thickness of the cadmium sulfide (CdS) window layer by adding intrinsic and doped zinc oxide (ZnO) layers, *etc.* Kamada *et al.* reported a new process to improve the open circuit voltage (V_{OC}) by depositing potassium fluoride between the CIGS and buffer layers (Zn, Mg)O/Zn(O, S, OH) in an effort to assemble a cadmium-free solar cell. They reported an efficiency of 22.8%.[98]

The typical configuration for solar cells based on CIGS (Figure 2.5) is a stacked layer formation consisting of a molybdenum back contact, the CIGS absorber, a thin CdS or ZnS or In_2S_3 buffer layer, and a bilayer of intrinsic and aluminium-doped zinc oxide as the transparent conductive oxide front contact. For high-efficiency solar cells, the sputtering technique is used to deposit molybdenum (Mo) and ZnO. Vacuum and non-vacuum (CBD) methods are used to deposit the CIGS absorber layer and the CdS window layer, respectively. The Mo film also acts as a reflector and reflects unused light back into the absorber. Significant improvement in reflection from the rear side has been reported when a Mo–Cu alloy is used as the rear contact.[99] The highest solar cell efficiency of 19.9% was achieved by depositing a CdS buffer layer using a CBD technique.[100] With a band gap of 2.5 eV, however, a CdS buffer layer limits the performance of the CIGS-based solar cells in the shorter wavelength region of the solar spectrum. Taking into account concerns about the toxicity of Cd and environmental factors, researchers have developed alternative buffer layers. Various materials and deposition techniques have been explored in this regard and materials like In_2S_3, ZnS, $Zn_{1-x}Mg_xO$ that have larger (or in range; 2.0–2.8 eV) band gap energies than that of CdS have been employed as buffer layers. High production costs and complex vacuum processes have led to the development of simple and low-cost non-vacuum solution methods such as atomic layer deposition, ion-layer gas reaction, sputtering and evaporation (PVD), and ultrasonic spray pyrolysis; many of these methods have produced promising results.[101–109] The non-vacuum technique is basically a two-step process, which involves precursor deposition at low temperature followed by high-temperature annealing in a chalcogen atmosphere. The main advantages of non-vacuum techniques over vacuum-based techniques (listed in full in Table 2.1) are reduced material costs and better material utilisation, better stoichiometric control, low energy input and high-throughput, and roll-to-roll compatible processing.

Nakada *et al.* reported an efficiency of 18.1% by fabricating cadmium-free CIGS-based thin-film solar cells.[111] They used a chemical bath deposition technique to deposit a wide band gap material layer (ZnS). This improved the quantum efficiency at short wavelengths, resulting in an increase in the short circuit current of fabricated solar cells. Tiwari *et al.* described a low-cost method to deposit a film of CIGS precursor paste using a "doctor blade" technique, which was subsequently selenised under selenium vapour in a quartz tube.[104] They reported an efficiency of 6.7% with ~0.5 μm thick absorber layers. Kaufmann *et al.* reported a method where they deposited the CIGSe absorber layer by co-evaporation with an *in situ* laser light scattering process control and reported an efficiency of 16.7%.[112] The performance of the absorber layer on Ti substrates was optimised through adjustment of the sodium content. The incorporation of sodium enhances the solar cell efficiency mainly through a higher open circuit voltage (V_{OC}), fill factor (FF), an increased hole density and film conductivity. During the absorber growth on soda-lime glass substrates at high temperature (*ca.* 550 °C), Na diffuses through the Mo layer from the substrates to the CIGS layer by thermal activation. The presence of Na in the CIGS absorber passivates the grain boundaries and surfaces. A moderate level (0.1%, hole concentration = 10^{20} cm^{-3}) of Na in the CIGS layer improves the cell efficiency.[113–115] When flexible substrates are used, Na doping at low temperature is accomplished by depositing a 10 nm NaF layer prior to CIGS layer deposition. Commercially available flexible (polyimide) substrates can withstand temperatures of up to 450 °C. Caballero *et.al.* further studied the effect of sodium incorporation in CIGS absorber layers and fabricated solar cells on polyamide substrates.[116] The researchers found that the NaF precursor layer evaporated onto the substrate modifies the CIGS growth kinetics and observed a stronger Ga-gradient and a decrease of the grain size when the Na content was increased. Se is included in the film to prevent the formation of a selenium donor vacancy (V_{Se}) and Na diffusion inside the CIGS grains passivates non-radiative centres located at the grain boundaries. Also, Na incorporation passivates shallow donor-type InCu anti-site defects by forming NaCu defects, and reduces charge compensation.

Schulz *et al.* reported an efficiency of 4.6% by using a spray deposition technique to deposit a CIGS nanocrystal thin film.[109] They prepared an amorphous CIGS nanoparticle colloidal solution which was sprayed onto heated molybdenum-coated soda-lime glass substrates to form the precursor films. After thermal annealing of the precursor films under a selenium atmosphere, CIGS solar cells were fabricated. The CIGS films prepared this way were comprised of a smaller grained intermediate layer and a larger grained top layer. In earlier reports using nanoparticles to make thin films

Table 2.1 List of various deposition techniques of CIGS films, together with their advantages and disadvantages.

Deposition technique	Advantages	Disadvantages
Co-evaporation	Well established technique for lab-scale devices	Simultaneous control of sources is difficult and results in variation in stoichiometry Poor reproducibility Poor uniformity over large area
Sputtering	Controlled deposition rate Better crystallinity	High operational cost Produces multiple phases, *viz.* CIS, CGS
Selenisation/sulfurisation	Large area deposition Established technique can be used for precursor deposition and reaction	Toxic gases (H_2Se) Additional processing step is required Poor adhesion with back contact
Screen printing	Little material wastage High packing density High throughput	Control of Ga profile is not possible while drying Phase segregation and inhomogeneities occur
Spray coating	Scalable High throughput Low cost	Precursor material wastage during spray coating
Spin-coating	Film uniformity (lab-scale) Low cost of equipment Easy operation	No uniformity over a large area material wastage Roll-to-roll incompatible
Doctor blade	Less material wastage Roll-to-roll compatible Better stoichiometric control	Solvent evaporation is slow, which leads to accumulation
MOCVD	Useful for basic studies Growth rate is faster than MBE	Not suitable for industrial processes Large area growth is not reported Process is not as abrupt as MBE
Electron beam deposition	Good stoichiometry and film purity	Large area deposition not reported Not compatible with industrial processes
Pulsed laser deposition	Target composition can be transferred to films and yield good stoichiometry in the film CuSe binary phase can be avoided	Not suitable for large area
Inkjet printing	Mask-less patterning simplifies processing steps Compatible for roll-to-roll technology	Large area stoichiometry has not been reported Low efficiency

and consequently solar cells, Korgel *et al.* reported the synthesis of different types of nanocrystals such as CIS, CISe, CIGS for solar cell fabrication.[117] In their paper, they reported very low efficiencies and suggested different ways to improve it by using nanocrystals with shorter chain surface ligands, incorporating Ga into the films, and using various chemical or thermal treatments of the nanocrystal layers to increase their conductivity. Guo *et al.* reported an efficiency of 3.2% by a facile solution synthesis of chalcopyrite $CuInSe_2$ nanocrystals.[118] Mitzi *et al.* reported an efficiency of 10% in a CIGS-based solar cell (glass/Mo/CIGS/CdS/i-ZnO/ITO) that involved the incorporation of all metal and chalcogenide components into a single hydrazine-based solution.[106,107] The solution was spin-coated to form a precursor film, and then heated at temperatures below 500 °C in an inert atmosphere, to form the desired CIGS film with up to μm-scale film thickness and grain size. Yoon *et al.* reported a method for a low-cost non-vacuum fabrication technique where the CIGS material was directly coated and the selenisation of the precursor was done *in situ*.[119] The double-layered films with an upper layer of chalcopyrite CIS and a carbon residue bottom layer were generated when this selenisation step was carried out at elevated temperatures. They showed an efficiency of 2% in solar cells fabricated using this method and argued that the high series resistance resulting from the residual carbon layer causes this low efficiency.

A number of factors affect CIGS solar cell efficiencies and performance. CIGS is a quaternary compound and film uniformity (thickness and composition) across the large area substrates is an issue. Uniformity is the main requirement for electronic and optical properties of high-efficiency solar cells and both reproducibility and uniformity over a large area directly influence the yield during production. Sputter deposition is suitable for large area substrates. Due to the scarcity and price volatility of In and Ga used in CIGS, other Earth-abundant and environmentally sustainable materials with related structures (*e.g.*, CZTS) have also attracted attention as absorbers in solar cells.

2.2.4 I_2–II–IV–VI_4 Semiconductor Compounds

Copper chalcogenide-based CIGS absorber films have shown great potential as absorber layers in solar cells, but the presence of rather expensive and rare earth metals like indium and gallium has caused issues with production and environmental regulations. The kesterite type $Cu_2ZnSn(S,Se)_4$ (CZTS) absorber has started to replace the CIGS layer in thin-film solar cells. By replacing In and Ga with Earth-abundant and non-toxic metals, such as zinc and tin, a more attractive system in terms of cost and low toxicity has emerged. CZTS and CIGS exhibit closely related characteristics due to their isovalency and similarities in crystal structure (Figure 2.6). CZTS offers a high absorption coefficient of $>10^4$ cm^{-1} and a tuneable band gap from 1.5 eV to 1.0 eV by introducing different ratios of selenium to sulfur.[120–122]

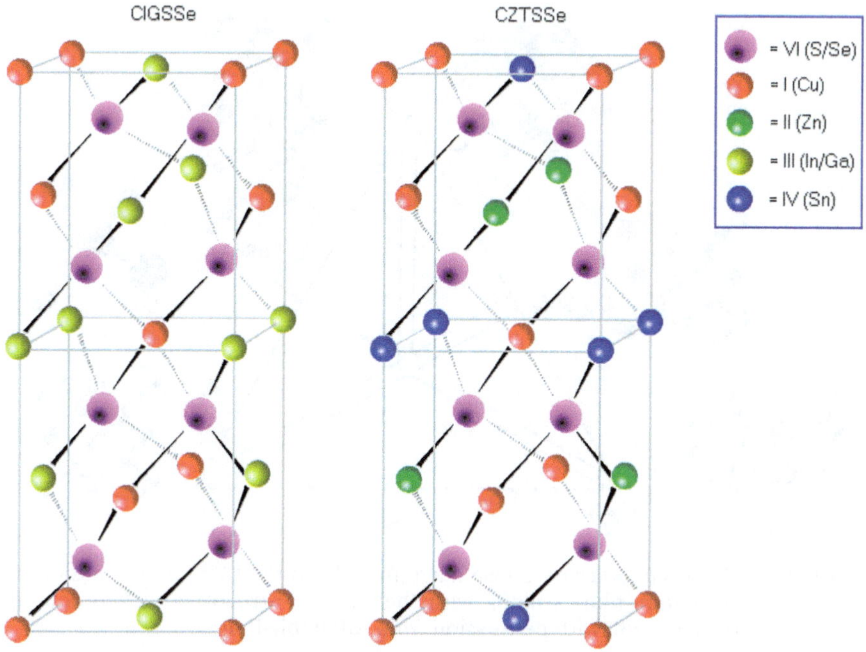

Figure 2.6 Chalcopyrite and kesterite crystal structures.[123] Adapted from ref. 123 with permission from the Royal Society of Chemistry.

The first CZTS-based solar cell was fabricated by Katagiri *et al.* in 1997, with a conversion efficiency of 0.66%.[124] With improvements made to the growth techniques and optimisation of the material fabrication, the conversion efficiency was improved to 6.7% by Guha *et al.*,[125] then to 10.1% by Mitzi *et al.*[126,127] Using a hydrazine pure-solution approach, targeting a Cu-poor and Zn-rich condition, Mitzi *et al.* also reported the highest efficiency for a CZTS device of 12.6%, which has been certified by the Newport Technology and Application Center.[128] A small area device with an efficiency of 13.8% was reported in 2016.[129] Despite the apparent advantages of CZTS absorber solar cells over CIGS, and the structural similarities, the gap in performance has remained an unresolved question for researchers. The difficulties in achieving an efficient CZTS system have arisen due to the quaternary nature of the material and key remaining challenges include the development of impurity-free and phase-controlled $Cu_2ZnSn(S_xSe_{1-x})_4$ thin films. The defects and secondary phases formed mainly during the growth and post-growth processes are due to a narrow stability region (Figure 2.7).

Figure 2.7 demonstrates an experimentally determined pseudoternary phase diagram from binary Cu_2S, ZnS and SnS_2 phases to form the CZTS phase, along with their crystal structures. The complexity of this quaternary system increases many-fold because of competition between CZTS growth

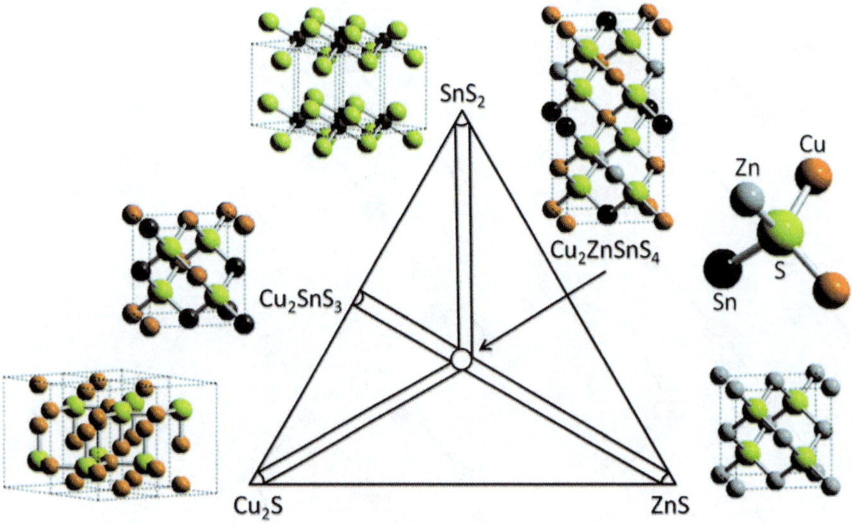

Figure 2.7 The pseudoternary phase diagram to form the CZTS phase and other secondary phases along with their crystal structures.[130] Reproduced from ref. 130 with permission from AIP Publishing.

and the growth of binary [Cu_2S, ZnS, SnS, and SnS_2] and ternary [Cu_2SnS_3] phases.[121,130–133] The formation of secondary phases, with different vacancies, anti-sites and self-compensating defect complexes in CZTS, together with carrier recombination at different interfaces, affects the device efficiency and may be the main reason for the performance gap between CZTS and CIGS solar cells.[125,131,134] A suitable polycrystalline semiconductor for high-efficiency photovoltaic devices requires benign surfaces and grain boundaries (GBs) that should not contribute significantly to the charge recombination in the device. Devices from polycrystalline CIGSSe consistently outperform monocrystalline films, implying that the GBs are beneficial to the device. However, the origins of this benign behaviour are still a topic of intense research, and significant work is still necessary to understand the nature of GBs in both CIGSSe and CZTSSe.

Researchers are developing various different deposition approaches for CZTS, such as co-evaporation,[135] sulfurisation of magnetron-sputtered precursors,[136] and solution processing.[126,137] Cu_2ZnSnS_4 solar cells fabricated on a molybdenum soda-lime glass (SLG) substrate with the structure Al/ZnO:Al/CdS/CZTS/Mo–SLG were reported by Katagiri *et al.*[138] The devices were assembled by sequential deposition of the precursors (*e.g.*, ZnS, Sn, and Cu) using thermal evaporation assisted by vapour-phase sulfurisation, and attained an efficiency of 2.62%. Noufi and co-workers reported a CZTSe solar cell with 9.15% efficiency using the co-evaporation method.[139] The device structure consisted of a Mo-coated SLG, an e-beam-evaporated NaF precursor (150 Å thick), a chemical bath deposited CdS layer, a sputtered resistive/

conductive ZnO bilayer, a Ni/Al grid by e-beam evaporation, and MgF_2 as an anti-reflective layer. In 2008, Motohiro *et al.* demonstrated the deposition of a ~1.3 μm CZTS film on 1 μm thick Mo coated on SLG using RF-magnetron co-sputtering of Cu, SnS, and ZnS targets, and obtained a solar cell efficiency of >6%.[136]

Non-vacuum methods of depositing kesterite films include solution processing of films *via* spin-coating of a CZTS precursor solution, chemically synthesising CZTS nanocrystal solution, and electrochemical deposition. Kamoun *et al.* reacted $CuCl_2$, $ZnCl_2$, and $SnCl_2$ and vulcanised them in a thiourea solution using spray pyrolysis.[140] The substances were reacted for 1 hour at the substrate temperature of 340 °C and were annealed for 120 minutes at 550 °C to give CZTS thin films with a band gap of 1.5 eV. According to experimental results, the CZTS thin film will have better optical properties if the substrate temperature is controlled within the range of 500°C–650 °C during pyrolysis. Moholkar *et al.* obtained a Cu_2S, ZnS, and SnS_2 powder using a grinding method, and made a CZTS target from the powder through a solid–state reaction. An excimer laser beam was used to bombard the target, and a CZTS thin film was deposited in a vacuum chamber, followed by annealing in a $N_2 + H_2S$ gas environment.[141] The thin-film cells based on this method have open circuit voltage of 585 mV, a short circuit current density of 6.74 mA cm^{-2} and a fill factor of 0.51, conversion efficiency of 2.02%, and band gap of 1.52 eV.

Scragg *et al.* made an early attempt to fabricate CZTS thin films using an electrodeposition technique.[142] CZTS thin films were formed on Mo-coated SLG substrates using sequential electrodeposition of constituent metals, Cu, Zn, Sn, followed by annealing in a sulfur atmosphere at 550 °C for 2 h. The X-ray diffraction (XRD) pattern showed some secondary peaks of SnS_2. The solar cell fabricated using these CZTS thin films showed an efficiency of 0.8%. Araki *et al.* reported the preparation of CZTS thin films using electroplating followed by the sulfurisation of metallic, stacked precursors.[143]

Agrawal and co-workers reported the fabrication of CZTSSe solar cells with an efficiency of 7.2%, by selenisation of CZTS nanocrystals.[137] CZTS nanocrystals were synthesised using the hot-injection method, wherein the precursor elements, Cu, Zn, and Sn, were mixed together in a solvent, oleylamine. The devices were made on ~1 μm thick molybdenum, which was coated on the SLG substrate by knife-coating. The dense and uniform CZTS film of 1 μm thickness was achieved by annealing at 500 °C for 20 minutes under selenium vapour. Ennaoui *et al.* reported a non-vacuum electroplating deposition technique for CZTS absorber layer formation.[144] This technique was chosen to prepare large area films using cheaper precursors and a simpler setup. Wang *et al.* reported a CZTS–Se thin-film solar cell with 12.6% efficiency, where films were made using a slurry-based deposition technique.[128] This deposition method overcame many challenges that occurred during solution processing and suspension of constituent elements. The slurry used for deposition consisted of Cu–Sn chalcogenide (S or S–Se) mixed in a hydrazine (N_2H_4) solution.

For CZTS to reach the efficiency levels of CIGS, a considerable improvement of deposition processes and process controls to accurately control composition and to avoid the occurrence of secondary phases will be required.

2.2.5 Other Compound Semiconductor Materials

2.2.5.1 *Copper Sulfides*

Copper sulfides (Cu_xS) have attracted a great deal of scientific attention due to their tuneable semiconducting properties and approximate metallic behaviour, depending on the stoichiometry of the mineral phase. Cu_xS represents a promising p-type semiconductor used in a variety of technologies such as nanoscale electronic devices, as a cathode material in lithium ion batteries, chemical sensors, and especially photovoltaic cells, because of its unique optical and electronic characteristics. Cu_xS is a family of compounds occurring in a large number of different crystallographic phases, *e.g.*, cupric sulfide (covellite) at $x = 1$ through diginite ($x \sim 1.8$) and djurleite ($x = 1.96$) to cuprous sulfide (chalcocite) at $x = 2$.[145–149] Because of these different phases, the energy band gap of Cu_xS thin films varies between 1.26 and 2.54 eV.[150–153] The photovoltaic effect is observed only for the chalcocite phase (Cu_2S). This is a p-type semiconductor, which is usually degenerate, or nearly so, because of copper vacancies, and which has an energy gap of 1.21 eV, close to the optimum for solar energy conversion. Its beneficial electrical properties and its composition of Earth-abundant and environmentally benign materials make it a promising candidate for an absorber layer in thin-film solar cells.[154] The optical absorption and the electron diffusion length both depend on crystallographic direction and, for efficient solar cells, the *c*-axis of the chalcocite must be normal to the film surface. The electron diffusion length decreases as the copper vacancy concentration increases, and, for an efficient cell, a composition with $x > 1.99$ is needed. As $x = 2$, the electrochemical potential needed for decomposition decreases, reducing the resistance of the solar cells to degradation, and so a composition of $x \sim 1.997$ is about optimum.

Solar cells of copper sulfide (Cu_2S) have been investigated intensively. The first efficient thin-film solar cells were Cu_2S/CdS heterostructures, with the Cu_2S absorber layer grown topotaxially into high-quality CdS films either by solution chemistry with Cu ions or by solid–state reaction of CuCl. These cells utilising a $Cu_2S/Cd_{1-x}Zn_xS$ heterojunction reached efficiencies exceeding 10% as well as open circuit voltages of 0.6 V.[155] Despite their good performance, research has been abandoned due to long-term stability issues, which were related to the degradation of the Cu_2S as well as to the inter-diffusion of copper ions into the CdS layer.[156] The most common method to produce Cu_2S for photovoltaic purposes was the chemiplating technique, a wet-chemical process where cadmium is replaced by copper in the outer layers of a thermally evaporated CdS film.[157,158] A variety of other deposition methods have been carried out to produce thin layers of Cu_2S for thin-film solar cells, including the dry method,[159,160] reactive magnetron sputtering,[161] thermal evaporation,[162] CVD,[163,164] spray pyrolysis,[165,166] and CBD[167–171] as well as spin-coating of Cu_2S nanocrystals.[172] Wu *et al.* have shown that Cu_2S nanocrystal-based photovoltaic devices yield a power conversion efficiency exceeding 1.6%.[172] In a review article by R. Hill in the late 1970s, he mentioned a set of formidable criteria for successful solar cell application of Cu_xS materials.[173] He states that the cadmium sulfide films must be free of pin-holes and of grain boundaries between the outer and inner surfaces. The electron mobility should be about 50 $cm^2\ V^{-1}\ s^{-1}$, or better, to give resistivities of about 10 Ω cm with the minimum donor densities, and the minority-carrier lifetime should be at least 10 ns, and preferably an order of magnitude higher, to give diffusion lengths about equal to the layer thickness. The layers should be monocrystalline. The CuS layer should be topotaxial to the CdS and should have a composition of about $Cu_{1.997}S$. The thickness of this layer should be about 0.1–0.15 μm with a minority-carrier diffusion length of about 0.5 μm.

2.2.5.2 *Iron Pyrite*

Iron pyrite (FeS_2) has attracted considerable attention as a potential candidate for photovoltaic and photoelectrochemical devices. Iron sulfide-based minerals are common in nature (iron is the fourth most abundant element in the Earth's crust, and is also the cheapest metallic element to extract from nature) and can exist in a variety of forms with varying stoichiometries that ranges from the sulfur-deficient mackinawite (FeS_{1-x}) through iron-deficient pyrrhotites ($Fe_{1-x}S$) to pyrite (FeS_2).[174] In the lower-temperature environments of the Earth's surface the iron sulfide group is particularly complex and a number of metastable phases become significant in biogeochemical processes. Fiechter *et al.* mentioned in their report that the strong d-character of the valence band contributes to the observed high stability of FeS_2, especially when used as an electrode in electrochemical cells.[175] They prepared FeS_2 thin films by sulfurisation of fresh iron deposited under vacuum and the obtained samples were p-type with 10^{-2} Ω cm resistivity.

The very high optical absorption coefficient ($\alpha \leq 5 \times 10^5$ cm, $\lambda \leq 700$ nm) and an energy band gap (E_g) of around 0.95 eV, along with rich abundance of the constituent elements in the Earth, make iron pyrite a promising candidate as an absorber material in thin-film solar cells.[176,177] This high absorption coefficient would make it possible to construct ultrathin solar cells with absorber thicknesses below 100 nm.[178] In addition, they would be cheap and non-toxic. Therefore, pyrite has been of interest as an alternative to silicon for thin-film solar cells.

There have been a number of investigations on the preparation of pyrite films with a variety of methods used, such as sulfurisation of the electrodeposited[179] or evaporated metallic iron films,[180] ion beam sputtering,[181] spray pyrolysis,[182,183] electrodeposition,[184,185] and MOCVD.[183,186] Birkholz *et al.* first showed that pyrite films could be prepared by reactive ion beam sputtering in H_2S discharges.[181] Because of the low H_2S partial pressure of about 0.01 Pa, the substrate temperature for growing stoichiometric pyrite was in this case limited to values below 100 °C which led to a small grain size of 10–15 nm. Reactive magnetron sputtering, however, offers the possibility of growing pyrite films at much higher H_2S partial pressures and, therefore, at higher substrate temperatures, since with this technique the deposition occurs at higher pressures. The use of Fe, in the form of FeS_2 pyrite, in solar cells was proposed approximately three decades ago, but the low photovoltage, sulfur deficiencies, and poor structure uniformity have limited its PV applications.[176,187,188] The highest efficiency of 2.8% has been obtained in a photoelectrochemical cell.[178,189–191] The limiting factor for high efficiency is the high dark current, which leads to small open circuit voltages (less than 0.2 V) that are caused by iron pyrite phase impurities and, more importantly, large density surface states.[189] The large density surface states originate from thermodynamically unstable particle surfaces that are terminated with sulfur dimers.[192] The sulfur dimers tend to segregate to the surface and cause vanishing of the hybridised band at the top of the Fe layer, making the surface effectively similar to a 1S termination. An iron monosulfide (FeS) structure is formed on the surface of the pyrite crystals. This results in a large density of defect states in the middle of the pyrite band gap that reduces the band gap at the surface to nearly zero.

In the last decade or so, researchers have started to use FeS_2 nanocrystals for thin-film preparation for applications in solar cells. Lin *et al.* reported a hybrid solar cell comprising poly(3-hexylthiophene) (P3HT) and FeS_2 NCs.[193] They demonstrated that adding FeS_2 NCs to P3HT could contribute to the extended photovoltaic response in the red light region, which can act as a potential candidate for applications in polymer/inorganic hybrid solar cells. A power conversion efficiency of 0.16% was reported. Huang *et al.* reported the synthesis of uniform, phase pure, and air stable pyrite FeS_2 NCs and deposited 400 nm thick films sandwiched between indium-tin-oxide (ITO) and Al contacts.[194] They reported the Hall effect mobility of the pristine pyrite NC films, which was around 80 $cm^2\ V^{-1}\ s^{-1}$ with p-type behaviour. Later on, Ren and co-workers reported the synthesis of FeS_2 nanocubes and fabrication of FeS_2 : CdS bulk heterojunction (BHJ) solar cells, enabling a large photovoltage of 0.79 V and power conversion efficiency (PCE) of 1.1% under AM 1.5 illumination to be attained.[195] Korgel *et al.* concluded in their review that, whereas pyrite has suitable optical properties for PVs and extremely low raw material costs, its electrical properties may be difficult to control properly due to the formation of surface defects related to sulfur vacancies.[196] This problem might be exacerbated in nanocrystal devices due to the very high concentration of interfaces.

2.2.5.3 Transition Metal Chalcogenides

Transition metal dichalcogenides (TMDs) are known as true two-dimensional (2D) materials with excellent semiconducting properties and strong mechanical flexibility, in addition to being atomically thin.[197–200] TMDs, *i.e.*, MX_2 where M is a transition metal (Mo, W, Ti, *etc.*) and X is a chalcogen (S, Se, Te), such as MoS_2, WS_2, $MoSe_2$, and WSe_2, form 2D-layered structures and are abundantly available in nature. In an MX_2 monolayer structure (X–M–X), an atomic layer of a transition metal (M) is sandwiched between two chalcogen (X) atomic layers, where transition metal atoms (M) such as Mo and W are covalently bonded with chalcogen atoms (X) such as S, Se, Te. Each layer is 6–7 Å thick in the TMD-layered structures. Weak van der Waals interactions occur between two adjacent MX_2 layers of TMDs, whereas the intra-layer M–X bonds are covalent.[201–204] From an electrical point of view, MoS_2, $MoSe_2$, WS_2, and WSe_2 are semiconductors, while WTe_2 and $TiSe_2$ are semimetals.[201–203,205]

TMDs possess intra-layer metal planes that are held between the atomic planes of a group VI element by covalent bonding, while the inter-layers are weakly positioned by van der Waals forces. Mechanical exfoliation using scotch tape techniques,[206] fast sonication,[207] and shear exfoliation in aqueous surfactant solution using a kitchen blender[208] *etc.*, allow convenient fabrication of high-quality monolayers that display a range of band gaps within the visible spectrum. Monolayers of well-known TMDs (MoS_2, $MoSe_2$, WS_2, and WSe_2) possess direct band gaps, unlike their bulk counterparts which display indirect band gaps.[198] Bulk MoS_2 has an indirect band gap of 1.29 eV and is optically active in the visible range, and the monolayer form has a direct band gap of 1.8 eV. For WSe_2, indirect band gaps ranging from 0.98 to 1.4 eV, and direct band gaps from 1.37 to 1.77 eV, have been reported.[209] TMDs with thicknesses of less than 1 nm show strong light–matter interactions, affording absorption of incident sunlight of as much as 5–10%, which is one order of magnitude higher than that of common semiconductors such as Si and GaAs.[210] These features make TMDs among the most attractive materials for high-performance, semitransparent, and flexible solar cells. However, due to the difficulties in fabricating large-scale defect-free 2D TMD materials, the development of 2D TMD-based solar devices is limited.

TMDs have traditionally been used as power conversion efficiency enhancers in GaAs- and Si-based solar cells. For example, Kis and co-workers demonstrated the fabrication of solar cells based on heterojunctions composed of monolayer MoS_2 and p-type silicon.[211] This choice of materials combines the advantages of the direct band gap and small thickness of 2D MoS_2 with the established silicon-based fabrication processes, and all the semiconducting materials used in this type of device can be considered Earth-abundant and non-toxic. They reported an external quantum efficiency exceeding 4% for their p–n diodes that also work as solar cells. Tsai *et al.* also fabricated a type-II heterojunction of MoS_2 monolayers with p-type Si.[212] The MoS_2 monolayer introduced a built-in electric field near the interface between MoS_2 and p-type Si to help photogenerated carrier separation. This heterojunction photovoltaic device achieved a PCE of 5.23%.

In the other type of fabrication, only TMDs are used as the photoactive layers. Because the first type (heterojunctions of TMDs with other materials) deteriorates the original high transparency and mechanical flexibility of TMDs, the second type (using TMD only) is necessary for the application of semitransparent and flexible solar cells. Balicas *et al.* studied the photovoltaic response of electrostatically generated p–n junctions composed of approximately 10 atomic layers of $MoSe_2$ stacked onto the dielectric h-BN.[213] They reported that these junctions could yield, under AM1.5 illumination, photovoltaic efficiencies exceeding 14%. Wi *et al.* also fabricated and characterised photovoltaic devices with a vertically stacked indium-tin-oxide electrode/multilayer MoS_2/metal electrode structure.[214] Utilising a plasma-induced p-doping approach, they were able to form p–n junctions in MoS_2 layers that facilitate the collection of photogenerated carriers, enhance the photovoltages, and decrease reverse dark currents. They exhibited power conversion efficiencies up to 2.8% under AM 1.5 illumination, as well as high external quantum efficiencies.

2.3 Conclusion

In summary, this chapter has reviewed the compound semiconductor material properties, current cell design, various growth techniques, and the current progress for fabricating low-cost, highly efficient PV devices. In the coming years, compound semiconductor materials will be competing with traditional silicon technologies. Meanwhile, fresh competition from new solution-processable materials, such as lead halide perovskites, is arriving due to their simple fabrication methods. There is still so much room for further improvement in these materials, including a better understanding of band alignments, the role of grain boundaries, and defect analysis. In Chapter 5, a theoretical approach is presented, to identify materials for high-efficiency solar cells, related to the compounds presented in this chapter. This approach could help to resolve some of the challenges described here and could be of benefit for future solar cell design and fabrication.

References

1. C.-H. Lee, G.-H. Lee, A. M. Van Der Zande, W. Chen, Y. Li, M. Han, X. Cui, G. Arefe, C. Nuckolls, T. F. Heinz and J. Guo, *Nat. Nanotechnol.*, 2014, **9**, 676.
2. R. W. Birkmire and E. Eser, *Annu. Rev. Mater. Sci.*, 1997, **27**, 625.
3. A. Polman, M. Knight, E. C. Garnett, B. Ehrler and W. C. Sinke, *Science*, 2016, **352**, 4424.
4. M. A. Green, Y. Hishikawa, W. Warta, E. D. Dunlop, D. H. Levi, J. Hohl-Ebinger and A. W. H. Ho-Baillie, *Prog. Photovoltaics*, 2017, **25**, 668–676, NREL/JA-5J00-68932.
5. C. Ferekides, U. Balasubramanian, R. Mamazza, V. Viswanathan, H. Zhao and D. Morel, *Sol. Energy*, 2004, **77**, 823.
6. X. Wu, J. C. Keane, R. G. Dhere, C. Dehart, D. S. Albin, A. Duda, T. A. Gessert, S. Asher, D. H. Levi and P. Sheldon, *Proceedings of the 17th European Photovoltaic Solar Energy Conference*, Munich, Germany, 2001, vol. 2, p. 995.
7. A. Luque and S. Hegedus, *Handbook of Photovoltaic Science and Engineering*, John Wiley & Sons, 2011.
8. N. Romeo, A. Bosio, R. Tedeschi and V. Canevari, *Mater. Chem. Phys.*, 2000, **66**, 201.
9. C. S. Ferekides, D. Marinskiy, V. Viswanathan, B. Tetali, V. Palekis, P. Selvaraj and D. L. Morel, *Thin Solid Films*, 2000, **361**, 520.
10. M. Burgelman, J. Verschraegen, S. Degrave and P. Nollet, *Thin Solid Films*, 2005, **480**, 392.
11. M. Afzaal and P. O'Brien, *J. Mater. Chem.*, 2006, **16**, 1597.
12. G. Y. Chung, S. C. Park, K. Cho and B. T. Ahn, *J. Appl. Phys.*, 1995, **78**, 5493.
13. S. Kumazawa, S. Shibutani, T. Nishio, T. Aramoto, H. Higuchi, T. Arita, A. Hanafusa, K. Omura, M. Murozono and H. Takakura, *Sol. Energy Mater. Sol. Cells*, 1997, **49**, 205.
14. B. E. McCandless and R. W. Birkmire, *Photovoltaic Specialists Conference, 2000 Conference Record of the Twenty-Eighth IEEE*, IEEE, 2000, p. 491.
15. K. Chopra, P. Paulson and V. Dutta, *Prog. Photovoltaics*, 2004, **12**, 69.
16. A. Niemegeers and M. Burgelman, *J. Appl. Phys.*, 1997, **81**, 2881.
17. O. Rotlevi, K. D. Dobson, D. Rose and G. Hodes, *Thin Solid Films*, 2001, **387**, 155.
18. T. Gessert, P. Sheldon, X. Li, D. Dunlavy, D. Niles, R. Sasala, S. Albright and B. Zadler, *Photovoltaic Specialists Conference, Conference Record of the Twenty-Sixth IEEE*, IEEE, 1997, p. 419.
19. D. Levi, L. Woods, D. Albin, T. Gessert, D. Niles, A. Swartziender, D. H. Rose, R. K. Ahrenkiel and P. Sheldon, *Photovoltaic Specialists Conference, Conference Record of the Twenty-Sixth IEEE*, 1997, p. 351.

20. B. E. McCandless and K. D. Dobson, *Sol. Energy*, 2004, **77**, 839.
21. S. S. Hegedus and B. E. McCandless, *Sol. Energy Mater. Sol. Cells*, 2005, **88**, 75.
22. C. Corwine, A. Pudov, M. Gloeckler, S. Demtsu and J. Sites, *Sol. Energy Mater. Sol. Cells*, 2004, **82**, 481.
23. B. E. McCandless and J. R. Sites, *Handbook of Photovoltaic Science and Engineering*, John Wiley & Sons, Ltd, 2003, vol. 2, p. 617.
24. G. D. Scholes, *ACS Nano*, 2008, **2**, 523.
25. D. R. Baker and P. V. Kamat, *Adv. Funct. Mater.*, 2009, **19**, 805.
26. D. Ginger and N. Greenham, *Phys. Rev. B*, 1999, **59**, 10622.
27. I. Gur, N. A. Fromer, C.-P. Chen, A. G. Kanaras and A. P. Alivisatos, *Nano Lett.*, 2007, **7**, 409.
28. L. Wang, Y. Liu, X. Jiang, D. Qin and Y. Cao, *J. Phys. Chem. C*, 2007, **111**, 9538.
29. J. Bouclé, S. Chyla, M. S. Shaffer, J. R. Durrant, D. D. Bradley and J. Nelson, *C. R. Phys.*, 2008, **9**, 110.
30. K. Ernst, A. Belaidi and R. Könenkamp, *Semicond. Sci. Technol.*, 2003, **18**, 475.
31. Z. Chen, X. Du, Q. Zeng and B. Yang, *Mater. Chem. Front.*, 2017, **1**, 1502.
32. A. Nozik, *Phys. E*, 2002, **14**, 115.
33. A. Kongkanand, K. Tvrdy, K. Takechi, M. Kuno and P. V. Kamat, *J. Am. Chem. Soc.*, 2008, **130**, 4007.
34. H. J. Lee, J.-H. Yum, H. C. Leventis, S. M. Zakeeruddin, S. A. Haque, P. Chen, S. I. Seok, M. Grätzel and M. K. Nazeeruddin, *J. Phys. Chem. C*, 2008, **112**, 11600.
35. I. Mora-Sero, J. Bisquert, T. Dittrich, A. Belaidi, A. Susha and A. Rogach, *J. Phys. Chem. C*, 2007, **111**, 14889.
36. O. Niitsoo, S. K. Sarkar, C. Pejoux, S. Rühle, D. Cahen and G. Hodes, *J. Photochem. Photobiol., A*, 2006, **181**, 306.
37. K. S. Leschkies, R. Divakar, J. Basu, E. Enache-Pommer, J. E. Boercker, C. B. Carter, U. R. Kortshagen, D. J. Norris and E. S. Aydil, *Nano Lett.*, 2007, **7**, 1793.
38. Z. Yang, C.-Y. Chen, P. Roy and H.-T. Chang, *Chem. Commun.*, 2011, **47**, 9561.
39. K. Wijayantha, L. M. Peter and L. Otley, *Sol. Energy Mater. Sol. Cells*, 2004, **83**, 363.
40. S.-C. Lin, Y.-L. Lee, C.-H. Chang, Y.-J. Shen and Y.-M. Yang, *Appl. Phys. Lett.*, 2007, **90**, 143517.
41. I. Robel, V. Subramanian, M. Kuno and P. V. Kamat, *J. Am. Chem. Soc.*, 2006, **128**, 2385.
42. I. Robel, M. Kuno and P. V. Kamat, *J. Am. Chem. Soc.*, 2007, **129**, 4136.
43. W.-T. Sun, Y. Yu, H.-Y. Pan, X.-F. Gao, Q. Chen and L.-M. Peng, *J. Am. Chem. Soc.*, 2008, **130**, 1124.
44. L. J. Diguna, Q. Shen, J. Kobayashi and T. Toyoda, *Appl. Phys. Lett.*, 2007, **91**, 023116.

45. C.-H. Chang and Y.-L. Lee, *Appl. Phys. Lett.*, 2007, **91**, 053503.
46. Y. L. Lee and Y. S. Lo, *Adv. Funct. Mater.*, 2009, **19**, 604.
47. J. H. Bang and P. V. Kamat, *ACS Nano*, 2009, **3**, 467.
48. G.-Y. Lan, Z. Yang, Y.-W. Lin, Z.-H. Lin, H.-Y. Liao and H.-T. Chang, *J. Mater. Chem.*, 2009, **19**, 2349.
49. X.-Y. Yu, B.-X. Lei, D.-B. Kuang and C.-Y. Su, *Chem. Sci.*, 2011, **2**, 1396.
50. I. Gur, N. A. Fromer, M. L. Geier and A. P. Alivisatos, *Science*, 2005, **310**, 462.
51. J. Olson, Y. Rodriguez, L. Yang, G. Alers and S. Carter, *Appl. Phys. Lett.*, 2010, **96**, 242103.
52. J. Jasieniak, B. I. MacDonald, S. E. Watkins and P. Mulvaney, *Nano Lett.*, 2011, **11**, 2856.
53. B. I. MacDonald, A. Martucci, S. Rubanov, S. E. Watkins, P. Mulvaney and J. J. Jasieniak, *ACS Nano*, 2012, **6**, 5995.
54. M. G. Panthani, J. M. Kurley, R. W. Crisp, T. C. Dietz, T. Ezzyat, J. M. Luther and D. V. Talapin, *Nano Lett.*, 2014, **14**, 670.
55. Z. I. Alferov, V. Andreev, M. Kagan, I. Protasov and V. Trofim, *Sov. Phys. Semicond.*, 1971, **4**.
56. V. Andreev, *Photovoltaic and Photoactive Materials—Properties, Technology and Applications*, Springer, Dordrecht, 2002, p. 131.
57. A. Bett, F. Dimroth, G. Stollwerck and O. Sulima, *Appl. Phys. A*, 1999, **69**, 119.
58. H. Cotal, D. Lillington, J. Ermer, R. King, N. Karam, S. Kurtz, D. J. Friedman, J. M. Olson, J. S. Ward, A. Duda and K. A. Emery, *Photovoltaic Specialists Conference, 2000 Conference Record of the Twenty-Eighth IEEE*, 2000, p. 955.
59. R. King, D. Law, K. Edmondson, C. Fetzer, G. Kinsey, H. Yoon, R. A. Sherif and N. H. Karam, *Appl. Phys. Lett.*, 2007, **9**, 183516.
60. R. King, D. Bhusari, D. Larrabee, X. Q. Liu, E. Rehder, K. Edmondson, H. Cotal, R. K. Jones, J. H. Ermer, C. M. Fetzer and D. C. Law, *Prog. Photovoltaics*, 2012, **20**, 801.
61. R. R. King, D. C. Law, K. M. Edmondson, C. M. Fetzer, G. S. Kinsey, H. Yoon, D. D. Krut, J. H. Ermer, R. A. Sherif and N. H. Karam, *4th International Conference on Solar Concentrators (ICSC-4)*, El Escorial, Spain, 2007, vol. 1, p. 7.
62. J. Woodall and H. Hovel, *Appl. Phys. Lett.*, 1972, **21**, 379.
63. L. James and R. Moon, *J. Vac. Sci. Technol.*, 1975, **12**, 1000.
64. J. Woodall and H. Hovel, *J. Vac. Sci. Technol.*, 1975, **12**, 1000.
65. J. Woodall and H. Hovel, *Appl. Phys. Lett.*, 1977, **30**, 492.
66. J. Batey and S. Wright, *J. Appl. Phys.*, 1986, **59**, 200.
67. N. Nelson, K. Johnson, R. Moon, H. A. Vander Plas and L. W. James, *Appl. Phys. Lett.*, 1978, **33**, 26.
68. R. Tatavarti, G. Hillier, A. Dzankovic, G. Martin, F. Tuminello, R. Navaratnarajah, G. Du, D. P. Vu and N. Pan, *Photovoltaic Specialists Conference, PVSC'08 33rd IEEE*, 2008, p. 1.

69. S. P. Tobin, S. Vernon, C. Bajgar, S. Wojtczuk, M. R. Melloch, A. Keshavarzi, T. B. Stellwag, S. Venkatensan, M. Lundstrom and K. A. Emery, *IEEE Trans. Electron Devices*, 1990, **37**, 469.
70. S. Moon, K. Kim, Y. Kim, J. Heo and J. Lee, *Sci. Rep.*, 2016, **6**, 30107.
71. G. T. Nelson, B.-C. Juang, M. A. Slocum, Z. S. Bittner, R. B. Laghumavarapu, D. L. Huffaker and S. M. Hubbard, *Appl. Phys. Lett.*, 2017, **111**, 231104.
72. H. Cotal, C. Fetzer, J. Boisvert, G. Kinsey, R. King, P. Hebert, H. Yoon and N. Karam, *Science*, 2009, **2**, 174.
73. F. Dimroth, U. Schubert and A. Bett, *IEEE Electron Device Lett.*, 2000, **21**, 209.
74. R. King, M. Haddad, T. Isshiki, P. Colter, J. Ermer, H. Yoon, H. L. Cotal, D. E. Joslin, D. D. Krut, R. Sudharsanan, K. Edmondson, B. T. Cavicchi and D. R. Lillington, *Proceedings of the 28th IEEE Photovoltaic Specialists Conference*, IEEE New York, Anchorage, Alaska, 2000, p. 15.
75. J. Geisz, D. Friedman, J. Ward, A. Duda, W. Olavarria, T. Moriarty, J. T. Kiehl, M. J. Romero, A. G. Norman and K. M. Jones, *Appl. Phys. Lett.*, 2008, **93**, 123505.
76. K. Sasaki, T. Agui, K. Nakaido, N. Takahashi, R. Onitsuka and T. Takamoto, *AIP Conf. Proc.*, 2013, **1556**, 22.
77. S. Essig, E. Stammler, S. Ronsch, E. Oliva, M. Schachtner, G. Siefer, A. W. Bett and F. Dimroth, *9th European Space Power Conference*, 2011, vol. 690.
78. M. Stan, D. Aiken, B. Cho, A. Cornfeld, V. Ley, P. Patel, P. Sharps and T. Varghese, *J. Cryst. Growth*, 2010, **312**, 1370.
79. F. Dimroth, M. Grave, P. Beutel, U. Fiedeler, C. Karcher, T. N. Tibbits, E. Oliva, G. Siefer, M. Schachtner, A. Wekkeli and A. W. Bett, *Prog. Photovoltaics*, 2014, **22**, 277.
80. K. Tanabe, D. J. Aiken, M. W. Wanlass, A. F. I. Morral and H. A. Atwater, *Conference Record of the IEEE Photovoltaic Specialists Conference*, 2006, vol. 1, p. 768.
81. D. Bhusari, D. Law, R. Woo, J. Boisvert, S. Mesropian, D. Larrabee, W. Hong and N. Karam, in *37th IEEE Photovoltaic Specialists Conference*, 2011.
82. K. Derendorf, S. Essig, E. Oliva, V. Klinger, T. Roesener, S. P. Philipps, J. Benick, M. Hermle, M. Schachtner, G. Siefer and W. Jäger, *IEEE J. Photovoltaics*, 2013, **3**, 1423.
83. I. Åberg, G. Vescovi, D. Asoli, U. Naseem, J. P. Gilboy, C. Sundvall, A. Dahlgren, K. E. Svensson, N. Anttu, M. T. Björk and L. A. Samuelson, *IEEE J. Photovoltaics*, 2016, **6**, 185.
84. J. L. Shay and J. H. Wernick, *Ternary Chalcopyrite Semiconductors: Growth, Electronic Properties, and Applications*, International Series in the Science of the Solid State, Elsevier, 2017, vol. 7.
85. A. Morales-Acevedo, *Energy Procedia*, 2010, **2**, 169.
86. U. Rau, K. Taretto and S. Siebentritt, *Appl. Phys. A*, 2009, **96**, 221.
87. S. Zhang, S.-H. Wei, A. Zunger and H. Katayama-Yoshida, *Phys. Rev. B*, 1998, **57**, 9642.

88. M. Nichterwitz, D. Abou-Ras, K. Sakurai, J. Bundesmann, T. Unold, R. Scheer and H. W. Schock, *Thin Solid Films*, 2009, **517**, 2554.
89. P. Migliorato, J. Shay, H. Kasper and S. Wagner, *J. Appl. Phys.*, 1975, **46**, 1777.
90. R. Noufi, R. Axton, C. Herrington and S. Deb, *Appl. Phys. Lett.*, 1984, **45**, 668.
91. H. Hahn, G. Frank, W. Klingler, A. D. Meyer and G. Störger, *Z. Anorg. Allg. Chem.*, 1953, **271**, 153.
92. S. Wagner, J. Shay, P. Migliorato and H. Kasper, *Appl. Phys. Lett.*, 1974, **25**, 434.
93. J. L. Shay, S. Wagner, K. Bachmann, E. Buehler and H. M. Kasper, *Proceeding 11th IEEE Photovoltaics Specification Conference Phoenix*, 1975, vol. 4, p. 503.
94. L. Kazmerski, F. White and G. Morgan, *Appl. Phys. Lett.*, 1976, **29**, 268.
95. W. E. Devaney, R. A. Mickelsen and W. S. Chen, *Conference Record of the IEEE Photovoltaic Specialists Conference*, 1985, vol. 18, p. 1733.
96. L. Kazmerski and G. Sanborn, *J. Appl. Phys.*, 1977, **48**, 3178.
97. R. Mickelsen and W. Chen, *15th Photovoltaic Specialists Conference*, 1981, p. 800.
98. R. Kamada, T. Yagioka, S. Adachi, A. Handa, K. F. Tai, T. Kato and H. Sugimoto, *IEEE 43rd Photovoltaic Specialists Conference (PVSC)*, 2016, p. 1287.
99. A. Krasnov, *US Pat.*, 8389852, Guardian Industries Corp, 2013.
100. I. Repins, M. A. Contreras, B. Egaas, C. DeHart, J. Scharf, C. L. Perkins, B. To and R. Noufi, *Prog. Photovoltaics*, 2008, **16**, 235.
101. V. K. Kapur, A. Bansal, P. Le, O. Asensio and N. Shigeoka, *Photovoltaic Energy Conversion, 2003. Proceedings of 3rd World Conference*, 2003, vol. 1, p. 465.
102. C. Eberspacher, K. Pauls and J. Serra, *Photovoltaic Specialists Conference, 2002. Conference Record of the Twenty-Ninth IEEE*, 2002, p. 684.
103. C. Eberspacher, C. Fredric, K. Pauls and J. Serra, *Thin Solid Films*, 2001, **387**, 18.
104. M. Kaelin, D. Rudmann, F. Kurdesau, H. Zogg, T. Meyer and A. N. Tiwari, *Thin Solid Films*, 2005, **480**, 486.
105. J. W. Park, Y. W. Choi, E. Lee, O. S. Joo, S. Yoon and B. K. Min, *J. Cryst. Growth*, 2009, **311**, 2621.
106. D. B. Mitzi, M. Yuan, W. Liu, A. J. Kellock, S. J. Chey, V. Deline and A. G. Schrott, *Adv. Mater.*, 2008, **20**, 3657.
107. D. B. Mitzi, M. Yuan, W. Liu, A. J. Kellock, S. J. Chey, L. Gignac and A. G. Schrott, *Thin Solid Films*, 2009, **517**, 2158.
108. V. K. Kapur, A. Bansal, P. Le and O. I. Asensio, *Thin Solid Films*, 2003, **431**, 53.
109. D. L. Schulz, C. J. Curtis, R. A. Flitton, H. Wiesner, J. Keane, R. J. Matson, K. M. Jones, P. A. Parilla, R. Noufi and D. S. Ginley, *J. Electron. Mater.*, 1998, **27**, 433.

110. J. Ramanujam and U. P. Singh, *Energy Environ. Sci.*, 2017, **10**, 1306.
111. T. Nakada and M. Mizutani, *Jpn. J. Appl. Phys.*, 2002, **41**, L165.
112. C. Kaufmann, A. Neisser, R. Klenk and R. Scheer, *Thin Solid Films*, 2005, **480**, 515.
113. D. Shin, J. Kim, T. Gershon, R. Mankad, M. Hopstaken, S. Guha, B. T. Ahn and B. Shin, *Sol. Energy Mater. Sol. Cells*, 2016, **157**, 695.
114. L. Oikkonen, M. Ganchenkova, A. Seitsonen and R. M. Nieminen, *J. Appl. Phys.*, 2013, **114**, 083503.
115. D. W. Niles, K. Ramanathan, J. Granata, F. Hasoon, R. Noufi, B. J. Tielsch and J. E. Fulghum, *MRS Online Proc. Libr.*, 1997, 485.
116. R. Caballero, C. Kaufmann, T. Eisenbarth, M. Cancela, R. Hesse, T. Unold, A. Eicke, R. Klenk and H. W. Schock, *Thin Solid Films*, 2009, **517**, 2187.
117. M. G. Panthani, V. Akhavan, B. Goodfellow, J. P. Schmidtke, L. Dunn, A. Dodabalapur, P. F. Barbara and B. A. Korgel, *J. Am. Chem. Soc.*, 2008, **130**, 16770.
118. Q. Guo, S. J. Kim, M. Kar, W. N. Shafarman, R. W. Birkmire, E. A. Stach, R. Agrawal and H. W. Hillhouse, *Nano Lett.*, 2008, **8**, 892982.
119. S. Ahn, C. Kim, J. H. Yun, J. Gwak, S. Jeong, B.-H. Ryu and K. Yoon, *J. Phys. Chem. C*, 2010, **114**, 8108.
120. S. C. Riha, B. A. Parkinson and A. L. Prieto, *J. Am. Chem. Soc.*, 2011, **133**, 15272.
121. S. Siebentritt and S. Schorr, *Prog. Photovoltaics*, 2012, **20**, 512.
122. K. Ramasamy, M. A. Malik and P. O'Brien, *Chem. Commun.*, 2012, **48**, 5703.
123. A. Polizzotti, I. L. Repins, R. Noufi, S.-H. Wei and D. B. Mitzi, *Energy Environ. Sci.*, 2013, **6**, 3171.
124. H. Katagiri, N. Sasaguchi, S. Hando, S. Hoshino, J. Ohashi and T. Yokota, *Sol. Energy Mater. Sol. Cells*, 1997, **49**, 407.
125. K. Wang, O. Gunawan, T. Todorov, B. Shin, S. Chey, N. Bojarczuk, D. B. Mitzi and S. Guha, *Appl. Phys. Lett.*, 2010, **97**, 143508.
126. T. K. Todorov, K. B. Reuter and D. B. Mitzi, *Adv. Mater.*, 2010, **22**, E156.
127. D. A. R. Barkhouse, O. Gunawan, T. Gokmen, T. K. Todorov and D. B. Mitzi, *Prog. Photovoltaics*, 2012, **20**, 6.
128. W. Wang, M. T. Winkler, O. Gunawan, T. Gokmen, T. K. Todorov, Y. Zhu and D. B. Mitzi, *Adv. Energy Mater.*, 2014, **4**, 1301.
129. Reported at PVSEC-36 by a research team led at DGIST in South Korea. A 0.181 cm^2 solar cell was certified at 13.80% by KIER.
130. H. Du, F. Yan, M. Young, B. To, C.-S. Jiang, P. Dippo, D. Kuciauskas, Z. Chi, E. A. Lund, C. Hancock, W. M. Hlaing Oo, M. A. Scarpulla and G. Teeter, *J. Appl. Phys.*, 2014, **115**, 173502.
131. S. Chen, A. Walsh, X. G. Gong and S. H. Wei, *Adv. Mater.*, 2013, **25**, 1522.
132. B. Shin, O. Gunawan, Y. Zhu, N. A. Bojarczuk, S. J. Chey and S. Guha, *Prog. Photovoltaics*, 2013, **21**, 72.

133. P. Salomé, J. Malaquias, P. Fernandes, M. Ferreira, J. Leitao, A. Da Cunha, J. C. González, F. N. Matinaga, G. M. Ribeiro and E. R. Viana, *Sol. Energy Mater. Sol. Cells*, 2011, **95**, 3482.
134. S. Ahmed, K. B. Reuter, O. Gunawan, L. Guo, L. T. Romankiw and H. Deligianni, *Adv. Energy Mater.*, 2012, **2**, 253.
135. B. A. Schubert, B. Marsen, S. Cinque, T. Unold, R. Klenk, S. Schorr and H.-W. Schock, *Prog. Photovoltaics*, 2011, **19**, 93.
136. H. Katagiri, K. Jimbo, S. Yamada, T. Kamimura, W. S. Maw, T. Fukano, T. Ito and T. Motohiro, *Appl. Phys. Express*, 2008, **1**, 041201.
137. Q. Guo, G. M. Ford, W.-C. Yang, B. C. Walker, E. A. Stach, H. W. Hillhouse and R. Agrawal, *J. Am. Chem. Soc.*, 2010, **132**, 17384.
138. H. Katagiri, K. Saitoh, T. Washio, H. Shinohara, T. Kurumadani and S. Miyajima, *Sol. Energy Mater. Sol. Cells*, 2001, **65**, 141.
139. I. Repins, C. Beall, N. Vora, C. DeHart, D. Kuciauskas, P. Dippo, B. To, J. Mann, W. C. Hsu, A. Goodrich and R. Noufi, *Sol. Energy Mater. Sol. Cells*, 2012, **101**, 154.
140. N. Kamoun, H. Bouzouita and B. Rezig, *Thin Solid Films*, 2007, **15**, 5949.
141. A. Moholkar, S. Shinde, A. Babar, K.-U. Sim, Y.-B. Kwon, K. Y. Rajpure, P. S. Patil, C. H. Bhosale and J. H. Kim, *Sol. Energy*, 2011, **85**, 1354.
142. J. J. Scragg, P. J. Dale, L. M. Peter, G. Zoppi and I. Forbes, *Phys. Status Solidi B*, 2008, **245**, 1772.
143. H. Araki, Y. Kubo, A. Mikaduki, K. Jimbo, W. S. Maw, H. Katagiri, M. Yamazaki, K. Oishi and A. Takeuchi, *Sol. Energy Mater. Sol. Cells*, 2009, **93**, 996.
144. A. Ennaoui, M. Lux-Steiner, A. Weber, D. Abou-Ras, I. Kötschau, H.-W. Schock, R. Schurr, A. Hölzing, S. Jost, R. Hock and T. Voß, *Thin Solid Films*, 2009, **517**, 2511.
145. R. Hill, *IEEE J. Solid-State Electron Devices*, 1978, **2**, S55.
146. T. J. Coutts, *Active and Passive Thin Film Devices*, Academic Press, 1978.
147. E. H. Roseboom, *Econ. Geol.*, 1966, **61**, 641.
148. D. Chakrabarti and D. Laughlin, *Bull. Alloy Phase Diagrams*, 1983, **4**, 254.
149. R. J. Goble, *Can. Mineral.*, 1985, **23**, 61.
150. A. Raevskaya, A. Stroyuk, S. Y. Kuchmii and A. Kryukov, *J. Mol. Catal. A: Chem.*, 2004, **212**, 259.
151. M. Behboudnia and B. Khanbabaee, *J. Cryst. Growth*, 2007, **304**, 158.
152. Y. Rodriguez-Lazcano, H. Martinez, M. Calixto-Rodriguez and A. N. Rodriguez, *Thin Solid Films*, 2009, **517**, 5951.
153. A. A. Sagade and R. Sharma, *Sens. Actuators, B*, 2008, **133**, 135.
154. C. Wadia, A. P. Alivisatos and D. M. Kammen, *Environ. Sci. Technol.*, 2009, **43**, 2072.
155. R. Hall, R. Birkmire, J. Phillips and J. Meakin, *Appl. Phys. Lett.*, 1981, **38**, 925.
156. A. Al-Dhafiri, G. Russell and J. Woods, *Semicond. Sci. Technol.*, 1992, **7**, 1052.
157. S. Martinuzzi, *Sol. Cells*, 1982, **5**, 243.

158. F. Z. Nataren, B. Bouchikhi, S. Chandrasekhar, J. Crest and S. Martinuzzi, *J. Cryst. Growth*, 1982, **59**, 409.
159. M. Islam and M. Haque, *Thin Solid Films*, 1982, **94**, 15.
160. A. Goldenblum, G. Popovici, E. Elena, A. Oprea and C. Nae, *Thin Solid Films*, 1986, **141**, 215.
161. E. Vanhoecke, M. Burgelman and L. Anaf, *Thin Solid Films*, 1986, **144**, 223.
162. E. Aperathitis, F. Bryant and C. Scott, *Sol. Energy Mater.*, 1990, **20**, 15.
163. S. Schneider, J. R. Ireland, M. C. Hersam and T. J. Marks, *Chem. Mater.*, 2007, **19**, 2780.
164. S. Schneider, A. Dzudza, G. Raudaschl-Sieber and T. J. Marks, *Chem. Mater.*, 2007, **19**, 2768.
165. L. Isac, A. Duta, A. Kriza, I. Enesca and M. Nanu, *J. Phys.: Conf. Ser.*, 2007, **61**, 477.
166. S. Schneider, Y. Yang and T. J. Marks, *Chem. Mater.*, 2005, **17**, 4286.
167. N. Mukherjee, A. Sinha, G. G. Khan, D. Chandra, A. Bhaumik and A. Mondal, *Mater. Res. Bull.*, 2011, **46**, 6.
168. M. Nair and P. Nair, *Semicond. Sci. Technol.*, 1989, **4**, 191.
169. P. Sebastian, O. Gomez-Daza, J. Campos, L. Baños and P. Nair, *Sol. Energy Mater. Sol. Cells*, 1994, **32**, 159.
170. S. Gadgil, R. Thangaraj, J. Iyer, A. Sharma, B. Gupta and O. Agnihotri, *Sol. Energy Mater.*, 1981, **5**, 129.
171. K. Xu and W. Ding, *Mater. Lett.*, 2008, **62**, 4437.
172. Y. Wu, C. Wadia, W. Ma, B. Sadtler and A. P. Alivisatos, *Nano Lett.*, 2008, **8**, 2551.
173. R. Hill, *IEEE J. Solid-State Electron Devices*, 1978, **2**, S49.
174. D. Rickard and G. W. Luther, *Chem. Rev.*, 2007, **107**, 514.
175. S. Fiechter, J. Mai, A. Ennaoui and W. Szacki, *J. Cryst. Growth*, 1986, **78**, 438.
176. A. Ennaoui, S. Fiechter, H. Goslowsky and H. Tributsch, *J. Electrochem. Soc.*, 1985, **132**, 1579.
177. K. Sato, *Prog. Cryst. Growth Charact.*, 1985, **11**, 109.
178. A. Ennaoui, S. Fiechter, C. Pettenkofer, N. Alonso-Vante, K. Büker, M. Bronold, C. Höpfner and H. Tributsch, *Sol. Energy Mater. Sol. Cells*, 1993, **29**, 289.
179. G. Pimenta, V. Schröder and W. Kautek, *Ber. Bunsenges. Phys. Chem.*, 1991, **95**, 1470.
180. B. Rezig, H. Dahman and M. Kenzari, *Renewable Energy*, 1992, **2**, 125.
181. M. Birkholz, D. Lichtenberger, C. Höpfner and S. Fiechter, *Sol. Energy Mater. Sol. Cells*, 1992, **27**, 243.
182. A. Yamamoto, M. Nakamura, A. Seki, E. Li, A. Hashimoto and S. Nakamura, *Sol. Energy Mater. Sol. Cells*, 2003, **75**, 451.
183. A. Luque, G. Sala, W. Palz, G. dos Santos and P. Helm, *Tenth EC Photovoltaic Solar Energy Conference: Proceedings of the International Conference*, Springer, Edition 2012, 1991.

184. A. Arico, V. Antonucci, P. Antonucci, D. Cocke and N. Giordano, *Electrochim. Acta*, 1991, **36**, 581.
185. S. Nakamura and A. Yamamoto, *Sol. Energy Mater. Sol. Cells*, 2001, **65**, 79.
186. B. Thomas, C. Höpfner, K. Ellmer, S. Fiechter and H. Tributsch, *J. Cryst. Growth*, 1995, **146**, 630.
187. W. Jaegermann and H. Tributsch, *J. Appl. Electrochem.*, 1983, **13**, 743.
188. C. Lin, D. Wang, Y. Tai, Y. Jiang, M. C. Chen, C. C. Chen, Y. J. Yang and Y. F. Chen, *J. Phys. D: Appl. Phys.*, 2011, **44**, 292002.
189. K. Büker, N. Alonso-Vante and H. Tributsch, *J. Appl. Phys.*, 1992, **72**, 5721.
190. A. Ennaoui, S. Fiechter, G. Smestad and H. Tributsch, *Energy and the Environment into the 1990s*, Pergamon Press, Oxford, 1990, p. 458.
191. R. Dasbach, G. Willeke and O. Blenk, *MRS Bull.*, 1993, **18**, 56.
192. R. Murphy and D. R. Strongin, *Surf. Sci. Rep.*, 2009, **64**, 1.
193. C. W. Lin, D. Y. Wang, Y. Tai, Y. T. Jiang, M. C. Chen, C. C. Chen, Y. J. Yang and Y. F. Chen, *J. Phys. D: Appl. Phys.*, 2011, **44**, 292002.
194. Y. Bi, Y. Yuan, C. L. Exstrom, S. A. Darveau and J. Huang, *Nano Lett.*, 2011, **11**, 4953.
195. A. Kirkeminde, R. Scott and S. Ren, *Nanoscale*, 2012, **4**, 7649.
196. C. Steinhagen, T. B. Harvey, C. J. Stolle, J. Harris and B. A. Korgel, *J. Phys. Chem. Lett.*, 2012, **3**, 2352.
197. A. Splendiani, L. Sun, Y. Zhang, T. Li, J. Kim, C.-Y. Chim, G. Galli and F. Wang, *Nano Lett.*, 2010, **10**, 1271.
198. K. F. Mak, C. Lee, J. Hone, J. Shan and T. F. Heinz, *Phys. Rev. Lett.*, 2010, **105**, 136805.
199. Z. Yin, H. Li, H. Li, L. Jiang, Y. Shi, Y. Sun, G. Lu, Q. Zhang, X. Chen and H. Zhang, *ACS Nano*, 2011, **6**, 74.
200. B. Radisavljevic, M. B. Whitwick and A. Kis, *ACS Nano*, 2011, **5**, 9934.
201. J. A. Wilson and A. Yoffe, *Adv. Phys.*, 1969, **18**, 193.
202. X. Huang, Z. Zeng and H. Zhang, *Chem. Soc. Rev.*, 2013, **42**, 1934.
203. M. Chhowalla, H. S. Shin, G. Eda, L.-J. Li, K. P. Loh and H. Zhang, *Nat. Chem.*, 2013, **5**, 263.
204. C. Ataca, H. Sahin and S. Ciraci, *J. Phys. Chem. C*, 2012, **116**, 8983.
205. E. Singh and H. S. Nalwa, *J. Nanosci. Nanotechnol.*, 2015, **15**, 6237.
206. K. S. Novoselov, D. Jiang, F. Schedin, T. J. Booth, V. V. Khotkevich, S. V. Morozov and A. K. Geim, *Proc. Natl. Acad. Sci. U.S.A.*, 2005, **102**, 10451.
207. X. Fan, P. Xu, D. Zhou, Y. Sun, Y. C. Li, M. A. T. Nguyen, M. Terrones and T. E. Mallouk, *Nano Lett.*, 2015, **15**, 5956.
208. E. Varrla, C. Backes, K. R. Paton, A. Harvey, Z. Gholamvand, J. McCauley and J. N. Coleman, *Chem. Mater.*, 2015, **27**, 1129.
209. D. Dholakia, G. Solanki, S. Patel and M. Agarwal, *Sci. Iran.*, 2003, **10**, 373.
210. M. Bernardi, M. Palummo and J. C. Grossman, *Nano Lett.*, 2013, **13**, 3664.

211. O. Lopez-Sanchez, E. Alarcon Llado, V. Koman, A. Fontcuberta i Morral, A. Radenovic and A. Kis, *ACS Nano*, 2014, **8**, 3042.
212. M.-L. Tsai, S.-H. Su, J.-K. Chang, D.-S. Tsai, C.-H. Chen, C.-I. Wu, L.-J. Li, L.-J. Chen and J.-H. He, *ACS Nano*, 2014, **8**, 8317.
213. S. Memaran, N. R. Pradhan, Z. Lu, D. Rhodes, J. Ludwig, Q. Zhou, O. Ogunsolu, P. M. Ajayan, D. Smirnov, A. I. Fernández-Domínguez, F. J. García-Vidal and L. Balicas, *Nano Lett.*, 2015, **15**, 7532.
214. S. Wi, H. Kim, M. Chen, H. Nam, L. J. Guo, E. Meyhofer and X. Liang, *ACS Nano*, 2014, **8**, 5270.

CHAPTER 3

Dye-sensitised Solar Cells

HAINING TIAN*[a], JAMES GARDNER*[b], TOMAS EDVINSSON[c], PALAS B. PATI[a], JIAYAN CONG[b], BO XU[a], MARIA ABRAHAMSSON[d], UTE B. CAPPEL[b] AND EVA M. BAREA[e]

[a]Department of Chemistry-Ångström Laboratory, Uppsala University, Box 523, SE 75120 Uppsala, Sweden; [b]Applied Physical Chemistry, Department of Chemistry, KTH Royal Institute of Technology, Teknikringen 30, SE-10044 Stockholm, Sweden; [c]Department of Engineering Sciences, Solid State Physics, Uppsala University, Box 534, SE 75121 Uppsala, Sweden; [d]Chalmers University of Technology, Department of Chemistry and Chemical Engineering, Gothenburg, Sweden; [e]Institute of Advanced Materials (INAM), Universitat Jaume I, 12006 Castelló, Spain
*E-mail: haining.tian@kemi.uu.se, jgardner@kth.se

3.1 Introduction

The concept of dye sensitisation was first applied in photography by Moser and others in 1887.[1] A hundred years later, during the 1950s and 1960s, Gerischer, Memming and others started exploring photoinduced reactions at planar semiconductor electrodes.[2–4] At that time, the main focus was on the photoinduced reactions of molecules such as erythrosine at semiconductor electrodes, and not on fully operational energy conversion devices. Of course, these pioneering works greatly inspired the subsequent application of dye-sensitised semiconductor electrodes in photoelectrochemical solar cells; however, the solar energy-to-electricity

Inorganic Materials Series No. 3
Solar Energy Capture Materials
Edited by Elizabeth A. Gibson

Published by the Royal Society of Chemistry, www.rsc.org

conversion efficiency of these original solar cells was unsatisfactory, showing a maximum efficiency of *ca.* 1%. In 1991, Grätzel and O'Regan reported a breakthrough in performance in an article in the journal *Nature*.[5] In their paper they demonstrated that sensitising a mesoporous, rather than a single crystal, TiO_2 photoelectrode with a ruthenium coordination complex could increase light-harvesting efficiency immensely, resulting in an impressive efficiency of up to 7%. The higher surface area of the mesoporous TiO_2 film meant that it could load many more dye molecules than the planar single crystals that were previously used, and the result was a significant improvement in photocurrent. Such devices are now known as dye-sensitised solar cells (DSSCs) or Grätzel cells. Instead of aiming for high purity materials, like conventional silicon solar cells, DSSCs use simple processing techniques that separate functions in the solar cell; dyes that absorb light, wide band gap semiconductors and electrolytes to transport charge. The individual components in a DSSC are quite inexpensive, easy to fabricate, and every component in the device is tuneable to optimise the performance and appearance. The flexibility and endless combinations of components have since attracted intense interest from scientists. Since the pioneering work of Grätzel and O'Regan, DSSCs have been extensively investigated and the technology has been developed by researchers in both academia and industry.[6]

The conventional DSSC device relies on a mesoporous, n-type, wide band gap semiconductor, such as TiO_2 (although several other materials have been used, see Section 3.2) to transport negative charge carriers (electrons); these are typically referred to as 'n-type' DSSCs (n-DSSC). A mirroring complementary device called a p-type DSSC (p-DSSC) uses a dye-sensitised, p-type, wide band gap semiconductor, such as NiO, to transport positive charge carriers (holes). The first reported p-DSSC was by Lindquist and co-workers in 1999.[7] Interest in p-type devices has intensified in the last decade, as they may be coupled to n-DSSCs to make tandem DSSCs[8,9] and dye-sensitised solar fuel devices.[10–13]

Figures 3.1 and 3.2 show the basic components and working principles of conventional dye-sensitised solar cells (n-type) and the inverted dye-sensitised solar cells (p-type), respectively. A DSSC consists of several basic components: a mesoporous semiconductor electrode, a dye/photosensitiser, a liquid/solid redox mediator and a counter electrode/back contact.

As seen in Figure 3.1, in an n-DSSC a dye sensitises a mesoporous n-type semiconductor such as TiO_2. The dye absorbs a photon of appropriate energy to excite it to a higher electronic state (i), the excited dye then injects an electron into the conduction band (CB) of the mesoporous semiconductor (ii). Upon electron injection, the dye becomes oxidised. The electrons are transported in the semiconductor and across grain boundaries between semiconductor particles to reach the front electrical contact (iii). The oxidised dye accepts an electron from the redox mediator (iv, dye regeneration step) and the oxidised redox mediator (either

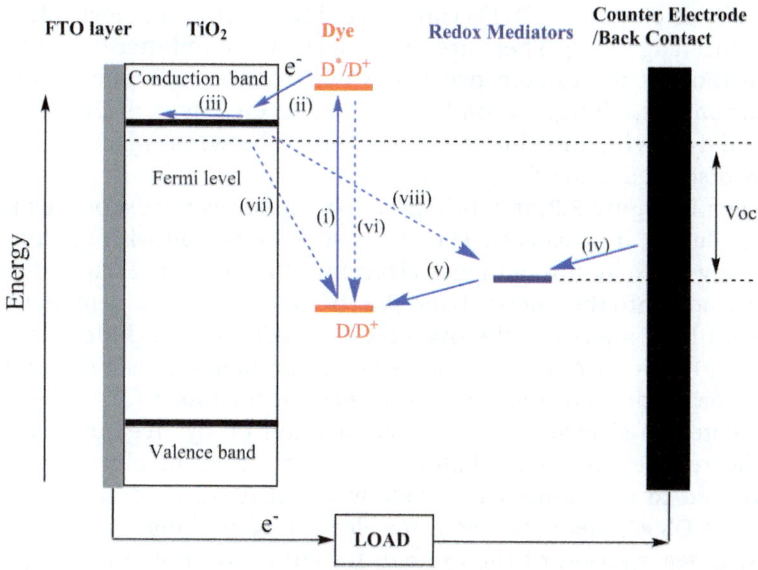

Figure 3.1 The basic components and working principles of a conventional n-type dye-sensitised solar cell (n-DSSC). Arrows display the working mechanisms (solid lines) and recombination mechanisms (dashed lines) in the device, which are further explained in the text.

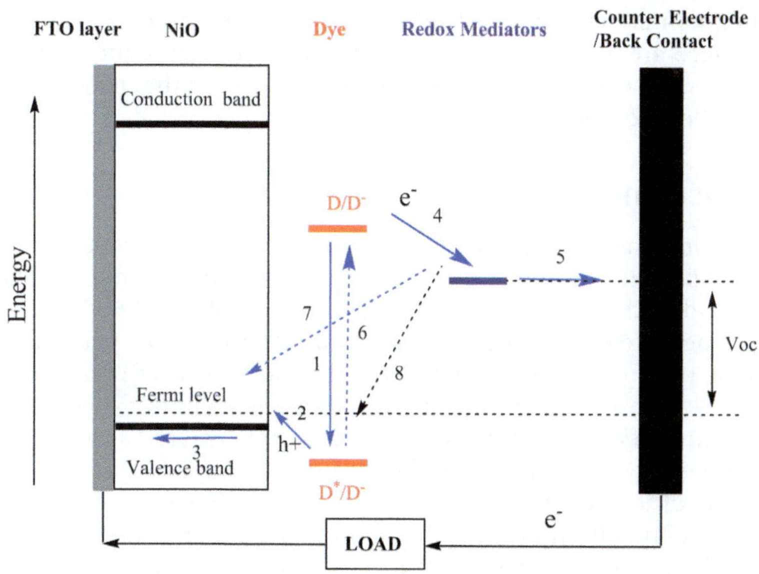

Figure 3.2 The basic components and working principles of a p-type dye-sensitised solar cell (p-DSSC). Arrows display the working mechanisms (solid lines) and recombination mechanisms (dashed lines) in the device, which are further explained in the text.

physically or electronically) is transported to the counter electrode where it is reduced again (v). There are three electron recombination processes: deactivation of the excited dye either by emission of a photon or thermalisation (vi), charge recombination between injected electrons and oxidised dyes (vii), and charge recombination between injected electrons and oxidised redox mediators (viii).

As seen in Figure 3.2, in a p-DSSC a dye sensitises a mesoporous p-type semiconductor such as NiO. The dye absorbs a photon with an appropriate energy to excite it to a higher electronic state (1), the excited dye then injects a hole into the valence band (VB) of the mesoporous semiconductor (2). Upon hole injection, the dye becomes reduced. The holes are transported in the semiconductor and across grain boundaries between semiconductor particles to reach the front electrical contact (3). The reduced dye donates an electron to the redox mediator (4, dye regeneration step) and the reduced redox mediator (either physically or electronically) is transported to the counter electrode where it is oxidised again (5). Like the n-type DSSC, there are also three electron recombination processes in p-DSSCs: deactivation of the excited dye either by emission of a photon or thermalisation (6), charge recombination between injected holes and reduced dyes (7), and charge recombination between injected holes and the reduced redox mediator (8).

Since both n- and p-type DSSCs are constructed from different components, changing one or several of the components can have a significant influence on the device performance. Optimisation of different components, the study of interfacial charge transfer processes, and improvement of device performance are, therefore, attractive topics for scientists and industry. In this chapter, we will address the individual components that make up DSSCs and how they are studied.

3.2 Electrodes

As previously mentioned, DSSCs first emerged as a concept at the end of the 1960s, with derivatives of Ru(II) tris-bipyridine complexes chemically bound to thin films of ZnO.[14–16] The use of TiO_2 and SnO_2 were also explored and a dye-TiO_2 solar cell with an electrolyte was patented by S. K. Deb in 1978.[17] A variety of dyes were bound (adsorbed) onto the electrodes.[17–20] Porous, micro- and nano-crystalline electrodes with dyes on porous sintered discs of ZnO, for example, were explored subsequently.[21] In all of these systems, the overall quantum yield was low and the power conversion efficiency (PCE) was very low, only a few percent for a close-to-monochromatic illumination.[21] A major breakthrough came with the utilisation of mesoporous TiO_2 electrodes, which increased the surface area by a factor of up to 200 and increased the external quantum efficiency (EQE) at the absorption maximum of the dye to 73%.[22] The porosity led to a substantial increase in efficiency by facilitating (1) an increase in surface area, which allowed more dye molecules to be absorbed and increased the light-harvesting efficiency, and (2) diffusion of the electrolyte through the semiconductor network to access (and regenerate) the dye. Further improvements to the accessible surface area, combined with optimising the electrolyte and applying a dye that displayed intramolecular charge separation, led to increases in performance, and 80% EQE and 8% (12% in diffuse sunlight) PCE were achieved by O'Regan and Grätzel in 1991 and 10% PCE was reported by Grätzel and co-workers in 1993.[5,23] As many groups in the world had a variety of different solar simulators and different approaches for masking stray light to the solar cells, it took until 1997 for a solar cell with 10% PCE to be certified, and in 2017 the certified PCE was 11.9%.[24] Crucial for the development of DSSCs is the synthesis and engineering of the nanocrystalline, mesoporous electrodes in parallel with dye and electrolyte development.

A typical n-type DSSC device consists of two electrodes, an anode where oxidation occurs and a cathode where reduction occurs. Normally, only the anode is photoactive and thus forms an n-type photoanode with an electronically excited and subsequently oxidised dye, while the cathode catalyses the regeneration of an electron transporter to mediate the reduction of the previously oxidised dye in a closed circuit.[25] The cathode can also be made into a photoactive electrode (p-type DSSC) and the system would then act as a serial interconnected system with voltage added from the n-type and p-type DSSCs. If spectral overlap is avoided and different parts of the solar spectrum are utilised, the tandem system can, theoretically, exceed the Shockley–Queisser limit.

The general operation mechanism in the conventional system is that the excited dye stimulates charge separation across the dye–semiconductor interface. Mesoporous semiconductor electrodes using some n-type semiconductors, such as TiO_2, ZnO or SnO_2, can, in principle, function as photoelectrodes on their own, without the dye; however, they cannot reach high efficiencies by themselves since their band gaps are in excess of 3.0 eV, which limits them to capturing UV light,[29,32,33] which is a very small portion of the solar irradiance. It is also possible to utilise a photoactive cathode, where the nanostructured metal oxide is a p-type semiconductor (NiO, CuO, $CuAlO_3$/$CuGaO_3$, *etc.*) and the highest occupied molecular orbital (HOMO) of the dye should be coupled to the valance band acceptor states to promote a hole transfer reaction. In the following sections, we will highlight the main materials used in photoanodes and photocathodes for the DSSC system and some of their intrinsic properties. The focus will be on the commonly used n-type photoanodes, but some p-type materials will also be included.

3.2.1 The Photoanode

A schematic image of a DSSC with a dye-sensitised nanostructured n-type photoanode and the I^-/I_3^- redox system is shown in Figure 3.3(a). In a typical device, a metal–organic or purely organic dye is attached on

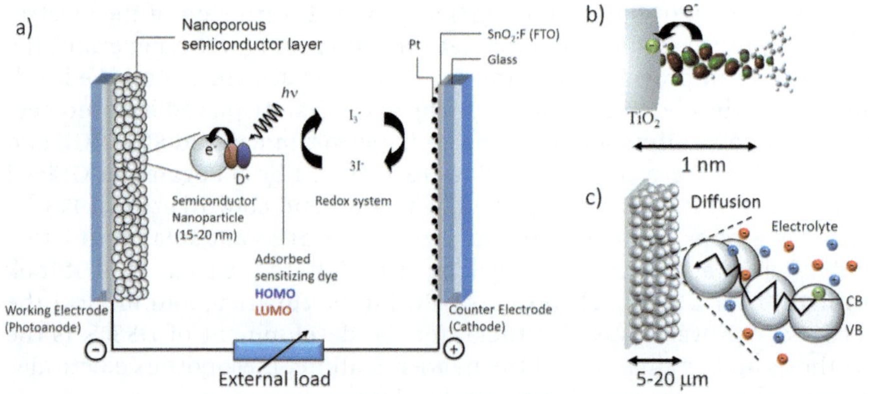

Figure 3.3 Schematic principle of (a) a DSSC with n-type photoanode, a FTO cathode with Pt nanoparticles, and I^-/I_3^- redox mediator. (b) The dimensions for a sensitiser dye attached on the surface of a TiO_2 nanoparticle with the LUMO frontier orbital and the charge injection process depicted. (c) The dimensions of the nanostructured electrode and a diffusion process of an electron throughout the interconnected TiO_2 particles depicted.

the surface of a porous electrode consisting of interconnected anatase TiO_2 nanoparticles *via* a carboxylic anchor group. In operation, light electronically excites the dye ($D \rightarrow D^*$) and if the excited-state orbitals couple to the acceptor states in the TiO_2 unoccupied states (Figure 3.4), electron transfer can occur, with subsequent oxidation of the dye ($D^* \rightarrow D^+$). For a dye with an intramolecular charge transfer (ICT) character oriented towards the binding groups, such as for the metal-to-ligand charge transfer in derivatives of tris(bipyridine)ruthenium(II) dyes or a strong ICT in an organic donor–π-acceptor system, an effective initial charge separation will occur that slows the kinetics of the charge recombination by many magnitudes (see Section 3.4, Figure 3.19).

The electron then diffuses to the back contact (Figure 3.3(c)) and can then be utilised for performing electrical work if there is a potential difference between the two electrodes. As the system is illuminated it reaches a state of quasi-equilibrium, wherein all kinetic processes of light absorption, charge injection, charge transport, and charge recombination result in a stable condition of constant incoming photon flux and a constant photocurrent and photovoltage that approximates a system in a steady-state condition. The Fermi level can then be described as a quasi-Fermi-level, which is valid under the steady-state conditions. The photovoltage in the system under illumination is defined as the potential difference between the quasi-Fermi levels in the anode and cathode. In n-type DSSCs, the cathode's quasi-Fermi-level is primarily determined by the redox potential of the electrolyte system. Thus, to increase the photovoltage for a DSSC, one must either change the redox mediator or shift the quasi-Fermi level in the photoanode.

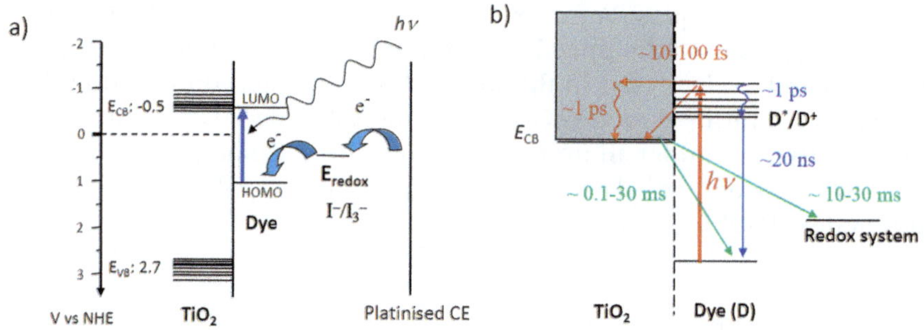

Figure 3.4 Overview of the energy levels involved at the electrode–dye and electrolyte interface.

This can be illustrated with a simple model of how the Fermi level in the metal oxide semiconductor (MOS) depends on the conduction band energy level and on the electron occupancy, where one can separate different contributions to the Fermi level shift and band edge shift. Starting with the Fermi–Dirac distribution, $f(E)$, the probability that an orbital at energy E will be occupied in an ideal electron gas at thermal equilibrium can be calculated.

$$f(E) = \frac{1}{1 + e^{(E-E_F)/k_BT}} \tag{3.1}$$

where k_B is the Boltzmann constant and T the temperature. The Fermi level, E_F, is thus defined as the energy where $f(E) = 1/2$. In the limit of $E - E_F >> k_BT$, the Fermi–Dirac distribution is reduced to the Boltzmann limit $f(E) = e^{-(E-E_F)/k_BT}$. The probability of occupancy of an orbital at the conduction band is thus $f(E_{CB}) = n_{CB}/N_{CB}$, where n_{CB} is the number of conduction band electrons and N_{CB} is the effective total density of CB states. For parabolic free-electron bands, the effective density of conduction band electrons can be expressed as

$$n_{CB} = N_{CB} e^{-(E_{CB}-E_F)/k_BT} \tag{3.2}$$

Taking the natural logarithm and rearranging, gives

$$E_F = E_{CB} + k_BT \ln \frac{n_{CB}}{N_{CB}} \tag{3.3}$$

In the dark, the Fermi energy level in the TiO_2 electrode is equilibrated with the Fermi level (electrochemical potential) in the electrolyte, $E_{F,redox}$. Under illumination and short circuit conditions, n_{CB} rises due to photoinduced electron transfer from the dye. Within milliseconds to seconds (depending on the light intensity), the system is adjusted into a steady-state condition where the increase in n_{CB} is balanced by losses in n_{CB} from recombination to the dye cations and to acceptors in the redox electrolyte which gives a quasi-equilibrium where the effective number of conduction band electrons is constant under the given illumination density. These conditions define a quasi-Fermi level, E_F^{light}, that obeys eqn (3.3). The photovoltage, V_{OC}, is the difference between the Fermi levels (electrochemical potentials) of the electrodes in the solar cell while under illumination. The anode (often fluorine-doped SnO_2 with nanocrystalline TiO_2) Fermi level is increased to E_F^{light} under illumination, while the cathode (often fluorine-doped SnO_2 coated with Pt nanoparticles) remains equilibrated with the redox system level $E_{F,\,redox}$ and so

$$V_{OC} = E_F^{light} - E_{F,redox} = E_{CB} + k_B T \ln \frac{n_{CB}}{N_{CB}} - E_{F,redox} \tag{3.4}$$

With the redox system levels fixed in the solar cell, the only contributions to the photovoltage are then the energies of the E_{CB} levels and their occupancy, n_{CB}. A linear increase in intensity of the light and subsequent linear increase of conduction band electrons will thus logarithmically increase the photovoltage, while any recombination or hindered injection will logarithmically decrease the photovoltage. From eqn (3.4), one observes that conduction band edge shifts to higher energies will contribute to increased photovoltage and this is commonly done by additives such as strong Lewis bases, for example, 4-(*tert*-butyl)pyridine, shifting the conduction band to higher energies and also suppressing recombination to the electrolyte. The commonly used n-type MOS materials (TiO_2, ZnO, and SnO_2) have different properties in their highly crystalline state (Table 3.1) where, for example, the dispersive nature of the s- and p-type orbitals at the conduction band edge contributes to a much higher electron conductivity in ZnO and SnO_2 compared to TiO_2. However, in an electrode with sintered nanoparticles, different grain orientations and grain boundaries strongly affect the situation where the electron transport has been found to be very similar in electrodes with sintered nanocrystals of ZnO and TiO_2.[26]

3.2.2 Cathode and Photocathode

The function of the cathode in an n-type DSSC is to supply electrons to the redox mediator *via* electron transfer from a catalyst (*e.g.*, nano-particulate Pt) to the redox system (*e.g.*, I^-/I_3^-). This is an essential function in the DSSC device as the recombination of electrons to the oxidised species of the redox couple from the TiO_2 surface in the anode is sluggish. However, if Pt particles fall off the electrode or make contact with the anode side or if a less selective redox mediator is utilised, such as Fe^{2+}/Fe^{3+}, the PCE for the DSSC rapidly approaches zero efficiency. Overall, it is desirable to have a catalyst on the cathode that has a low overpotential for the desired reduction reaction and a high overpotential for the oxidation. Depending on the selected redox system, other types of catalysts have been selected for the cathode, grouped broadly here as: transition metal carbides/nitrides,

Table 3.1 Selected material properties of n-type crystalline TiO_2, ZnO and SnO_2.

		TiO_2	ZnO	SnO_2	Unit	Reference
Crystal structure		Tetragonal, anatase	Hexagonal, wurtzite	Tetragonal, rutile		
Lattice constant	a	3.78	3.25	4.74	Å	27–29
Lattice constant	c	9.51	5.12	3.19	Å	27–29
Density	ρ	3.79	5.6	6.99	g cm^{-3}	27, 28, 30
Static dielectric constant	ε_s	31	7.9	9.6–13.2		29, 31, 32
Type of optical band gap		Indirect	Direct	Direct		
Optical band gap	E_{obg}	3.2	3.3	3.6	eV	29, 32, 33
Flat band potential[a]	E_{fb}	−0.5	−0.5	−0.2	V *vs.* NHE	33–35
Effective electron mass	m^*	1.0 m_e	0.24–0.30	0.23–0.33	$m_e = 9.11 \times 10^{-31}$ kg	29, 36–39
Electron mobility[b]	μ_e	30	200	100	cm^2 V^{-1} s^{-1}	40–43
Point of zero charge	pzc	5–6	8–9	3–4	pH	44–46

[a]The value from ref. 21 is recalculated to pH = 5 with a Nernstian shift of −59 mV/pH.
[b]Hall mobility for n-type materials at 300K.

transition metal chalgocenides, metals and polymeric p-type materials, and carbon allotropes.[47]

The cathode can also be sensitised into a photoactive part of the system, and thus function as a photocathode in a p-type or p/n tandem DSSC. Here, a mesoporous, wide band gap, p-type semiconductor is applied as the working electrode. The first and most commonly selected p-type semiconductor is NiO.[7–9] Other p-type semiconductors that have been tested in DSSC as photocathodes are CuO and $CuAlO_3/CuGaO_3$.[48,49]

The principle of the photocathode is to attach dyes with p-type character, with the LUMO close to the p-type electrode material and the HOMO far away to facilitate effective hole transfer into the valence band of the host electrode (Figure 3.5(a)). The simultaneous use of a photoanode and photocatode then makes the construction of a tandem device possible, as depicted in Figure 3.5(b). Proof of principle devices and different approaches to improve the initial charge separation in the dye *via* dyad structures have been published.[50–52]

3.3 Dyes

The dye is one of the crucial components of a DSSC; it is responsible for light absorption and electron/hole injection. In some cases, the dye also acts as a physical barrier (a blocking layer) to suppress charge

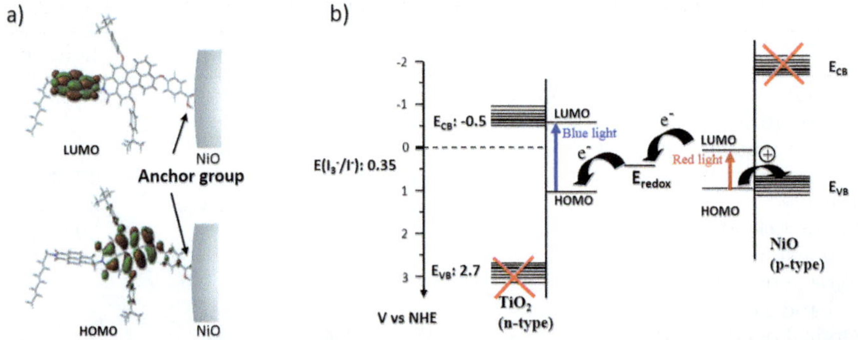

Figure 3.5 (a) Model of a p-type donor–acceptor dye bound to a p-type semiconductor (NiO). The electronic ground state for the dye has HOMO energy levels on the donor unit located spatially near the interface. In the electronic excited state, the electron density is located in LUMO frontier orbitals on a spatially distant acceptor unit. (b) Schematic principle of a DSSC with an n-type photoanode in combination with and a p-type photocathode separated by a redox mediator.

recombination between charges injected into the wide band gap semiconductor and redox mediators. When designing dyes, several points should be considered.

1. The spectral match between the absorption spectrum of the dye and the solar spectrum.
2. The photochemical and thermal stability of the dye.
3. The method of binding the dye to the surface of the wide band gap semiconductor, often through strong anchoring groups (–COOH, –SO_3H and –H_2PO_3).
4. The alignment of the energy levels of the dye should be appropriate for electron/hole injection into the semiconductor CB (for n-type) or VB (for p-type) and oxidation/reduction of redox mediators for dye regeneration.
5. The energy levels of the dye should not be substantially disturbed by interactions with other dyes (aggregation) on the semiconductor surface.

According to the type of devices that the dyes are used in, we categorise the dyes as either n-type or p-type for use in either n-type or p-type DSSCs, respectively.

3.3.1 n-type Dyes

Metal complexes, mainly ruthenium coordination complexes based on bipyridine ligands (Ru dyes), were explored as dyes intensively throughout the 1990s. Generally, metal complex photosensitisers consist of a central metal ion with ancillary ligands having anchoring groups such as –COOH. The broad absorption of these dyes comes from metal-to-ligand charge-transfer (MLCT) transitions. Among all the metal complexes, Ru dyes show the best performance and stability. Some ruthenium-based dyes such as N3, N719, and 'black dyes' were shown to have record efficiencies, up to 11% (Figure 3.6).[53,54] Many other attempts have been made to develop metal complex-based sensitisers with other metal ions such as Os, Re, Pt, Fe and Cu. Osmium-containing polypyridyl complexes have a prominent MLCT band like the more well-known Ru-based dyes (Figure 3.7). Lewis *et al.* introduced Os-containing polypyridine-based complexes as a promising class of dyes for DSSCs.[55,56] Rhenium(I)- and Fe(II)-based metal complexes have also been developed for DSSCs, although the device performance is low in comparison to Ru-based dyes.[57,58] Varieties of square-planar platinum(II) diimine dithiolato complexes were also explored as sensitisers for DSSCs; these dyes exhibit substantial solvatochromic shifts of their charge transfer band.[59,60] Due to the weak absorption of red and near-infrared (NIR) photons in the solar spectrum, these dyes have low power conversion efficiencies (PCE). Some Cu(I) coordination complexes have been tested in DSSCs as the full 3d shell prevents metal-centred d–d transitions that limit other first-row transition metals; these dyes have yet to show competitive photocurrents as obtained by Ru coordination complexes.[61,62] Due to their large extinction coefficients in the visible/NIR and stable redox chemistry, porphyrin and phthalocyanine dyes coordinated to Zn(II) have established themselves as an important class of sensitisers for DSSCs.[63,64] Devices containing porphyrin-derivative dyes are now some of the most efficient known.[65]

The use of ruthenium and other noble metal sensitisers has limited large-scale device fabrication due to innate costs and limited availability. Alternatively, organic dyes have several advantages over Ru coordination complexes, such as ease of synthesis and purification, high molar extinction coefficients, and potentially lower costs due to use of abundant elements. Organic dyes commonly consist of three components: an electron-donating functional group, a conjugated linking group, and an electron-accepting functional group (*i.e.*, the D–π–A molecular structure in Figure 3.8). For n-type DSSCs, the anchoring group is usually arranged in conjugation with the A unit to aid in injecting electrons from the excited dye into the semiconductor. The physical and chemical properties of organic dyes can be finely tuned by chemically modifying the individual functional groups in the D–π–A dyes. The acceptor group, A, is a functional group whose is electron density is deficient. The donor group, D, is a functional group whose electron density is enriched. The π-spacer moiety electronically connects the donor (D) and acceptor (A) groups and, by judicious selection of functional groups, can control the electron transfer rate between the D and A groups, as well as the degree of dye aggregation. Following the D–π–A molecular construction, designing new dye structures is facile. Electronic properties such as the absorption spectrum, ground- and excited-state

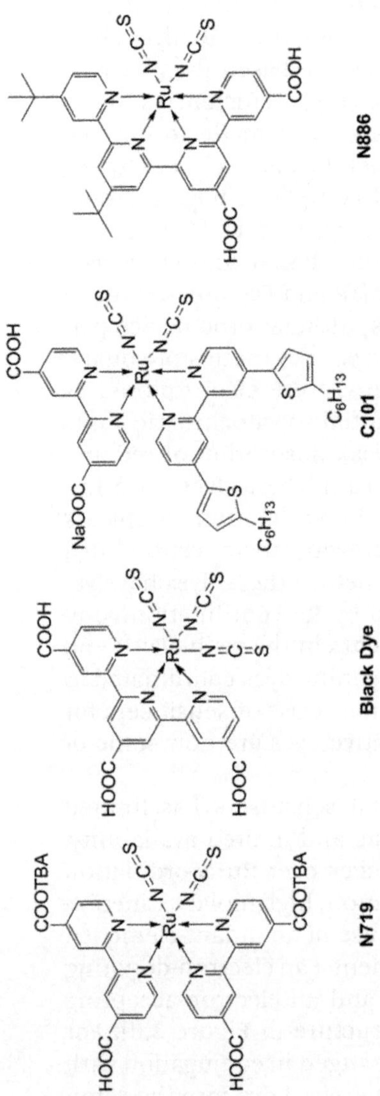

Figure 3.6 Some representative Ru-based dyes applied in DSSCs. TBA = tetrabutylammonium cation.

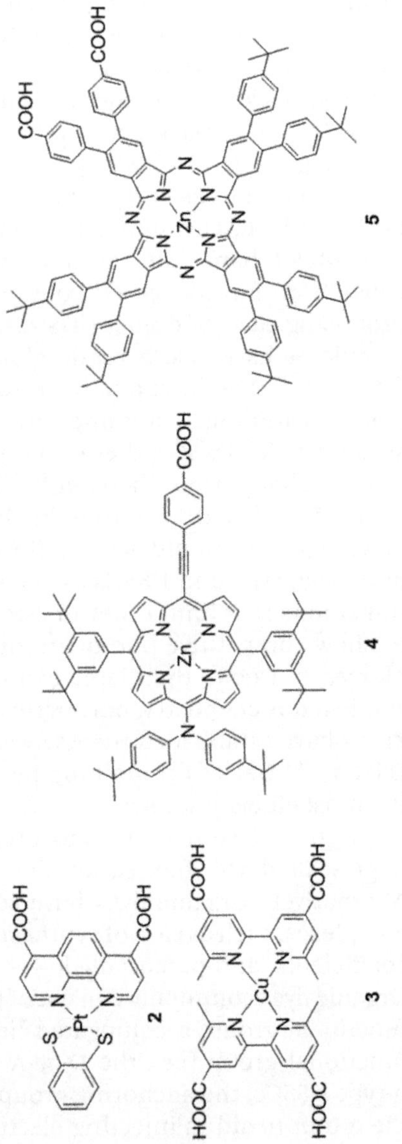

Figure 3.7 Some representative metal complexes used as photosensitisers in DSSCs.

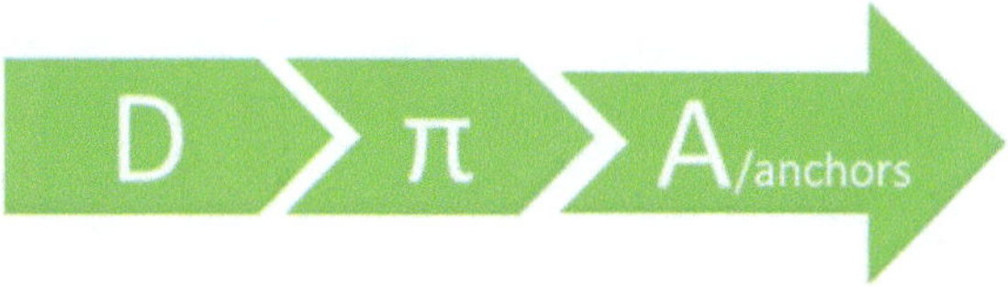

Figure 3.8 The classic structure of organic dyes in dye-sensitised solar cells. D: electron donor; π: conjugated linker; A: electron acceptor; anchors: anchoring group.

energy levels, and types of electronic transitions can be built into a dye. In D–π–A molecules, the presence of an ICT transition from the D to A state is essential for the dye function. When a dye absorbs light, an ICT transition is excited in the molecule, transferring charge from subunit D to A through the π-bridge. For n-type DSSCs, the excited dye injects one electron into the CB of the semiconductor *via* the electron acceptor group, A. As stated in Section 3.2, the wavefunction overlap between the anchoring group on A and the empty orbitals in the semiconductor facilitate charge injections. Therefore, the anchoring group functions best when in conjugation with the electron-accepting unit A.

Some of the most studied dyes are based on the triarylamine electron donor group, because of the group's stability as the oxidised radical, the ability to transport holes, and the relative ease of functionalisation (Figure 3.9). Yanagida and co-workers first introduced the triphenylamine unit as electron donor in organic dyes and obtained overall power conversion efficiencies of 3.3% and 5.3% for dye **6** and **7**, respectively.[66] Later, various spacer moieties were introduced to develop dyes with a broader absorption band in the visible region (**8–13**).[67,68] A series of dyes were introduced with an ethylene unit as the π-linker, but, due to the reactivity of the ethylene unit, its use in dyes has decreased (**14–16**).[69–71] A major concern when designing dyes is matching the absorption spectrum of the dye to the solar spectrum and increasing the spectral overlap can be achieved through extending the donor group (**17**).[72] Changing the electron-donating or -accepting nature of the π-linker group affects the light absorption of the dye (**18–21**).[73,74] To reduce dye aggregation on the semiconductor surface, extended donors and twisted structures have been introduced (**22**).[75]

Organic dyes based on the coumarin functional group are an important class of sensitiser, as shown in Figure 3.10 for dyes **23** and **24** with cyanoacrylic acid as the acceptor unit. However, coumarin-based dyes are prone to aggregation, which can reduce power conversion efficiencies for solar cells using them with high surface coverage. With careful dye design, dye aggregation can be reduced and power conversion efficiencies can raised to 6.1–8.1% (**23**).[76] Horiuchi *et al.* first reported indoline-based dyes as possible low-cost, efficient materials for DSSCs (Figure 3.10).[77] Further structural optimisation, including the incorporation of the rhodanine framework (**27**) increased device efficiency to 9.5%.[78] Another important class of dyes

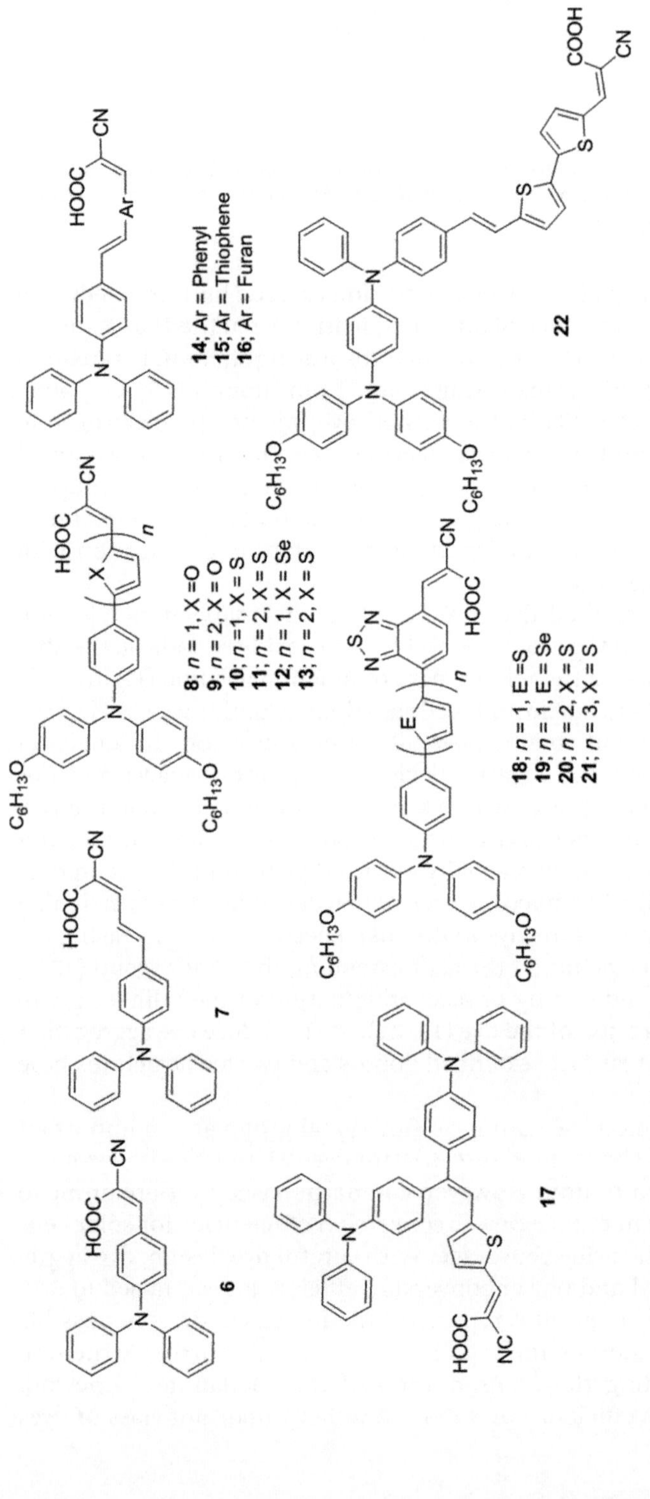

Figure 3.9 Representative examples of triphenylamine-based dyes.

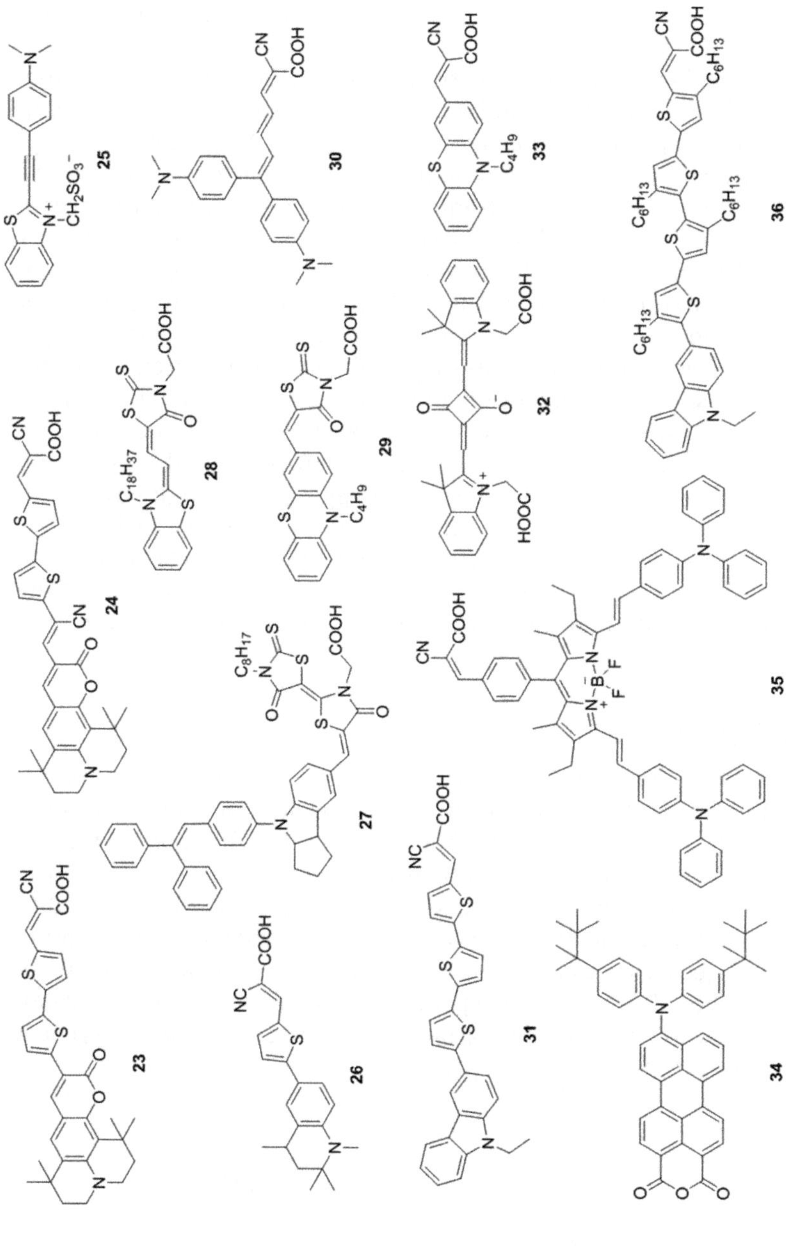

Figure 3.10 Representative examples of organic dyes containing alternative electron donors.

include *N,N*-dialkylaniline derivatives as the electron donor; molecule **30** is a prototypical example that has yielded power conversion efficiencies of 6.8% in optimised devices.[79] In heteroanthracene dyes, heteroatoms have been introduced into the anthracene framework; for example, xanthene, phenothiazine, and phenoxazine.[80] While these dyes have strong absorption in the visible spectrum, their transitions are narrow and do not capture a significant portion of the solar spectrum. In carbazole dyes, the carbazole functions as the electron donor. Solar cells based on carbazole dye, **36**, consisting *n*-hexyl-substituted quaterthiophene achieved 8.3% PCE.[81] Hemicyanine dyes are cationic dyes and usually consist of basic *p*-dialkylaniline groups as electron donors and cationic moieties as strong electron acceptors, linked together by methine (–CH=CH–) π-bridges. After enormous effort from Huang *et al.* the hemicyanine dye **25** produced an IPCE of nearly 100%.[82] Merocyanine-based dyes (**28**) have been well studied, but only achieve high power conversion efficiencies by suppressing aggregation on the surface.[83] Squaraine dyes (**32**) exhibit sharp and intense absorption bands in the visible and NIR regions that are often complementary to other dyes for co-sensitisation.[84] Perylene-based dyes have been selected for DSSCs due to their large molar extinction coefficients and outstanding chemical stability. However, it has been challenging to get high power conversion efficiencies due to poorly matched energetics and relatively narrow absorption bands. The boradiazaindacene (BODIPY) dye has been incorporated in a DSSC as a photosensitiser in a few cases mainly because of the excellent NIR absorption of the BODIPY moiety, as for example compound **35** in Figure 3.10.

3.3.2 p-type Dyes

From the working principle of a p-type DSSC, the excited state of the dye must inject holes into the valence band (VB) of a wide band gap p-type semiconductor, such as NiO. It is better that the donor (D) functional group of the dye is bound to the semiconductor surface to facilitate rapid hole injection. Therefore, the anchoring group(s) in p-type dyes should be linked to the electron-donating moiety. In order to suppress charge recombination between the injected holes and reduced dyes, the electron-accepting functional group should be spatially separated from the surface of the semiconductor. The classic p-type dyes arrangement features an anchor–D–π–A structure (Figure 3.11).

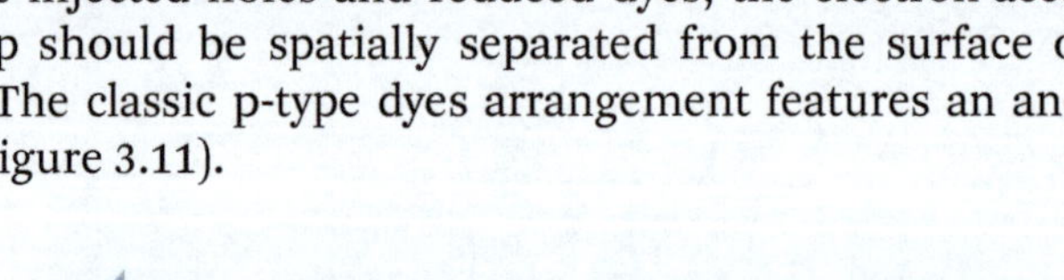

Figure 3.11 The classic structure of dyes for p-type DSSCs.

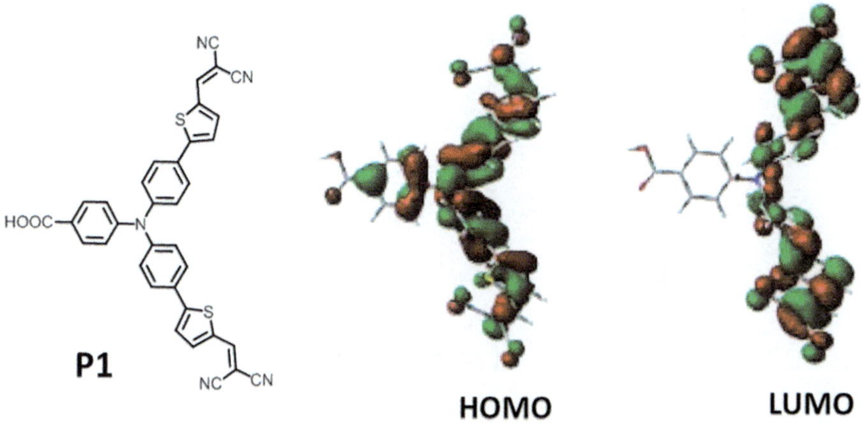

Figure 3.12 Molecular structure, and the calculated HOMO and LUMO of the P1 dye. HOMO and LUMO pictures are reproduced from ref. 86 with permission from American Chemical Society, Copyright 2010.

Figure 3.12 shows the state-of-the-art organic p-type dye, P1. In P1, the carboxylic acid acts as an anchoring group, which is linked to the electron-donating triphenylamine group. Thiophene acts as a π-spacer to connect the electron donor and acceptor units and also to extend the π-conjugated system, which broadens the absorption spectrum of the dye. The electron-accepting group is a malononitrile moiety. When a dye that is anchored on the semiconductor surface absorbs a photon, an electron will be moved from the donor to the acceptor. The donor can then oxidise the semiconductor (inject a hole) and the dye becomes reduced. On the basis of this design, the P1 dye rendered an overall efficiency of 0.15% with a J_{SC} of 5.48 mA cm^{-2}, a V_{OC} of 84 mV and a fill factor (FF) of 0.34 in a NiO-based p-type device.[85] The PCE is well below that from n-type devices. One of the primary reasons for the poor efficiency is that charge recombination between the reduced P1 dye and an injected hole in NiO was *ca.* 200 ps, which results in insufficient dye regeneration. The results imply that the acceptor unit in the P1 dye is not sufficiently far away from the semiconductor interface.

One strategy to solve the issue of charge recombination is increasing the linker unit length to make the acceptor group further away from the donor part, which is beneficial to push the electrons away from the semiconductor surface. Theoretically, this can prevent charge recombination and enhance the dye regeneration and thus improve overall efficiency. Bach and co-authors therefore developed a series of oligothiophene-based dyes for p-DSSCs and systematically investigated the effect of linker length on the charge separation lifetime as well as on the device performance (see Figure 3.13).[87]

According to transient absorption data, they conclude that the increase in the number of thiophene units between the donor and acceptor can

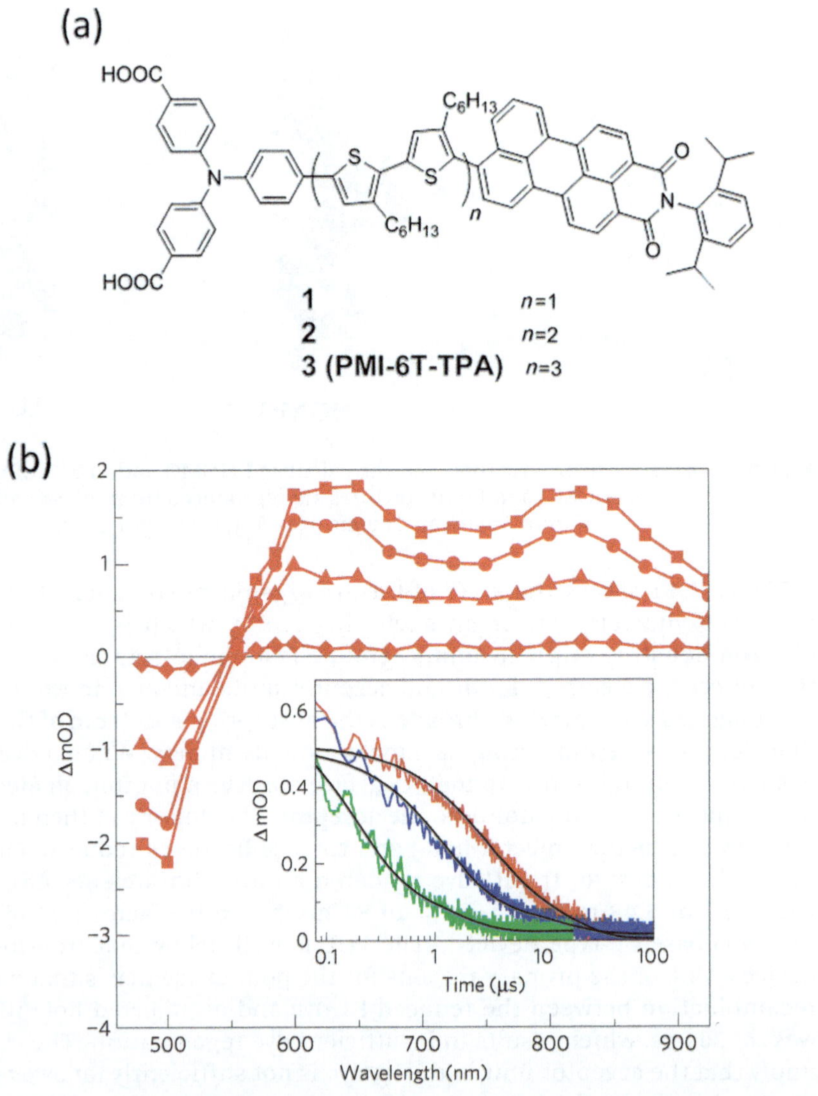

Figure 3.13 (a) Molecular structures of oligomer thiophene-based dyes. (b)Transient absorption decay signals of NiO films sensitised with 1 (green), 2 (blue) and 3 (red). Observation wavelength: 700 nm, laser intensity: 70 µJ cm^{-2} per pulse for 1 and 2 and 35 µJ cm^{-2} for 3. Reproduced from ref. 87 with permission from Springer Nature, Copyright 2009.

significantly prolong the lifetime of the reduced dye. The dyes with two, four and six thiophene units showed that the lifetime $\tau_1(\tau_2)$ of the reduced dye increased from 0.17 µs (1.34 µs) to 1.22 µs (12.17 µs), then to 2.54 µs (13.56 µs), as obtained from a double exponential model. The efficiencies of the p-type devices based on these dyes also improved along with the increase

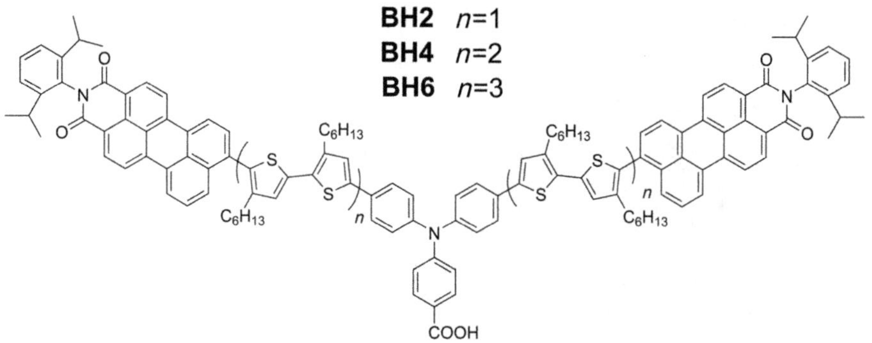

Figure 3.14 Triphenylamine dyes with double arms for p-DSSCs.

in thiophene linking groups and charge separation time. The dye featuring six thiophene units gave the highest efficiency among the three dyes, 0.41% with a J_{SC} of 5.35 mA cm^{-2}, a V_{OC} of 218 mV and an *FF* of 0.35.

A similar strategy has been further improved by Wu and co-workers.[88] Like the P1 dye, in the BH series of dyes there are two electron-accepting groups that are bound through linkers to a triphenylamine electron donor (Figure 3.14). The BH dyes have significantly increased extinction coefficients (99 980 M^{-1} cm^{-1}), which aids in light harvesting on the comparatively thin p-type semiconductor films. In addition, the BH series dyes can effectively block the semiconductor surface, which inhibits the contact between semiconductors and redox couples and suppresses charge recombination between them. However, with six thiophene units in each arm, BH6 did not yield conversion efficiency as a result of low dye loading from the large and bulky molecular structure. Therefore, BH4 dye showed the best efficiency of 0.28% with a J_{SC} of 7.4 mA cm^{-2}, a V_{OC} of 128 mV and an *FF* of 0.30.

By changing acceptor units in dyes, it is also possible to tune the performance of the device. The concept is that different electron acceptors have different electron withdrawing ability, which can influence the potential levels of excited dye and reduced dye and band gap, which in turn will affect the hole injection, dye regeneration, and the absorption spectrum. Gibson and co-workers used BODIPY and cationic 1-hexyl-2,3,3-3H indolium (CAD) as acceptors to build up the dyes (Figure 3.15).[89] Compared to the P1 dye, the dyes with BODIPY (GS1) and CAD (CAD3) acceptors show much more red-shifted absorption spectra. In addition, GS1 and CAD3 dyes have a smaller driving force for charge separation, −0.98 eV and −0.91 eV, respectively, compared to P1 dye, −1.21 eV, thus minimising the energy loss in solar cell devices. Consequently, CAD dye rendered an impressive J_{SC} of 8.21 mA cm^{-2} due to its broad light absorption, without sacrificing the charge separation efficiency.

Besides organic dyes, metal complexes have also been applied in p-type dye-sensitised solar cells; however, most of them have worked unsatisfactorily. One of the reasons is that recombination processes in metal

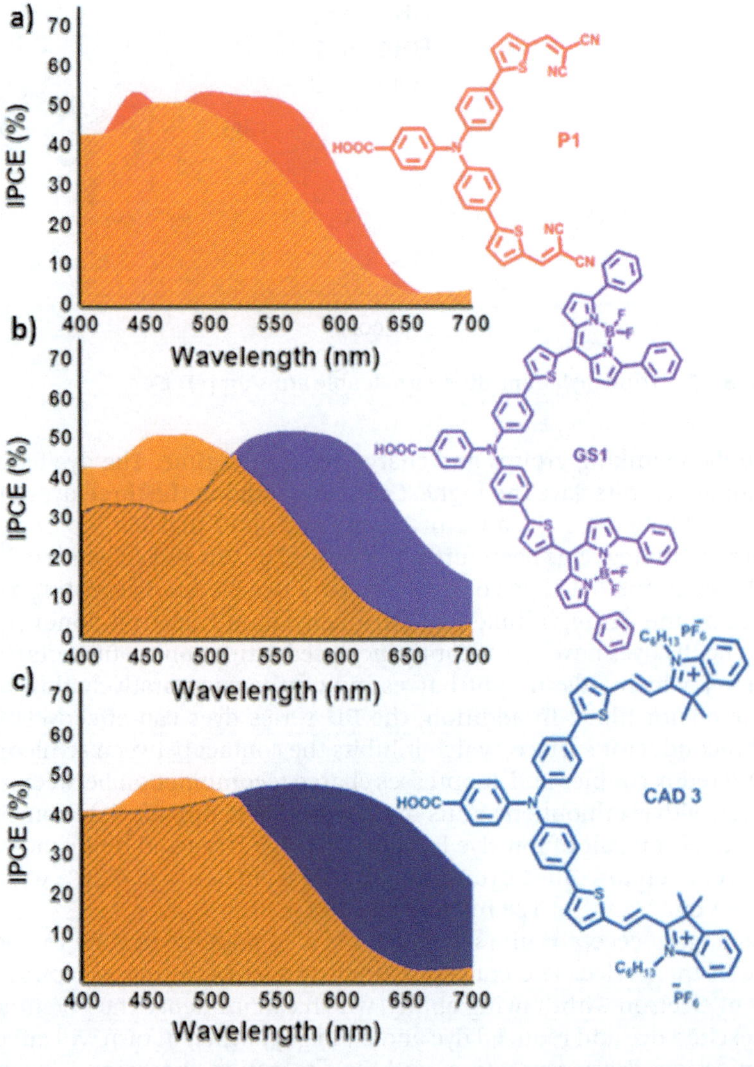

Figure 3.15 The structures and IPCE spectra of triphenylamine dyes with different double arms in p-DSSCs. In orange are the complementary IPCE spectra for the n-type dye D35 on TiO_2 used in a tandem solar cell. Adapted from ref. 89 with permission from the Royal Society of Chemistry.

complex-based p-type devices are rapid. Odobel and co-workers developed several iridium complexes that showed long-lived charge separation states up to a nanosecond timescale.[90] However, these dyes did not yield high photocurrent due to poor light-harvesting efficiency over the visible light region. Porphyrin dyes have excellent light harvesting of visible light and some Zn–porphyrins have shown outstanding performance in n-type devices. In 1999, Lindquist and co-workers[7] tried to use a simple porphyrin dye, meso-tetra(carboxyphenyl) porphyrin (TCPP) in a p-type device, but its performance was very poor. The poor performance could be due to the short intramolecular charge separation state of TCPP, which influences the hole injection. Also, the excited electrons are still concentrated on the porphyrin framework, which is close to the semiconductor surface, leading to fast charge recombination (Figure 3.16).

One strategy to solve the problem of interfacial charge recombination and poor intramolecular charge separation, was to introduce a secondary acceptor in TCPP; however, the synthetic work is usually complex for porphyrin dyes. Alternatively, Tian and co-workers adopted a zinc-coordinated TCPP (ZnTCPP) dye to sensitise a NiO electrode and the electrode was subsequently immersed in a fullerene derivative with a pyridine unit (named as C_{60}PPy) solution to form a supramolecule with ZnTCPP *in situ*.[52] The structure of the supramolecule formed is shown in Figure 3.17. The axially coordinated C_{60}PPy can work as a strong electron acceptor. The matched energy between ZnTCPP and C_{60}PPy is beneficial for fast intramolecular electron transfer from the porphyrin to the fullerene. Also, since the fullerene part in this case is located far away from the porphyrin framework as well as from the NiO surface, it can help with prolonging charge separation, and so is expected to improve dye regeneration and to suppress the charge recombination processes. The supramolecular system ZnTCPP–C_{60}PPy eventually gave an efficiency of 0.09%. In the absence of C_{60}PPy, the sole ZnTCPP-based device only gave an efficiency of 0.02%. The J_{SC} was significantly increased from 0.5 mA cm^{-2} to 1.5 mA cm^{-2} by simply adding C_{60}PPy as a secondary acceptor. From the IPCE spectrum, the ZnTCPP–C_{60}PPy supramolecule-based device showed the maximum value of 33% at 450 nm and 12% at 560 nm, which is responsible for the increased photocurrent.

As shown throughout this section, the structure of the dye plays a crucial role in solar cell performance, for both n-type and p-type DSSCs. Therefore, optimising the dye is the most common strategy applied to improve the performance and stability of DSSCs.

3.4 Charge Transport Mediators

3.4.1 Liquid Electrolyte

The electrolyte is one of the main components of liquid-junction dye-sensitised solar cells (DSSCs). Its function is to transport ions to the counter electrode and to regenerate the ground-state dye, which completes the circuit inside the DSSC. The electrolyte can greatly affect the performance and stability of the solar cells.

There are three parts in the electrolyte of the DSSC: (1) the redox couple, (2) additives, and (3) solvent. In this section, we will discuss the details of these three components.

Figure 3.16 The structure, HOMO and LUMO of TCPP. Reproduced from ref. 52 with permission from The Author.

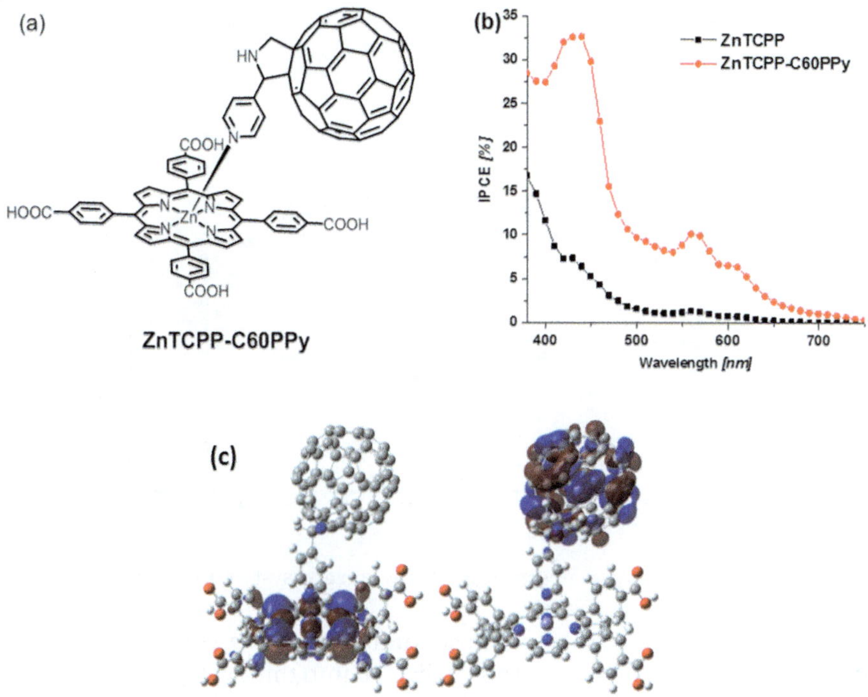

Figure 3.17 (a) The molecular structure of the ZnTCPP–C60PPy aggregate. (b) IPCE spectra of p-type DSSCs based on ZnTCPP alone and the supramolecular system ZnTCPP–C60PPy. (c) Electron distribution at HOMO (right) and LUMO (left) levels. Reproduced from ref. 52 with permission from The Author.

3.4.1.1 The Redox Couple

Redox couples and redox systems are the crucial parts in the electrolyte. They are the carriers of the electron, and fulfil the electron transfer process to ensure current flows through the solar cell. There are two main redox systems in liquid-type DSSCs, called the single redox system and the tandem redox system.

3.4.1.1.1 Single Redox System. In the single redox system there is only one redox couple, which is made up of oxidised species (Ox) and reduced species (Red). The single redox system works as follows in an n-type DSSC. When the dye absorbs the light, the dye injects electrons into the semiconductor forming Dye$^+$ (D$^+$), which is now a sufficiently strong oxidant to oxidise the Red in the redox couple and in so doing forms Ox. The Ox diffuses to the counter electrode of the solar cell device and collects the electron supplied by the circuit and is reduced back to Red. This process is, typically, catalysed by a thin layer of Pt deposited on the counter electrode.[91] The basic function cycle of a single redox system inside a DSSC is shown as a schematic description in Figure 3.18.

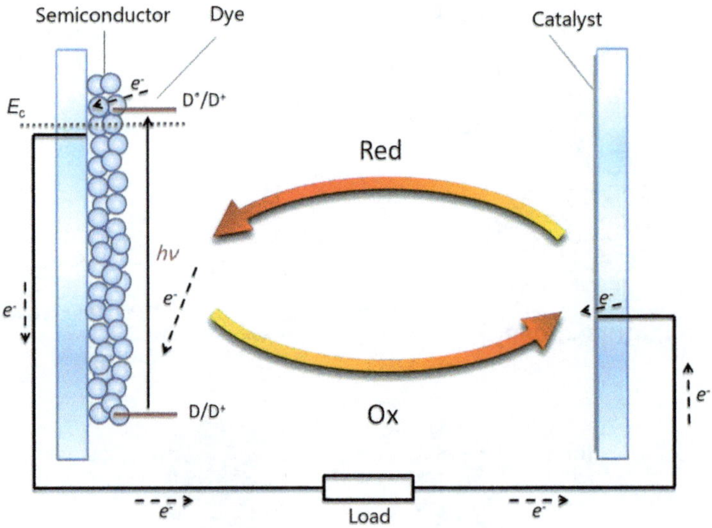

Figure 3.18 The schematic description of single redox system in n-type DSSCs.

The single redox system is the most studied redox system in DSSCs. The different types of redox couples can be divided into three categories: (1) inorganic redox couples, (2) metal complex redox couples, and (3) organic redox couples. Since most studies have been carried out on the I^-/I_3^- redox couple we discuss this separately.

I^-/I_3^- Redox Couple. The most frequently applied redox couple for dye-sensitised solar cells is I^-/I_3^-. It dominated high-efficiency DSSC devices for nearly 20 years, following the pioneering paper of Grätzel in 1991.[5] However, the oxidation of I^- to I_3^- is not a single-electron process, and polyiodide species like I_5^-, I_7^- and I_9^- exist in equilibrium with I^- and I_3^- in the electrolyte. Therefore, the complete operational mechanism inside a solar cell that includes I^-/I_3^- is not as straightforward as it is shown in the schematic description in Figure 3.19. Boschloo and Hagfeldt have studied the mechanism taking place in the typical high-efficiency solar cell with a $[Ru(bpy)_3]^{2+}$-based dye (known as N3) and I^-/I_3^--based electrolyte. They proposed that, rather than a simple outer-sphere electron transfer reaction directly between the oxidised Ru dye and I^-, instead, the Dye^+ coordinated with I^- to form a single complex that reacts with a second I^- species to form an $I_2^{-\bullet}$ radical. After this, two $I_2^{-\bullet}$ radicals react by disproportionation to form I_3^- and I^-.[92] The detailed kinetics for the processes can be found in eqn (3.5)–(3.7) and Figure 3.19.

$$D^+ + I^- \rightarrow (D\cdots I) \tag{3.5}$$

$$(D\cdots I) + I^- \rightarrow D + I_2^{-\bullet} \tag{3.6}$$

$$2I_2^{-\bullet} \rightarrow I_3^- + I^- \tag{3.7}$$

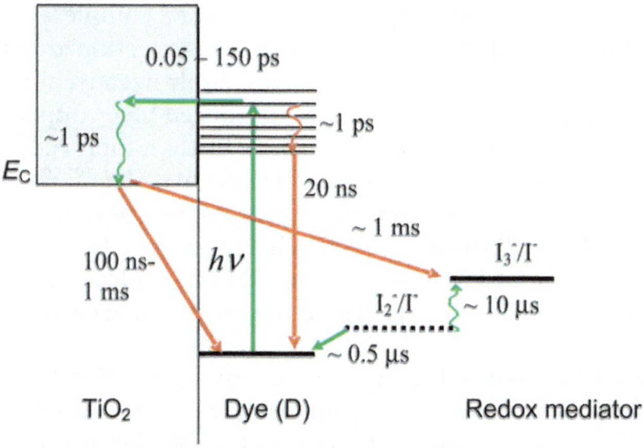

Figure 3.19 Kinetics of the N3-sensitised TiO_2-based solar cell with I^-/I_3^- redox mediator. Reproduced from ref. 92 with permission from American Chemical Society, Copyright 2009.

At present, the highest reported efficiency for a DSSC is not based on I^-/I_3^-. However, so far the highest certified recorded efficiency for a DSSC remains with this redox couple.[93] Moreover, most commercialised solar cells are still based on this electrolyte due to its high long-term stability in the device.

Metal Complex Redox Couples. For quite a long time the combination of the I^-/I_3^- redox couple and Ru dyes dominated in the field. However, the corrosive nature of the I_2 towards metal contacts and sealing materials, as well as the comparably fixed redox potential, limited the large-scale application of I^-/I_3^--containing devices.

To resolve these problems, many metal coordination complex redox couples were studied, in which the energy potentials can be adjusted by changing the ligand or metal in the complex. However, unlike with I^-/I_3^-, coordination complexes rely on a one-electron transfer system that resulted in high rates of charge recombination between the oxidised metal complex species and electrons in the conduction band of the wide band gap semiconductor (most often TiO_2). This "shunt" lowered the efficiency and initially hindered the use of these redox couples in solar cells.[94–96]

In 2010, Hagfeldt and Sun devised a new strategy to improve the efficiency of DSSCs with metal coordination complex redox couples, which combined an organic dye with long alkyl chains and a cobalt-based redox couple.[97] The long alkyl chains on the organic dye efficiently blocked the recombination between the oxidised metal complex and the electrons in TiO_2. In the following year, the Grätzel group at EPFL reported a new record efficiency of 12.3% for a DSSC device with this new strategy.[65] At present, the highest published efficiency for a DSSC is 14.3% for a device containing a cobalt(III/II) tris(1,10-phenanthroline) complex ($[Co(phen)_3]^{3+/2+}$).[154]

Besides cobalt complexes, copper complex-based complexes are also suitable redox couples in DSSCs, due to their faster electron transfer kinetics (particularly for dye regeneration) and comparably negative redox potential. In 2005, Fukuzumi and co-workers first introduced blue copper model complexes into DSSCs.[98] These devices have been further improved,[99,100] and currently, the record efficiency for Cu-based DSSCs is 10.3%.[101,102]

Other metals are also used to form the complexes for redox couples, such as Fe[103–106] and Ni.[107,108] However, solar cell devices based on these redox couples are not as efficient as those with Co and Cu complexes at present. More efforts are needed to show the potential of these complexes in the future.

Organic Redox Couples. Organic redox couples have also been applied in DSSCs, and can be divided into the following categories: (1) TEMPO and its derivatives,[109–111] (2) disulfide/thiourea and disulfide/thiolate-based redox couples,[112–118] (3) tetrathiafulvalene-based redox couples,[119,120] and (4) other organic redox couples such as quinone/hydroquinone,[121] phenothiazines,[122] and triarylamine.[123]

Inorganic Redox Couples. Inorganic redox couples other than the I^-/I_3^- redox are mainly Br^-/Br_3^-,[124–126] interhalogen,[127] and pseudohalide redox couples.[128–130] Due to efficiency limitations similar to I^-/I_3^- (most are two-electron systems) and stability problems, focus on these redox couples has diminished in the field.

3.4.1.1.2 Tandem Redox Systems. A perfect redox couple should undergo rapid reductive quenching of the oxidised dye (regeneration) and rapid reduction at the counter electrode, with the help of a catalyst. However, it is very difficult to find one perfect redox couple that has excellent properties for both aspects. Therefore, a new strategy of tandem redox systems has been invented, in which one redox couple is optimised for the working electrode, and the other is optimised for the counter electrode. This overcomes issues such as slow diffusion or electrode kinetics to improve the efficiency of the whole system.

In 2006, Bignozzi *et al.* studied the tandem $[Co(DTB)_3]^{2+}$/Fc system. (Fc = ferrocene, DTB = 4,4′-dimethyl-2,2′-bipyridine), and claimed that the photooxidised dye would be regenerated by Fc forming Fc^+ with Fc^+ then rapidly intercepted by the Co(II) in the $[Co(DTB)_3]^{2+}$/Fc system.[131] In 2010, Caramori *et al.* demonstrated that a combination of $[Fe(DMB)_3]^{2+}/[Fe(DMB)_3]^{3+}$ and $[Co(DTB)_3]^{2+}/[Co(DTB)_3]^{3+}$ redox couples could be used to improve the electron-collection efficiency in a DSSC sensitised by a (thienylterpyridine)ruthenium complex (DMB = 4,4′-dimethyl-2,2′-bipyridine). The Fe-based redox couple was assumed to work mainly on the TiO_2 working electrode side, whereas the cobalt-based redox system was thought to work at the counter electrode side.[103] However, the overall efficiencies of these solar cells was quite low (less than 1%). In 2012, Cong *et al.* reported the first high-efficient redox system S^{2-}/S_x^{2-}–I^- system, which reached an efficiency of 9.1%;[132] then the $Co(bpy)_3{}^{2+/3+}$–I_2, $[Co(bpy)_3]^{2+/3+}$–Fc^{+}[133] and TEMPO– $[Co(bpy)_3]^{2+/3+}$ [134] (TEMPO = 2,2,6,6-tetramethylpiperidine 1-oxyl) systems were also reported. At present, the most efficient system is TPAA–$[Co(bpy)_3]^{2+/3+}$ (TPAA = tris(*p*-anisyl)amine) which has an efficiency of 10.5%.[135]

The tandem redox system usually has two different redox couples with different redox potentials, in which, typically, one redox couple (Red1/Ox1) dominates the dye regeneration process on the working electrode, and the other redox couple (Red2/Ox2) works on the counter electrode. However, it is very difficult for all redox species of these two redox couples to exist at the same time. Usually, three redox species exist in the electrolyte, or we can say that there is an extremely small amount of the fourth species in the electrolyte. Therefore, the tandem system can be divided into two types: the electrolyte with Red1, Red2 and Ox2, and the electrolyte with Red1, Ox1 and Ox2. For the first type of electrolyte in an n-type DSSC, Red1 first reacts with oxidised dye forming Ox1-g. Ox1-g diffuses and has electron exchange with the existing Red2, generating Red1 and Ox2. Finally, Ox2 gets an electron from the counter electrode and changes to Red2 to finish the electron transfer cycle. TEMPO–$[Co(bpy)_3]^{2+/3+}$ and TPAA–$[Co(bpy)_3]^{2+/3+}$ systems work with this mechanism. The schematic description can be found in Figure 3.20(a). For the second kind of electrolyte, Red1 regenerates the oxidised dye and forms Ox1 as usual. However, on the counter electrode, Ox2 commonly has a faster speed collecting electrons from the counter electrode than that of Ox1. Therefore, mainly Ox2 gets electrons from the counter electrode, forming Red2-g, then Red2-g and Ox1 have an electron exchange and generate Ox2 and Red1 to finish the electron transfer cycle. $[Co(bpy)_3]^{2+/3+}$–I_2 and $[Co(bpy)_3]^{2+/3+}$–Fc^+ systems belong to this kind of mechanism. A schematic description can be found in Figure 3.20(b).

3.4.1.2 *Additives*

In addition to the redox couple, additives are added to the electrolyte to increase the efficiency and stability of the solar cells. These typically include inorganic salts, 4-(*tert*-butyl)pyridine (TBP), ionic liquids, and polymers.

3.4.1.2.1 Inorganic Salts. Inorganic salts are a standard additive in DSSC electrolytes. The cations of these salts can affect the conduction band of the semiconductor in solar cells and increase the conductivity of the electrolyte to improve the performance. For example, when electrons are accumulated in the TiO_2 particles, the cations start to adsorb on the surface or intercalate in the lattice to compensate the charge.[136] As a consequence, the conduction band moves in the positive direction, often leading to an increase in the photocurrent because electron transfer from the excited dye to the TiO_2 is more thermodynamically favourable. The magnitude of this shift is proportional to the ionic radius of the cation ($Li^+ < Na^+ < K^+ < Rb^+ < Cs^+$).[137] Consequently, Li salts are most frequently used, and are usually present in high efficiency.

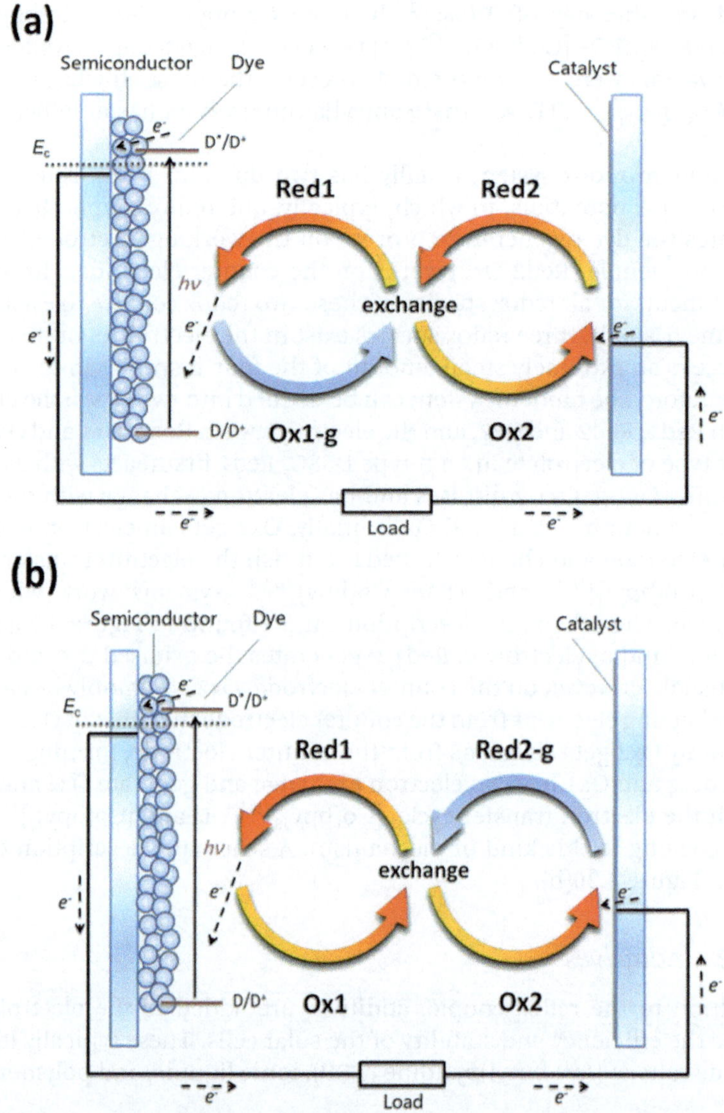

Figure 3.20 The schematic description of tandem redox systems for DSSCs with rapid regeneration at the photoanode (a) or cathode (b).

Another function of Li^+ in the electrolyte is charge-screening of the electrode surface. The Li^+ cations closely surrounding the TiO_2 can strongly affect the rate of electron recombination between the electrode and the electrolyte.[138] This reduces the energy loss and increases the fill factor of the DSSC. The commonly used Li salt for the I^-/I_3^- redox-based electrolyte is LiI, and in iodine-free electrolytes, $LiClO_4$ and lithium bis(trifluoromethane)sulfonamide (LiTFSI) are typically used.

3.4.1.2.2 4-(*tert*-Butyl)pyridine. Nitrogen-containing heterocyclic compounds, such as pyridine, benzimidazole, pyrazole, quinoline, and their derivatives, are typically added to the electrolyte in DSSCs. Among these chemicals, 4-*tert*-butylpyridine (TBP) gives the highest cell performance and is most commonly used. The addition of TBP can improve the photovoltage of the DSSC. The reason for this is that TBP acts as a Lewis base and adsorbs onto the surface of the TiO_2 particles,[139] leading to a negative shift in the conduction band. Furthermore, the bulky butyl groups help to block the TiO_2 surface from the redox couple, increasing the electron lifetime in the TiO_2.[140] Therefore, the photovoltage and fill factor of the solar cell device are increased.

3.4.1.2.3 Ionic Liquids. Ionic liquids are often used in I^-/I_3^- redox-based electrolytes. Ionic liquids can be used as both additives and solvents. When used as additives, they have a dual function of increasing the electrolyte conductivity and supplying the reduced species of the redox couple (*e.g.*, I^- or Br^-). Typical ionic liquid additives are 1-butyl-3-methylimidazolium iodide (BMII),[141] 1,2-dimethyl-3-propylimidazolium iodide (DMPII),[132,142] 1-methyl 3-propyl imidazolium iodide (PMII)[143] and dimethyl imidazolium iodide (DMII).[144] The role of ionic liquids as solvents will be discussed in the Section 3.4.1.2.

3.4.1.2.4 Polymer and Metal Oxide. Polymers and metal oxides can be added to form a quasi-solid-state electrolyte (*i.e.*, increasing the viscosity of the liquid so that it behaves almost as a solid to improve the stability of electrolyte). Two categories of polymers are typically used: (1) thermoplastic polymer electrolytes (TPPE) and (2) thermosetting polymer electrolytes (TSPE). The difference between these two kinds of electrolytes is that TPPEs are based on physical cross-linking and TSPEs are based on chemical cross-linking.[145] The polymers used in TPPE are always linear, such as poly(ethylene glycol) (PEG), poly(acrylonitrile) (PAN), poly(vinyl pyrrolidinone) (PVP), polystyrene (PS), poly(vinyl chloride) (PVC), poly(vinylidene ester) (PVE), poly(vinylidene fluoride) (PVDF), poly(methyl methacrylate) (PMMA), *etc.*[146] For TSPE, with a poly(oxyethylene)-*block*-imide[147] or P(VA-co-MMA)[148] copolymer-based quasi-solid-state electrolyte, the solar cell efficiencies have already exceeded 9.4%.

SiO_2,[149] Al_2O_3,[150] and ZnO[151] have also been added to form quasi-solid-state electrolytes. These metal oxide particles can also be added to the electrolyte with polymer-forming composite electrolytes.[152]

3.4.1.3 Solvents

The remaining component of the electrolyte is the solvent. The requirements for the perfect solvent for DSSCs are that it should (1) be chemically and physically stable, (2) have low viscosity to ensure the mass transport of the redox species, (3) completely dissolve/solvate the redox and additives, and (4) not dissolve the dye, electrodes, or sealing materials.[153] According to these requirements, three kinds of solvents are commonly applied to the electrolyte of the DSSC. They are organic solvents, water and ionic liquids.

3.4.1.3.1 Organic Solvents. Organic solvents are most frequently chosen as the electrolyte solvent for DSSCs. Many organic solvents have been tested, together with different kinds of redox couples, but one that is frequently found to fulfil the necessary criteria is acetonitrile. State-of-the-art solar cells, either with I^-/I_3^- or metal complex redox couples, have been assembled with acetonitrile-based electrolytes, with efficiencies of 11.9% and 14.3% separately.[93,154]

One disadvantage of acetonitrile (AN) is its high volatility, which limits its application in industry. To reduce the volatility, valeronitrile (VN) is sometimes mixed with the acetonitrile,[155] typically in a ratio of 85 : 15 (VN : AN; v/v) for high-efficiency solar cells.[156] Another disadvantage of acetonitrile is its relatively high toxicity. An alternative nitrile solvent, with a higher boiling point (164 °C) than acetonitrile (82 °C) and lower toxicity, is 3-methoxypropionitrile (MPN), which is often chosen for stable DSSCs. For example, a solar cell with an MPN-based electrolyte reached 10.0% efficiency, and kept 92% of its initial efficiency after a 1000-hour accelerated ageing test, under full Sun soaking conditions at 60 °C.[157] The physical parameters of some frequently used organic solvents for the electrolytes in DSSCs are listed in Table 3.2.

3.4.1.3.2 Water. Water is another preferred solvent for electrolytes, as it is inexpensive, abundant and environmentally friendly. During the early development stages of DSSC research, Ru-based dyes were often used.

Table 3.2 Physical parameters of some commonly used solvents for the electrolytes used in DSSCs[a] (ref. 6 and 145).

Name	Formula	Melting/ Boiling point (°C)	Viscosity (cP)	Dielectric constant	Donor number	$D(I_3^-)$ (cm^2 s^{-1})[b]
Water	H_2O	0/100	0.89	78	18.0	1.1×10^{-5}
Acetonitrile (AN)	CH_3CN	−44/82	0.33 (30 °C)	36	14.1	1.5×10^{-5}
Valeronitrile (VN)	$CH_3(CH_2)_3CN$	−96/139	0.78 (19 °C)	21		
3-Methoxy-propionitrile (MPN)	$CH_3O(CH_2)_2CN$	−63/164	2.5	36	16.1	$4\text{-}5 \times 10^{-6}$

[a]All data at 25 °C and 760 mmHg unless otherwise indicated.
[b]Apparent diffusion coefficient for triiodide in DSSC electrolyte, this also depends on electrolyte composition.

Since Ru dyes tend to be soluble in water, water-based electrolytes were avoided. However, according to the research of O'Regan *et al.*, Ru dye-based DSSCs with electrolytes containing up to 20% water and an I^-/I_3^- redox couple can maintain their efficiency without compromising their long-term stability.[158]

Examples of metal complex redox couples used in water-based electrolytes include iron complexes such as $K_4Ni^{II}[Fe^{II}(CN)_6]/K_3Ni^{II}[Fe^{III}(CN)_6]$[104] and $Fe(CN_6)^{4-/3-}$,[106] which have attained over 4% efficiency. Cobalt complexes are typically cationic (*e.g.*, $[Co(bpy)_3]^{2+/3+}$, $[Co(phen)_3]^{2+/3+}$ and $[Co(bpy\text{-}pz)_3]^{2+/3+}$, where bpy = 2,2′-bipyridine, phen = 1′,10-phenanthroline, bpy-pz = 6-(1*H*-pyrazol-1-yl)-2,2′-bipyridine) with PF_6^- or $[B(CN)_4]^-$ counter ions, and are insoluble in water. Chloride salts of the metal complexes are reasonably water-soluble and Ellis *et al.* reported an efficiency of 5.7% with this redox electrolyte.[159]

3.4.1.3.3 Ionic Liquids. Another class of solvent that is commonly used in DSSCs is ionic liquids. An ionic liquid is a salt that has a low melting point and relatively low viscosity, and is usually a free-flowing liquid at room temperature. Ionic liquids are composed of anions and cations. The cations are generally bulky ammonium or phosphonium salts or heteroaromatics, which have low symmetry, weak intermolecular interactions, and low charge densities. Some of the weakly coordinating anions may act as redox species.[160] The extremely low vapour pressure and high boiling point of ionic liquids make them a perfect solvent for electrolytes in stable DSSCs. However, ionic liquids tend to be hygroscopic and the high viscosity slows the diffusion of ions in the electrolyte, hindering their widespread application.

Ionic liquids have been used in a number of ways, such as (1) combined with polymers to form quasi-solid electrolytes, (2) or combined with different kinds of ionic liquids to reduce viscosity. For (1), TPPE[161] and TSPE[162,163] -based quasi-solid-state DSSCs have both been reported. However the polymers in the electrolyte slow down the ion diffusion in the device further. For (2), the $[B(CN)_4]^-$ anion has been shown to improve stability and reduce the viscosity.[164] However, the high toxicity of $[B(CN)_4]^-$ limits the application of this kind of electrolyte on a large scale. Therefore, an alternative anion needs to be found.

3.4.2 Solid-state Electrolyte

While liquid-junction DSSCs of the type developed by O'Regan and Grätzel in 1991,[5] and discussed in detail in the previous sections, have yielded the highest power conversion efficiencies (14%),[154] the volatile nature of the liquid electrolyte that is used in the DSSC significantly limits its future commercial application for this solar technology.[165] In order to address these issues, a solid-state hole transport material (HTM), 2,2′,7,7′-tetrakis-(*N*,*N*-di-*p*-methoxy-phenyl-amine) 9,9′-spirobifluorene (spiro-OMeTAD, Figure 3.21) was introduced as a replacement for the liquid electrolyte in an

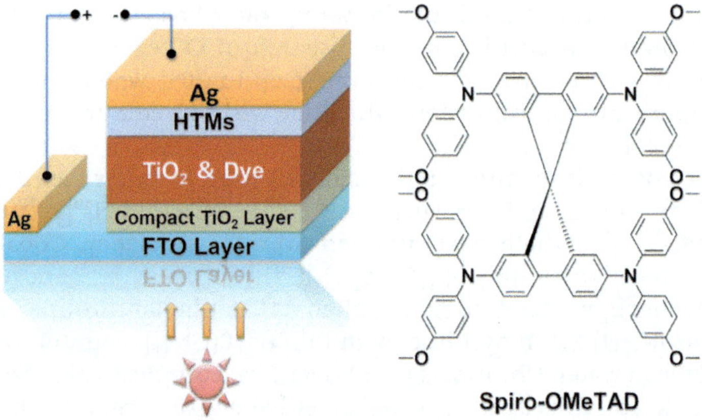

Figure 3.21 Device structure of an n-type ssDSSC and chemical structure of spiro-OMeTAD.

n-type solid-state dye-sensitised solar cells (ssDSSC, Figure 3.21) by Bach *et al.* in 1998.[166]

A typical device structure for a ssDSSC is shown in Figure 3.21. A dense, compact TiO_2 layer is deposited on the fluorine-doped tin oxide ($F:SnO_2$; FTO)-coated glass substrate by spray pyrolysis, which prevents short circuiting that would result from direct contact between the FTO and the HTM. The mesoporous TiO_2 film is applied on top of the blocking layer by screen printing, spin-coating or doctor blading a commercial TiO_2 paste, which is then sintered together with the compact layer *via* a heat treatment at 450–500 °C, to form an interconnected network of TiO_2. This network is subsequently decorated with a monolayer of dye molecules during the so-called dye sensitisation, which is typically achieved by immersing the substrate in a solution of the dye molecules. Then, the dyed mesoporous electrode is infiltrated, often by spin-coating, with a transparent organic HTM, typically spiro-OMeTAD. Assembly of the device is completed by evaporating a layer of metal as the counter electrode, for example, Ag or Au.[167]

A p-type ssDSSC was first proposed by Tian and co-workers in 2016.[168] A typical device structure of a p-type ssDSSC is shown in Figure 3.22. The p-type semiconductor NiO is used as the photocathode as well as as the electron-blocking layer, and the phenyl-C_{61}-butyric acid methyl ester ($P_{61}CBM$, Figure 3.22) is employed as the electron transport material (ETM). An impressive open circuit photovoltage of 620 mV was obtained in this p-type ssDSSC. Femtosecond- and nanosecond-resolution transient absorption spectroscopy has provided convincing evidence for hole injection from the excited p-type dye to NiO, followed by electron transfer from to $P_{61}CBM$. Although the PCE of p-type ssDSSCs is poor at present, there is much scope for research into this type of solar cell, due to its potential application in all-solid-state tandem DSSCs. In addition, materials development, fundamental understanding and device engineering of p-type ssDSSC are all attracting increasing research interest.

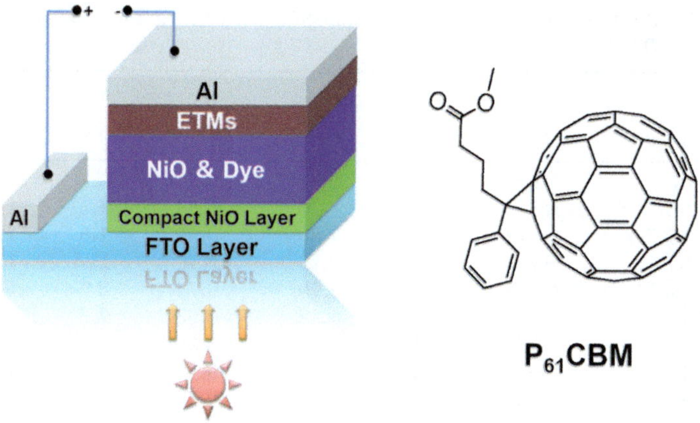

Figure 3.22 Device structure of a p-type ssDSSC and chemical structure of $P_{61}CBM$.

3.4.2.1 Organic Hole Transport Materials for ssDSSCs

To date, the record efficiency of 8.0% for an n-type ssDSSC has been achieved by using 1,1,2,2-tetrachloroethane-doped spiro-OMeTAD as the HTM and an organic D–A–π–A dye, AQ310, as the photosensitiser, which was reported by Sun and co-workers in 2017.[169] Although n-type ssDSSCs have been reported for quite some time, their efficiency is still less than 9%, which is far below the PCEs of liquid-junction DSSCs. There are two main limitations for further improving the photovoltaic performance of ssDSSCs compared with liquid-junction DSSCs. First, poor light-harvesting ability: the active mesoporous TiO_2 layer of a ssDSSC has been limited to less than 2.5 μm, resulting in incomplete light harvesting and consequently a lower photocurrent of the device.[170] Second, rapid interfacial electron–hole recombination decreases open circuit voltage.[165] The low conductivity and hole mobility of the HTM layer results in rapid electron–hole recombination in the device, thereby reducing the photovoltage of the solar cell and limiting the thickness of the electrode.[165] The solution to these problems is to develop new materials, such as photosensitisers with high molar extinction coefficients and HTMs with high hole mobility/conductivity.[171,172] Table 3.3 summarises the photovoltaic data of ssDSSCs with inorganic- and organic-based HTMs.

Organic HTMs play a vital role in the regeneration of the photooxidised dyes and transport of holes to the back contact in ssDSSCs. A high-performance HTM for ssDSSCs must meet several general requirements, which are as follows.[167]

1. The HTM must have high thermal and chemical stability.
2. The HTM must have redox potentials that are less energetic than the light absorber in its oxidised (or reduced state), for n-type and p-type DSSCs, respectively.

Table 3.3 Photovoltaic parameters of ssDSSCs with inorganic- and organic-based HTMs.

No	HTMs	Dyes	V_{OC} [V]	J_{SC} [mA cm^{-2}]	*FF*	η [%]	Reference
1	CuI	N3	0.59	6.84	0.57	3.8	174
2	CuBr	N3	0.40	4.30	0.44	1.5	175
3	CuSCN	D149	0.54	4.80	0.60	1.5	176
4	NiO	N3	0.48	0.15	N/A	N/A	177
5	CsSnI3	N719	0.72	15.90	0.74	8.5	178
6	AS44	Z907	0.72	5.38	0.58	2.3	188
7	X3	LEG4	0.90	9.70	0.66	5.8	179
8	X51	LEG4	0.92	9.27	0.70	6.0	185
9	X60	LEG4	0.89	11.38	0.72	7.3	186
10	Spiro-OMeTAD	AQ310	0.87	12.30	0.75	8.0	169

3. The HTM must be solution-processable, so that it may be deposited by spin-coating or similar techniques.
4. The HTM should be facile to synthesise and, ideally, non-toxic.
5. The HTM should have high charge-carrier mobility and conductivity.
6. The HTM should have a low tendency towards crystallisation so that homogenous thin films can be deposited on large areas.

Initially, inorganic p-type semiconductors, copper(I) salts (*e.g.*, copper(I) iodide,[173,174] copper(I) bromide,[175] copper(I) thiocyanate,[176] and nickel oxide)[177] were used as the HTMs for ssDSSCs by scientists, and unsatisfactory efficiencies of about 2% were achieved. However, such inorganic HTMs suffer from a tendency to crystallise, poor solubility in organic solvents and problems of insufficient pore filling, which are not easy to control for the fabrication of high-performance ssDSSCs. In 2012, Kanatzidis *et al.* employed an inorganic p-type semiconductor, $CsSnI_3$, as the HTM and a ruthenium dye, N719, as the photosensitiser for ssDSSCs, and an inspiringly high PCE of 8.5% was achieved.[178] The $CsSnI_3$ exhibited a high hole mobility of 585 cm^2 V^{-1} s^{-1} and excellent solubility in organic solvents for solution processing, such as γ-butyrolactone (GBL) and *N*,*N*-dimethylformamide (DMF). However, the $CsSnI_3$ crystals are sensitive to oxygen and moisture, which will lead to poor stability and reproducibility of the ssDSSC.

Organic HTMs exhibit unique advantages compared with inorganic HTMs, such as solution-processability, ease of synthetic variation, and tuneability of electronic properties.[179,180] The triphenylamine (TPA)-based compounds are the most commonly investigated HTMs in ssDSSCs. Durrant and co-workers reported in 2006 on a series of TPA-based oligomers (Figure 3.23) with varying degrees of polymerisation and hole mobilities.[181] The authors further investigated the hole transfer at the dye/HTM interface in ssDSSC devices. The results did not show an obvious correlation between the hole

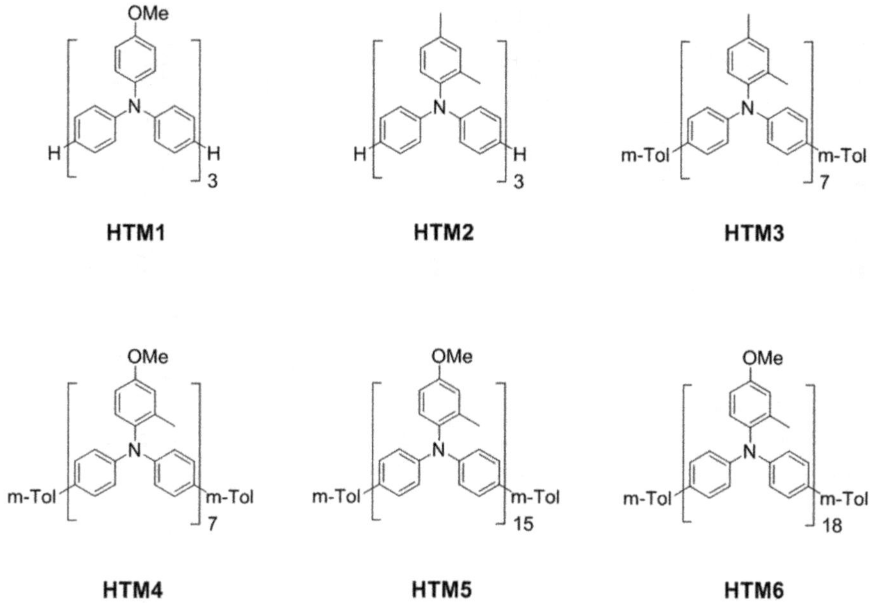

Figure 3.23 Chemical structures of the TPA-based oligomers.

mobility of the HTM and the overall device performance.[182] A variable that is difficult to control but has a strong effect on the overall device performance, including observed hole mobility, is how the HTM infiltrates the electrode (often called pore filling). In 2015 Sun *et al.* designed and synthesised a sequence of TPA-based HTMs named X1, X2, X3 and X35 (Figure 3.24(a)) with the addition of methoxy groups in order to further improve solubility and eventual pore filling of the HTM in the mesoporous TiO_2 film.[165,179,183] After systematically investigating the HTM in a ssDSSC by the transient photovoltage decay method, which can represent the charge recombination kinetics in a solar cell, they demonstrated that the V_{OC} was not only dependent on the oxidation potential of the HTM, but also proportional to the molecular size of the HTM (Figure 3.24(b) and (c)). The results showed that for smaller HTM molecules there was faster electron recombination in the device, and the authors reasoned that this was due to increased infiltration of the small HTM molecules into the mesoporous TiO_2, resulting in close contact between TiO_2 and the HTM.

Other small-molecule-based HTMs have been developed based on different core units, such as fluorene,[184] carbazole,[185] and spiro[fluorene-9,9′-xanthene] (SFX).[186,187] McGehee and co-workers introduced long alkyl chains into molecular HTMs (termed AS37 and AS44, Figure 3.25) to increase the solubility and the pore filling ability of the HTM.[188] Devices using the low-cost HTMs X51 and X60 (Figure 3.25) with carbazole and SFX as core units, developed by Sun and co-workers, achieved power conversion

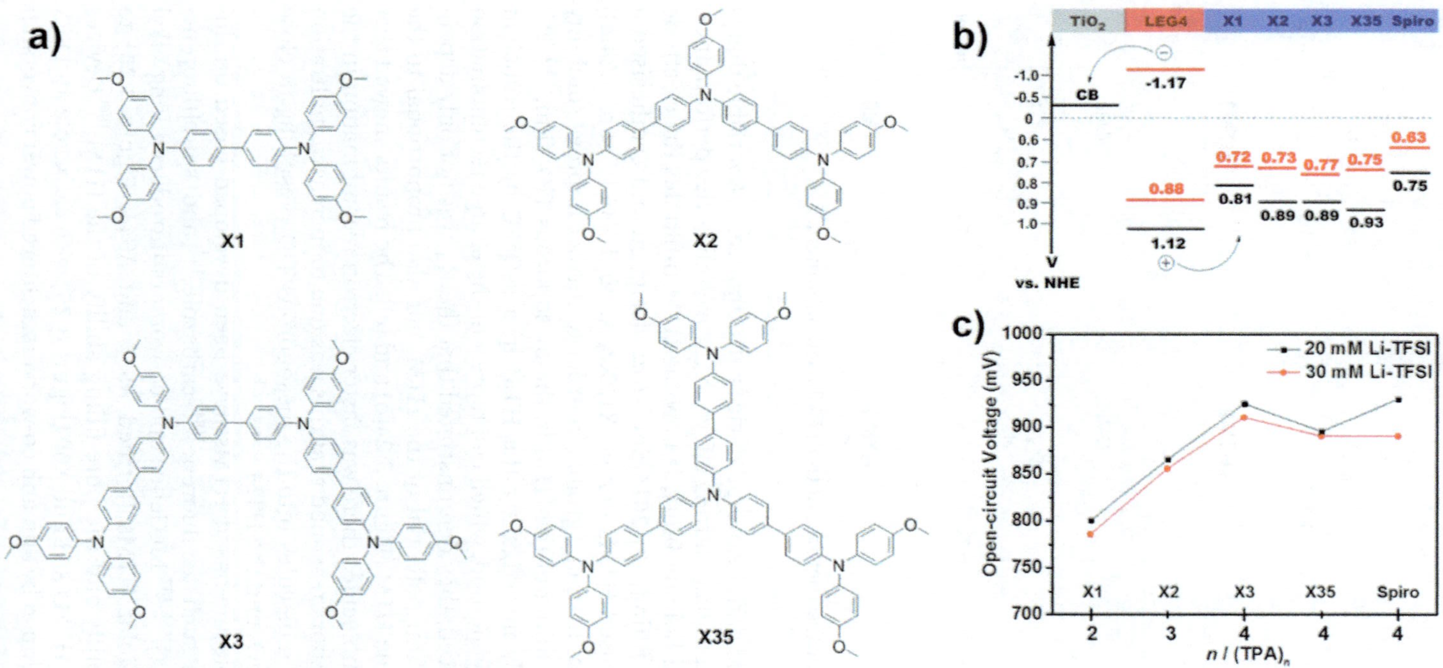

Figure 3.24 (a) Chemical structures of the TPA-based HTMs. (b) Energy level diagram of the TiO_2, dye and HTMs (red line measured in DCM, black line measured on TiO_2 electrode). (c) V_{OC} of the devices *vs.* oligomerisation number (n) of the HTM with different Li–TFSI doping concentrations. Reproduced from ref. 196 with permission from John Wiley and Sons, © 2014 WILEY-VCH Verlag GmbH & Co. KGaA, Weinheim.

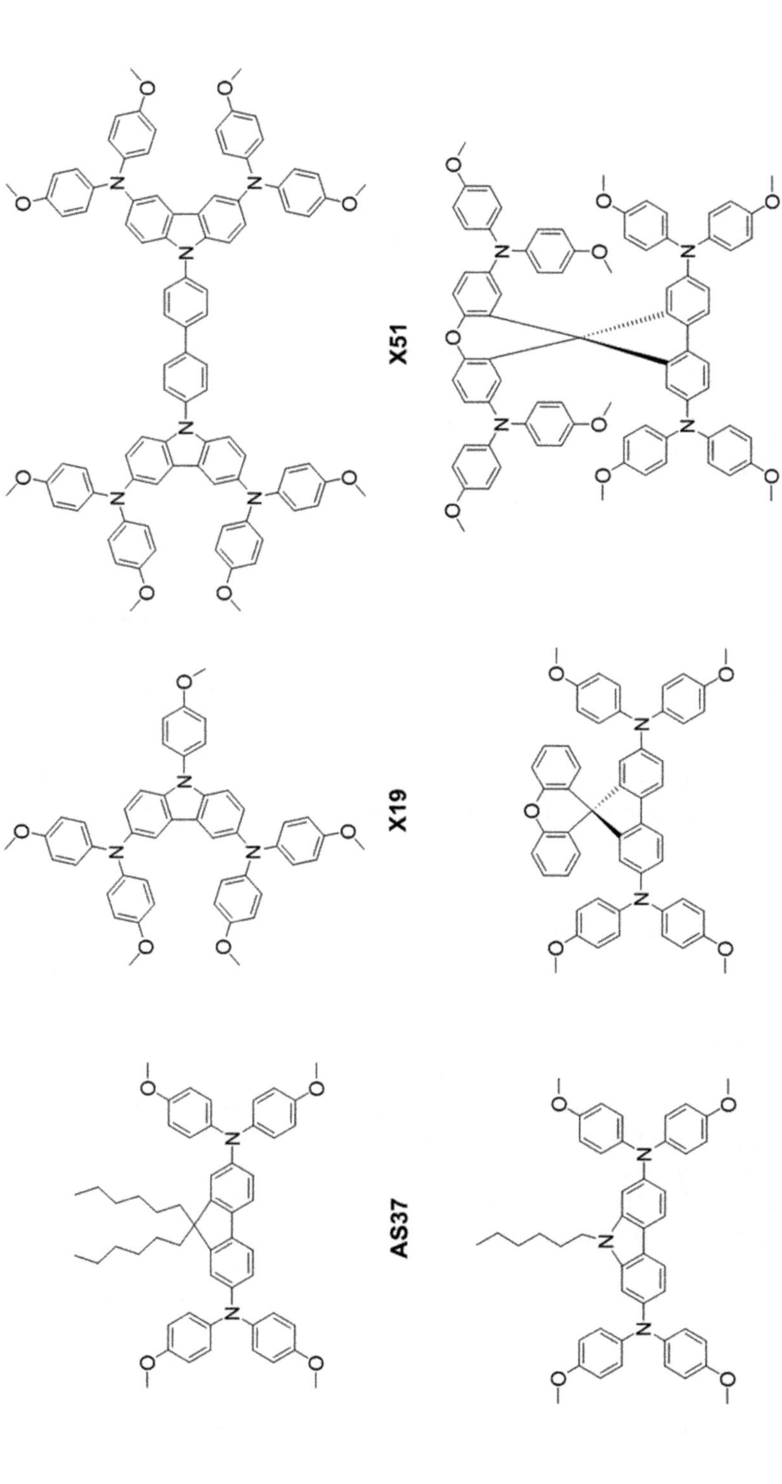

X51 X60 X19 X59 AS37 AS44

Figure 3.25 Chemical structures of the TPA-based HTMs with different core units.

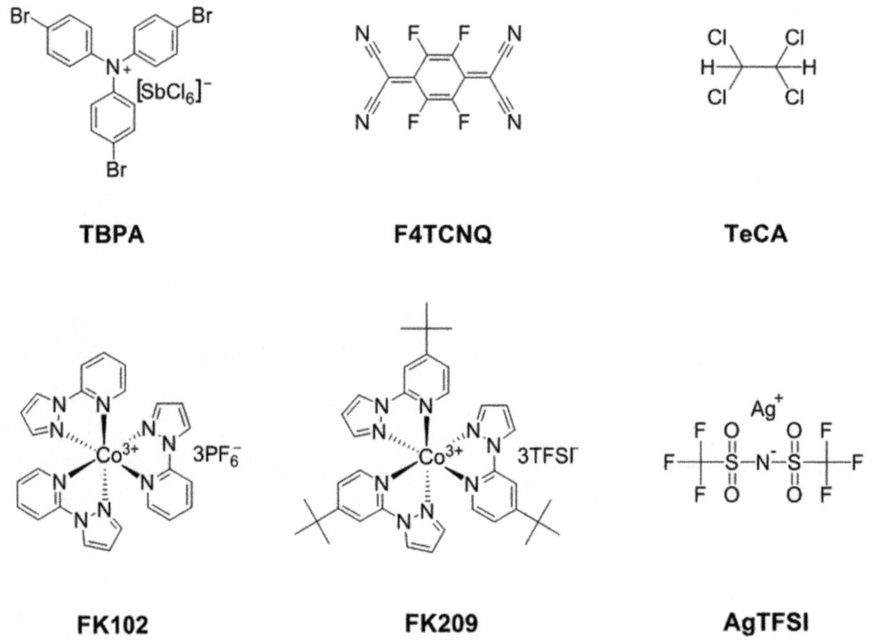

Figure 3.26 Chemical structures of different p-type dopants for HTMs.

efficiencies of over 6.0%.[185,186] Overall, the synthesis and development of new organic and inorganic HTMs is an area of active research that is continuously developing.

3.4.2.2 P-type Chemical Doping for HTMs

Compared to the inorganic HTMs, organic-based HTMs have suffered from low charge-carrier mobility that is obviated by chemical doping. Chemical doping of HTMs is used to control and increase the charge-carrier concentration and therefore the electrical conductivity of organic HTMs. The fundamental principle of chemical p-type doping in organic HTMs is similar to that applied to conventional semiconductors; small concentrations of an acceptor are used to oxidise a fraction of the HTM, increasing the hole density in the material. This method has been widely applied to p-type HTMs in ssDSSCs and related electronic devices.[189]

A series of chemical p-type dopants which have been utilised in spiro-OMeTAD-based ssDSSCs are shown in Figure 3.26. Tris(4-bromophenyl) ammoniumyl hexachloroantimonate ($(p\text{-BrC}_6H_4)_3N[SbCl_6]$, TBPA, Figure 3.26) was the first p-type dopant applied in a ssDSSC, and was reported by Bach *et al.* in 1998.[166] In 2011, Grätzel and co-workers applied a cobalt(III) complex (named FK102, Figure 3.26) as a p-type dopant,[190] which substantially increased the electrical conductivity of the spiro-OMeTAD layer, resulting in improved ssDSSC performances. A second-generation cobalt(III) complex, FK209 (Figure 3.26), was developed to function as a p-type dopant for spiro-OMeTAD.[191] This complex exhibited better solubility in the precursor solution used to deposit spiro-OMeTAD. Recently, scientists demonstrated that HTMs can become oxidised in the presence of lithium bis(trifluoromethanesulfonyl)imide (LiTFSI), a common additive to spiro-MeOTAD in ssDSSCs, and an oxygen atmosphere.[192,193] This implies that no additional chemical dopant is needed, however, The LiTFSI doping method is also reported to cause poor device stability and reproducibility. Alternative electron-accepting dopants have been reported (Figure 3.26), such as 2,3,5,6-tetrafluoro-7,7,8,8-tetracyanoquinodimethane (F4TCNQ)[194] and silver bis(trifluoromethanesulfonyl)imide (AgTFSI).[195] However, some of these strongly absorb visible light, have limited solubility in organic solvents (*e.g.* toluene or chlorobenzene), and/or are expensive to synthesise on a large scale. There is a need to develop solution-processable, visibly transparent, and low-cost p-type dopants for high-performance ssDSSC. In 2015, Sun and co-workers first demonstrated that a chlorinated, organic solvent, 1,1,2,2-tetrachloroethane (TeCA) (Figure. 3.26),[196] which can be used both as a co-solvent and as an effective additive for both the triarylamine-based organic HTM and spiro-OMeTAD. This additive significantly improves the conductivity of the HTM film and the corresponding ssDSSC device performance. While research into dopants for HTMs has led to improved device efficiency, further work is needed on developing stable, environmentally friendly additives for ssDSSCs.

3.5 Characterisation of DSSCs

While there are many different methods used to characterise DSSCs, this chapter will focus on three specific methods that report on different mechanistic aspects of a DSSC. These are: transient optical absorption spectroscopy, X-ray photoelectron spectroscopy, and electrochemical impedance spectroscopy. Transient absorption spectroscopy (see Section 3.5.1) is designed to provide insights into kinetic processes in DSSCs that take place on a femtosecond to microsecond timescale. In DSSCs, there are many interfaces between molecules and materials, and photoelectron spectroscopy (see Section 3.5.2) can directly probe the molecular interactions at these interfaces. Finally, since DSSCs are electronic devices, impedance spectroscopy (see Section 3.5.3) can be used to measure electronic components in DSSCs.

3.5.1 Transient Absorption Spectroscopy for DSSCs

Due to the shear breadth of studies that have been performed on DSSCs by transient absorption spectroscopy (TAS), this section does not aim to be a comprehensive overview. Instead, it highlights several examples of what

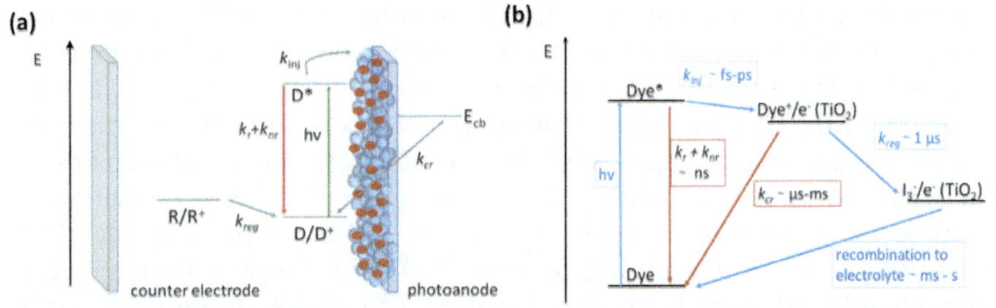

Figure 3.27 (a) A schematic drawing of the kinetic processes in an n-type DSSC photoanode. (b) A simplified energy diagram depicting the same processes; k_{inj} describes the electron injection process, k_{cr} the charge recombination from the conduction band electrons to the oxidised dye, k_{reg} is the rate constant for regeneration of the oxidised dye by the redox mediator, and k_{r} and k_{nr} represent the radiative and non-radiative decay of the oxidised dye back to the ground state.

transient absorption can be used for and how powerful it has been in showing the complexity of the interfacial processes occurring in DSSC materials of all kinds.

In a DSSC all of the desired chemical processes: charge injection, electron transport, and regeneration of the oxidised dye, compete with various undesired processes that decrease the overall efficiency of the cell. These parasitic processes include: radiative and non-radiative decay of the excited dye, which compete with the initial charge transfer; and charge recombination from the semiconductor to the dye, which competes both with electron transport to the outer circuit and with the dye regeneration. Tuning of the rates for both desired and undesired reaction pathways is crucial for a fundamental understanding of how the interfacial processes occur and for successfully optimising solar cells. Thus, the ability to distinguish between these processes and to measure their kinetics is of key importance, and is why transient absorption is a widely used technique for DSSC material characterisation.[197] The relevant charge transfer processes and the corresponding rate constants are schematically depicted in Figure 3.27.

In short, a transient absorption measurement is performed by photoexcitation of a sample with pulsed laser light, which excites the dye and thus initiates the primary charge separation, *i.e.*, electron injection. At a certain, controlled, time delay after excitation, the sample is subject to white light, in a process equivalent to ground-state absorption but that measures the absorption of the transient state(s). Thus, it is in principle possible to follow the formation and/or decay of any transient state in the process, be it the excited state of the dye, the charge injection, the dye being regenerated or the charge recombination. However, these processes occur on very different time scales,[197] as indicated in Figure 3.27.

Photon absorption and charge injection typically occur on timescales of the order of femtoseconds to picoseconds (10^{-15} s to 10^{-12} s), and thus require femtosecond transient absorption set-ups. However, it should be noted that electron injection on the nanosecond timescale (10^{-9} s), has been reported for dye molecules designed to attenuate the injection rate. Recombination between the reduced semiconductor and the oxidised dye has been observed on ultrafast time scales, but is typically measured in the nanosecond to millisecond time regime, while the dye regeneration process is usually complete within the shorter microsecond time regime.

The different time scales put high demands on the experimentalists that perform the measurements. Even though femtosecond and nanosecond transient absorption experiments share the same general characteristics, the experimental set-ups are quite different, and it is not always trivial to relate data from a femtosecond pump–probe experiment with those of a millisecond transient absorption experiment. The main difference between the different time regimes is the light sources. In a typical femtosecond pump–probe measurement the excitation light is a laser beam with very high peak power, a 1 mJ pulse with a 100 fs duration results in *ca.* 10 GW peak power. This should be compared to a 10 ns pulse of 1 mJ which results in a peak power of 0.1 MW. Furthermore, one must take into account the difference between laser-based excitation sources and the Sun. A full Sun provides *ca.* 100 mW cm^{-2}, which would correspond to only *ca.* 100 photons if it was focused on a normal sample spot size of around 10^{-3} cm^2, if the flux was the same as in a fs experiment. In contrast, a femtosecond laser pulse typically contains around 10^{11}–10^{14} photons.[198] Another important difference is that, depending on the type of detector used, the magnitude of the signals that can be measured differs significantly. Thus, a meaningful comparison between transient absorption in different published results requires detailed understanding of the experimental conditions.

One of the main reasons why the dynamics from excited states and the charge transfer products that result from them continue to attract interest is their ability to provide mechanistic insights. To determine reaction mechanisms from transient data requires detailed modelling of the kinetic and spectral data. In a non-viscous solution, the excited states of dyes typically exhibit decays that follow classic, mono-exponential behaviour, while charge separation (oxidation or reduction) and charge recombination can typically be sufficiently modelled by double- or multi-exponential functions. However, as soon the dye is immobilised on the photoanode; such that translational, rotational, and vibrational degrees of freedom are decreased, the kinetic behaviour becomes more complex. Even for dyed electrodes, multi-exponential functions are commonly needed to correctly represent the data. Alternatively, there are cases where other functions, like stretched exponential functions (*e.g.*, Kohlrausch–Williams–Watts) have been successfully used. It can be hard to intuitively understand the more complex decay functions, and comparing, for example, a rate constant for charge recombination based on a stretched exponential function and an average rate constant from a multi-exponential decay function is not always straightforward. Another reason for caution is large differences in time scales, where direct comparisons may not be valid. While the extracted decay rates can be very useful to understand a set of data, they should always be complemented with actual visual inspection to ensure that the rate constants obtained from modelling do provide a reliable representation of reality. Another complicating factor is that the number of charges that are injected into the semiconductor can significantly affect the observed kinetics. At low surface coverages and low excitation intensities, charge recombination is often close to mono-exponential, while at high surface coverages and/or high excitation intensities, the decays often appear more complex. Thus, care is necessary when comparing results from different studies.

Another common criticism of transient absorption measurements is that the experimental conditions are too far from operational conditions of DSSCs.[199] While this statement is true in itself, it is important to understand that by clever experimental design, transient absorption spectroscopy has been of tremendous importance to the field, and will continue to be important in the study of new materials, dyes, redox mediators, *etc.*

Transient absorption measurements have provided extremely useful insights into the mechanisms of DSSC materials and will continue to do so. One notable example is how the development of femtosecond pump–probe measurements can be used to study the long-standing question of which dye excited states (Franck–Condon state, an equilibrated singlet state, or a triplet state) inject electrons into the wide band gap semiconductor. For example, Sundström and co-workers have shown that the classic ruthenium bipyridine-derived dye, N3, injects electrons into TiO_2 from a singlet excited state (τ ~30 fs) that is complemented by an injection component on the ps time scale originating from a triplet excited state with metal-to-ligand charge transfer character. Similarly, they could show that when N3 is bound to a SnO_2 electrode, the electron injection from the singlet state is slower, while the injection occurring from the triplet excited state had the same kinetics regardless of the wide band gap semiconductor chosen. From the kinetic data it could be concluded that the injection efficiency was not responsible for the low overall PCE observed for N3-sensitised SnO_2 DSSCs.[200] Transient absorption has also been used to measure significantly slower charge injection rate constants. The groups of Galoppini and Meyer showed that by strategic dye design the injection rate could be slowed down to *ca.* $3.5 \times 10^7\ s^{-1}$ despite a thermodynamic driving force for injection on a par with many widely used Ru-based dyes, suggesting that the chromophore design and spatial arrangement of the dye relative to the semiconductor surface is key for good performance.[201]

Transient absorption studies have been instrumental to mechanistic studies of both dye regeneration by the redox couple and charge recombination between electrons in the conduction band of TiO_2 and the oxidised dye. The I_3^-/I^- redox couple has been extensively studied in combination with different dyes, and several factors have been shown to influence the kinetics of the regeneration reactions. This includes intuitive factors like I^- concentration and solvent viscosity, but also that the cation of the iodide salt has been shown to influence the kinetics of the process. Indeed, the available data even suggest that there may be a different mechanism in play for different dyes, which can be understood, at least partly, by the different steric hindrance for different dyes.[92,202]

Charge recombination studies are typically performed in the absence of a redox mediator to yield less complicated data. Transient absorption studies have proved that the charge recombination is typically very dependent on the electron density in the TiO_2 (the conduction band and/or in lower lying trap states).[57,203,204] This can sometimes complicate comparisons with working cells, but there are also examples where a trend measured with transient absorption can readily be related to cell performance – for example this has been established for open circuit voltages for operating cells and charge recombination rates within a series of model molecules.[205] Furthermore, electrolyte composition, for example viscosity, has been shown to influence the charge recombination kinetics, again at least for some dyes.[160,206] The dye structure and the electronic configuration appear important in this regard. In much the same manner it has also been shown that the spatial location of the oxidised dye HOMO relative to the metal oxide surface has an influence on the kinetics.[207,208] Transient absorption spectroscopy has helped establish that, while electron injection into the semiconductor appears to be a process with effectively no energy barrier, charge recombination typically occurs in the Marcus inverted region.[204,207] Another example of when transient absorption proved to be a very useful tool was when the groups of Gerald Meyer and Anders Hagfeldt, at more or less the same time, started understanding the details of the Stark effect formed after charge separation between dye and semiconductor in DSSC photoanodes.[209]

These examples are just a handful from an extensive literature concerning DSSCs and transient absorption spectroscopy. The overall efficiency is a measure of all processes at once, but for clever optimisation, the detailed mechanistic understanding provided by transient absorption measurements will continue to be important in the field.

3.5.2 X-ray Photoelectron Spectroscopy for DSSCs

Photoelectron spectroscopy (PES) has been applied to the study of DSSC components and to complete working electrodes to determine:

- the surface coverage and geometry of dye molecules and other adsorbates on the surfaces of semiconductors such as TiO_2; and
- the electronic structure of the frontier orbitals (HOMO or valence) and energy alignment between components.

This section will briefly outline how this can be achieved by going from model to real systems. More complete descriptions of photoelectron spectroscopy can be found in other textbooks, while longer descriptions of the application of this method to solar cells and DSSCs in particular can be found in the ref. 210 and 211.

Briefly, PES relies on the determination of the kinetic energy (E_K) of electrons emitted from a sample upon illumination with monochromatic radiation with energies in the X-ray or ultraviolet region. From known energies of the X-ray or UV radiation and the kinetic energy of the emitted electron, the binding energy (E_B) of electrons in a solid sample can be determined using the following equation:

$$E_B = h\nu - E_K - \Phi \tag{3.8}$$

where $h\nu$ is the photon energy and Φ is the work function of the analyser. In this case, the binding energy scale is set in such a way that the Fermi level lies at $E_B = 0$ eV. To ensure that electrons are detected without a loss of energy after exiting the sample, measurements usually have to be carried out under ultrahigh vacuum (UHV) conditions. The electron signal is typically divided into two regions: core and valence levels. Core levels can be clearly assigned to atomic orbitals within the material and occur with minor changes at specific (known) binding energies. The exact position of the core levels will depend on the chemical environment of an atom in the material. Measurements of core levels can therefore give both the elemental composition of a material as well as the chemical state and surroundings of specific atoms within the material. Valence levels at low binding energies are due to molecular orbitals in molecules and bands in semiconductors, and therefore relate to the chemical bonds in the studied material. For DSSCs, these can be used, for example, to determine the HOMO energy of a dye or a HTM.

The intensity of the detected photoelectrons from specific core and valence levels depends on several factors, which include the concentration of a particular element in the sample, the probability of X-ray absorption, and the probability of the photoelectron escaping the solid without a loss in energy. The probability of X-ray absorption is given by the absorption cross-section for specific orbitals and depends on the photon energy. Tabulated absorption cross-sections can be found in reference tables. The probability of an electron escaping the solid decays exponentially with sample depth and is related to the mean free path (λ). The mean free path varies with the kinetic energy of the photoelectron according to the universal curve, with small variations for different elements.[212] The minimum of λ is about 0.4–0.5 nm for electrons with a kinetic energy of *ca.* 50 eV and from there λ increases to values exceeding 10 nm at several 1000 eV kinetic energy. PES is therefore, in general, a very surface-sensitive technique, but the probing depth can be adjusted by changing the photon energy, allowing for depth profiling of the surface and sub-surface composition and for studies of buried interfaces. Varying the photon energy can therefore be an important tool in PES studies, as it allows for both a change in the probing depth, as well as a change in the sensitivity to different atomic orbitals due to the variations in cross-sections.

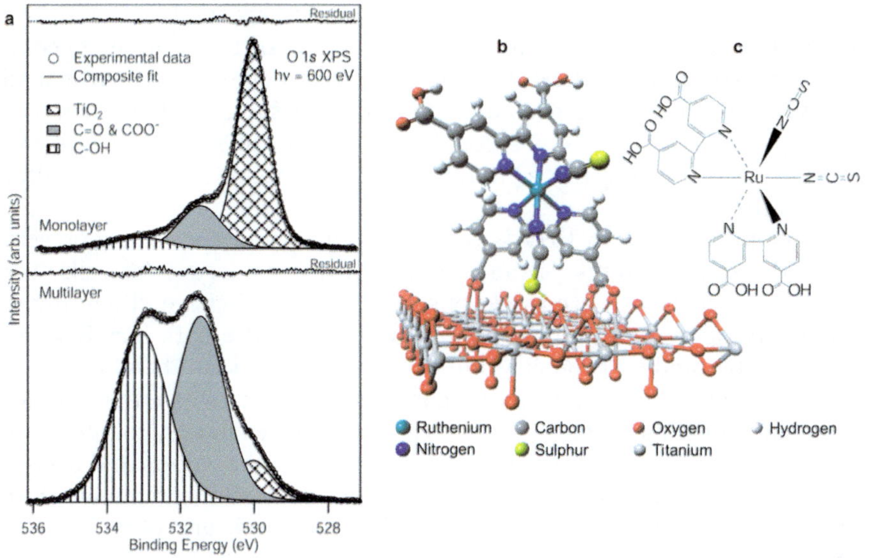

Figure 3.28 (a) O 1s core-level spectra of N3 deposited on a rutile surface by electrospray measured using a photon energy of 600 eV. (b) N3 molecule adsorbed on TiO_2, calculated using CASTEP at the DFT-GGA level. (c) Chemical structure of the N3 molecule. Reproduced from ref. 216 with permission from AIP Publishing.

Photoelectron spectroscopy has been used to assess the binding geometry of molecules on semiconductor surfaces. Studies have been carried out for molecules with anchoring groups of relevance in DSSCs such as bi-isotonic acid[213,214] and maleic anhydride[215] and even for complete dye molecules (N3).[216] Controlled and clean sample conditions are needed to obtain this type of information from direct measurements of the binding groups, as typical surface contaminants often contain oxygen and carbon. Such conditions can be achieved by preparing a clean surface of a metal oxide single crystal under UHV. The molecules can then be deposited on this clean surface, for example by evaporation, and the variation in core-level photoemission peaks when going from mono- to multi-layer coverage can be used to determine the typical binding geometry of a monolayer. Figure 3.28 shows an example of such a measurement of the N3 dye, which was deposited onto a rutile TiO_2 (110) surface in UHV using an electrospray method.[216] In the O 1s spectra of the dye on the surface, three different peaks can be observed, which vary in intensity for the mono- and the multi-layer. For the monolayer, the most intense peak stems from the TiO_2 surface itself and its intensity is decreased in relation to the other peaks for the multilayer due to the surface sensitivity of PES. Two peaks are observed for the dye itself, which have approximately equal intensity for the multilayer: the oxygen from OH groups and the oxygen from C=O and COO^- groups. In the multilayer, most dyes are not bound to the surface and a 1:1 ratio is expected for the unbound carboxylic acid groups. In the case of the monolayer, a 1 : 3.2 ratio between C–OH and C=O and COO^- is observed. As this is very near to a ratio of 1 : 3, a binding geometry as shown in Figure 3.28(b) can be inferred where two carboxylic acid groups bind to the surface in a bidentate fashion, resulting in both their oxygens appearing at the COO^- position. Measurements of S 2p core levels give evidence in a similar way that a binding interaction occurs through the sulfur of one of the NCS groups.

Sample preparation in UHV is time consuming and does not represent exactly the same systems as those used in dye-sensitised solar cells. For example, the surface of a mesoporous TiO_2 film may behave differently to a single crystal surface and the solvent used in the dye bath may have an impact on the binding geometry and surface composition. Therefore, studies are also carried out on *ex situ* prepared samples, *i.e.*, where the samples are prepared under ambient conditions and transferred to UHV for measurement. In this case, a detailed quantitative analysis of the carbon and oxygen core levels is difficult, as these elements are also present in common surface contaminants. However, the analysis of other core levels and of the valence band region can give relevant information for the DSSC. For example, the dye orientation has been determined from nitrogen core levels for triphenylamine dyes through the ratio of two chemically different nitrogens in the dye structure.[217,218] Furthermore, PES has been used to determine the variation in surface coverage between a series of sensitisers[219] and the surface structure in the presence of co-adsorbers[220] and co-sensitisers.[221] The impact of the dye bath solvent and of the presence of water in the dye bath on the electronic structure has also been assessed.[222]

Photoelectron spectroscopy can be used to map the frontier orbitals of dye molecules and even of HTMs for ssDSSCs.[223] From this, the character of the orbitals and the energy alignment in the solar cell can be determined. In contrast to other techniques, such as electrochemical methods, the different components can be measured in relation to each other: for example, for an adsorbed dye, core levels from the substrate (TiO_2) can be used to directly reference the HOMO level of the dye to the valence and conduction band of the TiO_2. In this way, energy level changes caused by interactions between the materials or surface dipoles are accounted for Figure 3.29 shows an example of determining the energy alignment for the perylene dye ID176 on TiO_2, where all spectra were referenced to the Ti 3p level.[224] Using this method, it was observed that the addition of $LiClO_4$ to the surface caused a change in the HOMO energy of the dye, which was accompanied by a change in the absorption onset of the dye. The measurements therefore suggested that the driving force for electron injection in a solid-state DSSC was not increased by the addition of Li ions to the surface, in contrast to what is usually expected.

In a similar way to dyes, HTMs can also be characterised by photoelectron spectroscopy under UHV conditions. For example, the doping mechanism of HTMs has been studied in a more bulk sensitive measurement with higher energy X-rays (>2000 eV).[225,226] For studies of a complete semiconductor/dye/electrolyte interface, new opportunities are becoming available with

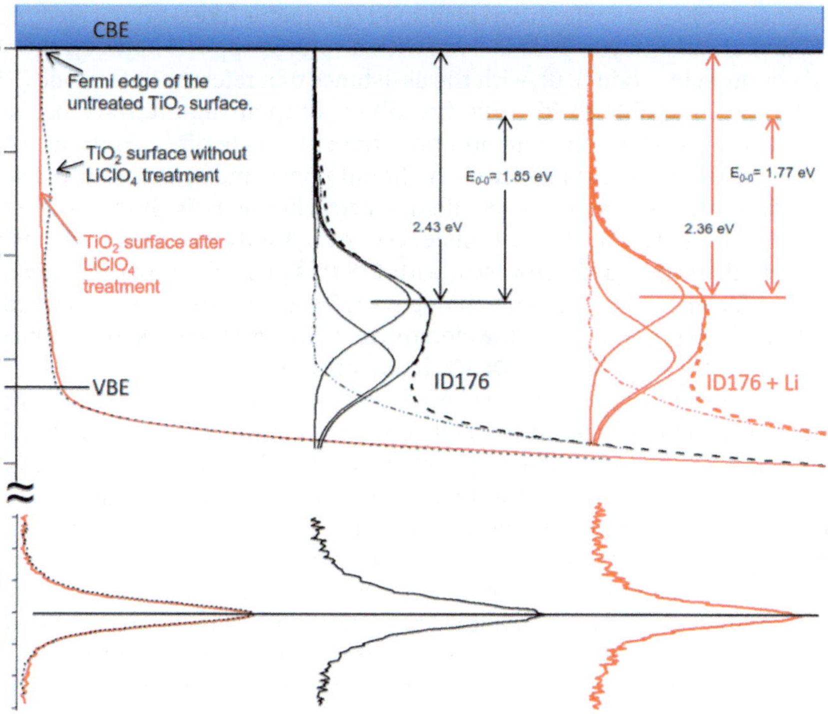

Figure 3.29 Valence band and Ti 3p spectra measured with photon energy of 100 eV resulting in an energy diagram of the TiO_2/ID176 system. Left: TiO_2 in absence (black dashed) and presence (red) of $LiClO_4$. Middle: ID176 on TiO_2. Right: ID176 on TiO_2 with $LiClO_4$. Dashed lines are measured data and solid lines are fits. The $E_{0\text{-}0}$ transition energies for the dye under corresponding conditions are indicated. Reproduced from ref. 224 with permission from American Chemical Society, Copyright 2011.

ambient pressure X-ray photoelectron spectroscopy (APXPS).[227] In these measurements, the pressure close to the sample surface can be increased up to approximately 100 mbar, enabling measurements of solid/vapour and solid/liquid/vapour interfaces. An APXPS study of relevance to DSSCs, which has recently been carried out, was the measurement of the Z907 dye at different water vapour pressures.[228] This represents a first step towards new opportunities for studying more realistic DSSC interfaces in the future.

3.5.3 Electrochemical Impedance Spectroscopy for DSSCs

Electrochemical impedance spectroscopy (EIS) is a non-destructive, alternating current electrical technique that is particularly sensitive to small systemic changes in an electrical device. EIS allows the electrical properties of materials to be characterised even in electrochemical systems with poorly conducting media. EIS is a traditional method that is central to electrochemical sciences and technology. Electrochemical methods usually investigate interfacial charge transfer between a solid conductor (the working electrode, WE) and an electrolyte. This is done with a voltage applied between the WE and counter electrode (CE), with the assistance of a reference electrode (RE), rendering it possible to identify the voltage drop at the interface between the WE and the electrolyte. In addition, the electrolyte often contains a salt that increases the conductivity of the liquid phase and removes limitations by drift transport in an electrical field. Electrochemistry is thus mostly concerned with interfacial charge transfer events, possibly governed by diffusion of reactants or products. However, with EIS it is possible to readily separate the interfacial capacitance and charge transfer resistance, as well as to identify diffusion components in the electrolyte. It is these features that make EIS a highly useful technique for DSSC characterisation.

The technique consists of the application of a small electrical bias (E) to an electrode and its response in current (I) is measured at varied alternating current frequencies. The impedance is defined as the quotient between the applied potential and the measured current at the output. In many materials the impedance varies with the frequency of the applied potential in a shape that is related to the properties of such materials. This is due to the physical structure of the material, to the electronic processes that take place, or to a combination of both. Therefore, if the impedance is measured in a suitable frequency range it is possible to relate the results with the physical and chemical properties of the materials and charge movements in a DSSC.

There are two ways to represent the impedance results from EIS analysis: Nyquist plots or Bode plots. In a Nyquist plot, the imaginary part of the fitting is multiplied by minus 1 ($-Z''$) and plotted *versus* the real part (Z'). Nyquist plots are the most widely used representation system and the information obtained from them is based on the form that the spectrum takes, usually representing the charge transfer resistance between interfaces (Figure 3.30(a)). A Bode plot is represented by the logarithm of the impedance module ($\log|Z|$) and the phase shift (φ) as a function of the

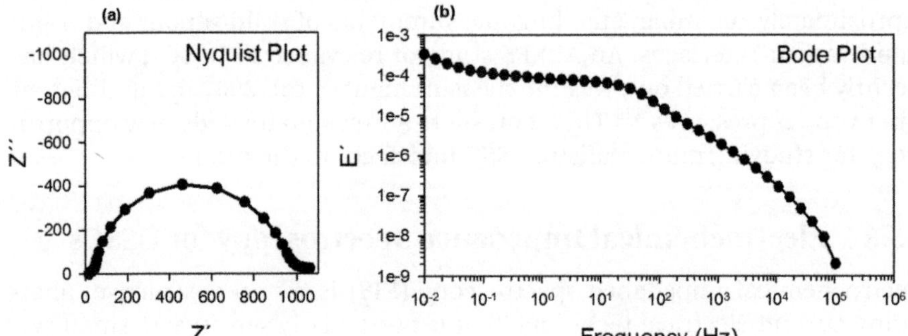

Figure 3.30 Typical representations for the EIS measurements carried out on a DSSC: (a) Nyquist plot and (b) Bode plot.

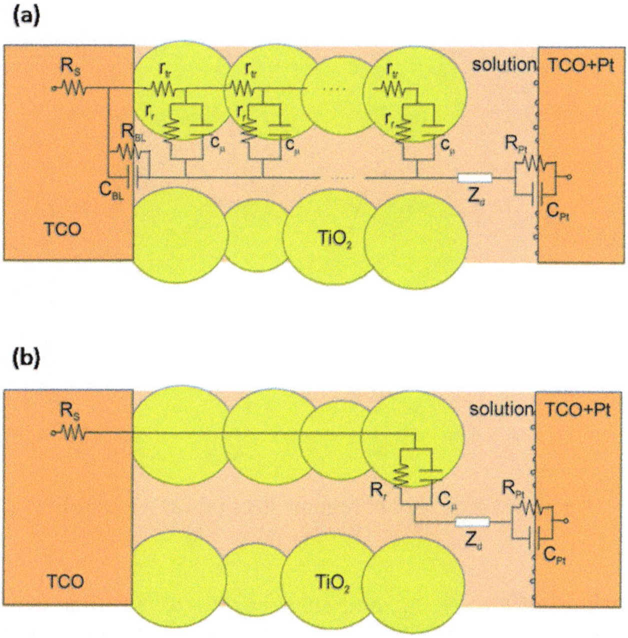

Figure 3.31 The complete equivalent circuit model to fit experimental data (a, top) and a simplified equivalent circuit model that is valid for impedance spectra measured at high voltage (b, top). Adapted from ref. 229 with permission from the PCCP Owner Societies.

logarithm of the frequency (log ω). The information obtained from this type of representation is directed mainly at the behaviour as a function of frequency (Figure 3.30(b)).

Interpretation of the impedance spectrum requires selection of an appropriate electrical model that fits the experimental data. Through the model, the measurements obtained using this technique provide information related to the resistance of the electrolyte, the resistance of polarisation, and the capacitance of the Helmholtz double layer. The resistance of the electrolyte is obtained at high frequencies and the data acquired at low frequencies gives information on the kinetics of the reaction. Depending on the model proposed, information can be obtained from the characteristic parameters. Following a large amount of time developing the most suitable model for DSSCs, a proposed and generally accepted model is shown in Figure 3.31.[229]

Figure 3.31(a) represents the complete transmission line model that may be used to fit the impedance data shown in Figure 3.30.[230–232] The model includes the effect of the counter electrode/electrolyte interface, described by the parallel combination of the charge transfer resistance, R_{Pt}, and the interfacial Helmholtz capacitance, C_{Pt}. The suffix Pt is used to stress the fact that a platinum-coated counter electrode is the most

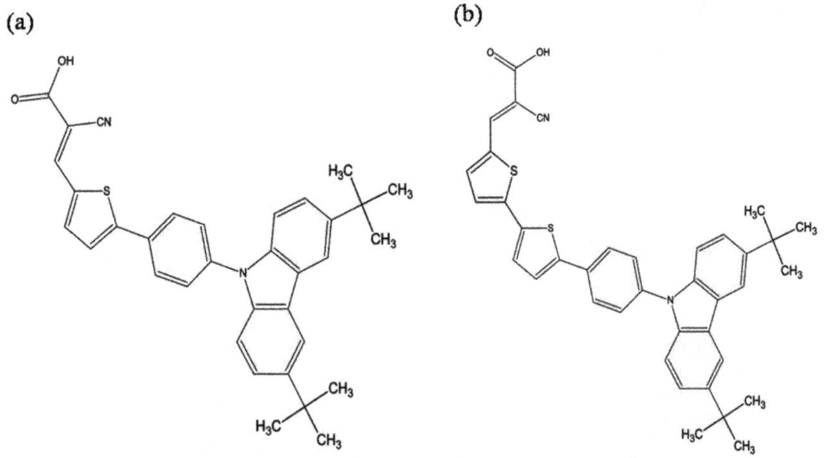

Figure 3.32 Chemical structure of organic dyes (a) BG-1 and (b) BG-2.

Table 3.4 Values of V_{OC}, FF, J_{SC}, and efficiency are obtained at steady-state measurement under 100 mW cm^{-2} light intensity and simulated AM 1.5 illumination.

Sample[a]	V_{OC} (V)	J_{SC} (mA cm^{-2})	FF	η (%)
BG-1	0.47	8.47	0.60	2.39
BG-2	0.40	10.9	0.57	2.48
N719	0.52	19.3	0.38	3.90

[a]Samples fabricated with non-optimum electrolyte solution for N719.

commonly employed for liquid-junction DSSCs. The model also contains R_{BL} and C_{BL} elements for the recombinations through the substrate that are not blocked by TiO_2 nanoparticles (or the blocking layer). The equivalent circuit shown in Figure 3.31(a) can then be simplified as shown in Figure 3.31(b), with the transmission line reduced to a simple R–C parallel combination.

In a real device, for example a DSSC based on *N*-substituted carbazole dyes[233] (Figure 3.32 and Table 3.4), the EIS characterisation combined with the model described above[229] lets us understand the observed experimental results from different DSSCs.

Comparing the performance of the two organic sensitisers in DSSCs, a trade-off is observed between an enhanced photocurrent and a reduced photovoltage when the number of conjugated units increases from BG-1 to BG-2. Under the circumstances where the electrolytes are identical, this observation should be related to the role that the organic dye plays in the charge recombination process. Different suggestions have been reported in the literature concerning the increase of recombination, including dye interaction with I_3^-,[234] lower reorganisation energy for organic dyes,[235] or possibly electrostatic effects induced by the dye structure.[236] Nevertheless, one effect that could significantly increase the overall recombination, is recombination through the dye cation. While this is usually negligible for N719 and related dyes, it may become a relevant factor in organic dye molecules functionalised with long, conjugated chains. It is clear that the family of dyes that we have investigated (Figure 3.32) follow exactly the behaviour described in the literature. From the chemical structure of the dyes, it is clear that the BG-2 dye (with two thiophene units) absorbs more broadly in the visible spectrum and this is correlated with higher photocurrents (Table 3.4).[233] Nevertheless, the overall conversion efficiency is similar to that of BG-1. However, BG-1 produces a larger photovoltage than BG-2, which compensates for the lower extent of photon absorption. A variety of causes may exist for the increased photovoltages, *i.e.*, an upward shift of the conduction band of TiO_2, E_c, or a decrease of the rate of charge recombination. EIS measurements are a suitable technique to answer these questions. The measurement conditions for such an experiment are as follows: illuminate the sample under simulated 1 Sun conditions (AM 1.5) and apply an electrical bias ranging from zero volts to open circuit potential. Under these conditions, the alternating current is varied between 1 MHz and 0.1 Hz. The data are analysed using the impedance model explained above so that the recombination resistance is isolated from other resistive contributions in the cell.[232,237] The obtained data (Figure 3.33(a) and (b)) indicate the measured capacitance, C_μ, and recombination resistance, R_{rec}, as a function of the potential for the different cells. Here, the potential means the Fermi level voltage V_F that corresponds to the bias voltage, without the effect of ohmic loss in the solar cell. The chemical capacitance describes the density of states in the band gap of the TiO_2[238] and the recombination resistance is related to the recombination current, J_{rec}, as follows:[239]

$$R_{rec} = \left(\frac{\partial J_{rec}}{\partial V}\right)^{-1} \tag{3.9}$$

The analysis of R_{rec} provides information about the extent of recombination in each solar cell. However, the change of dye can also produce a change of the energetics, *i.e.*, of the conduction band of the TiO_2 semiconductor, E_c, with respect to the redox energy in the electrolyte, E_{F0}. As observed in the chemical capacitance values reported in Figure 3.33(a), the conduction band edge of TiO_2 is lower for BG-2-sensitised films. The lower position of the conduction band facilitates the electron injection from the dye and tends to increase the photocurrent.

In Figure 3.33(c) the voltage scale has been shifted, so that chemical capacitances coincide. In this graph the voltage is obtained as:

$$V_{ecb} = V_F - \Delta E_c \tag{3.10}$$

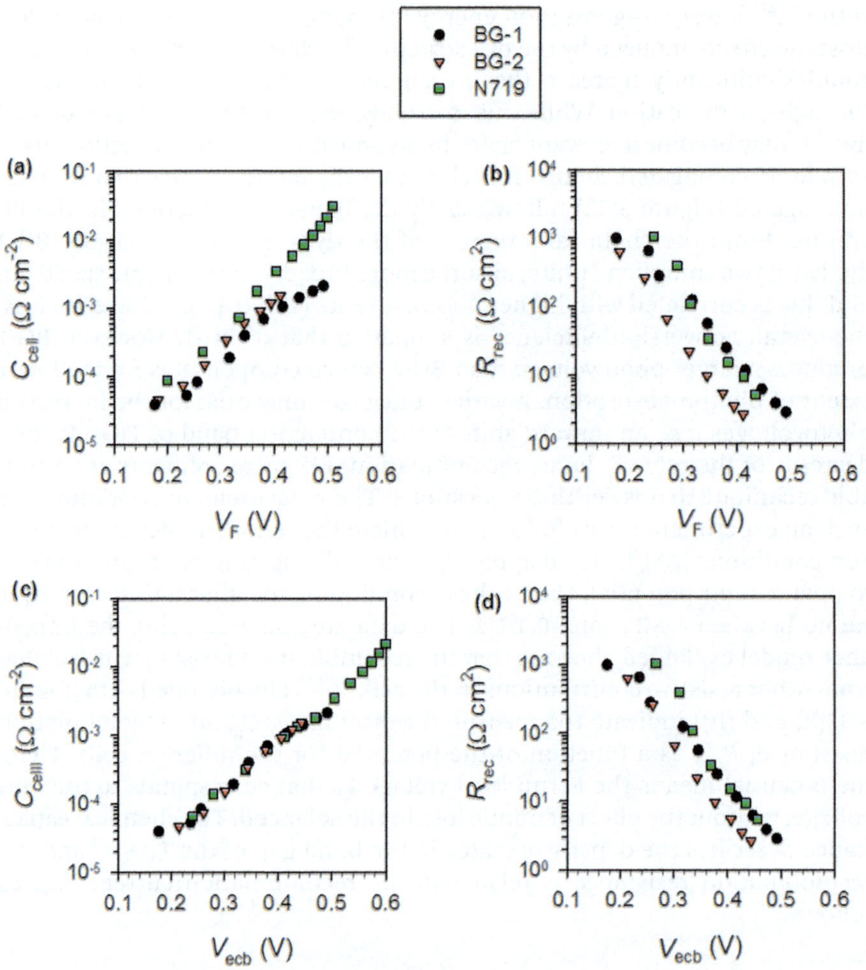

Figure 3.33 Capacitance (a) and recombination resistance (b) obtained from impedance spectroscopy of the solar cells with BG-1, BG-2, and commercial N719 as sensitisers under simulated AM 1.5 illumination, with respect to Fermi level voltage (without the effect of series resistance). In (c) and (d) capacitance and recombination resistance are shown replotted with respect to equivalent common conduction band voltage so that the distance between Fermi level and conduction band is the same in all cases. Reproduced from ref. 233 with permission from American Chemical Society, Copyright © 2010.

where "ecb" stands for "common equivalent conduction band" and ΔE_c is the shift of the conduction band with respect to a reference sample. In Figure 3.33(d) the recombination resistance is plotted on the same scale. In effect this procedure normalises the conduction band edge potential of TiO_2 for all the cells and allows the comparison of charge transfer rate. It is observed in Figure 3.33(d) that at low voltage all dyes show similar charge transfer resistance, although N719 has less recombination. When the forward voltage increases and we get into the maximum power point and close to open circuit voltage, the recombination resistance is similar for N719 and BG-1, and larger than in BG-2. Therefore, the higher photovoltage from BG-1 results from a smaller recombination rate (larger resistance) than for BG-2.

It can be concluded that EIS techniques are powerful tools to characterise DSSCs and, when applying the correct electrical model, it is possible to extract mechanistic information for charge transfer processes.

3.6 Conclusions

The dye-sensitised solar cell is a complex photovoltaic system and has been studied intensively over more than 25 years. Techniques such as EIS and TAS have enabled scientists to investigate the charge transfer processes at the interfaces in the devices. On the basis of these fundamental studies, the performance of DSSCs has been improved from less than 1% efficiency initially to more than 10% efficiency after systematically optimising every individual component. Currently, the technology is transitioning from fundamental research to commercialisation. Although it is often argued that the cost of DSSCs is competitive with silicon solar cells, compatibility between efficiency and stability is still a big challenge. More effort from both academia and industry is required to optimise the device, from solar cell engineering and new materials design to improving the stability without sacrificing the efficiency, in order to make DSSCs really competitive with established photovoltaic technology.

Acknowledgements

HT and JG would like to gratefully thank the Swedish Energy Agency (43237-1) for financial support. JG acknowledges support from the Swedish Government through the strategic research area STandUP for ENERGY.

References

1. J. Moser, *Monatsh. Chem.*, 1887, **8**, 373.
2. H. Gerischer, *J. Electrochem. Soc.*, 1978, **125**, 218C.
3. H. Gerischer and F. Willig, in *Topics in Current Chemistry, No. 61*, Springer, 1976, p. 31.
4. R. Memming, *Prog. Surf. Sci.*, 1984, **17**, 7.
5. B. O'Regan and M. Grätzel, *Nature*, 1991, **353**(6346), 737.
6. A. Hagfeldt, G. Boschloo, L. Sun, L. Kloo and H. Pettersson, *Chem. Rev.*, 2010, **110**(11), 6595.
7. J. He, H. Lindström, A. Hagfeldt and S.-E. Lindquist, *J. Phys. Chem. B*, 1999, **103**(42), 8940.

8. E. A. Gibson, A. L. Smeigh, L. Le Pleux, L. Hammarström, F. Odobel, G. Boschloo and A. Hagfeldt, *J. Phys. Chem.*, 2011, **115**(19), 9772.
9. Z. P. Shao, X. Pan, H. Chen, L. Tao, W. Wang, Y. Ding, B. Pan, S. Yang and S. Dai, *Energy Environ. Sci.*, 2014, **7**, 2647.
10. H. Tian, *ChemSusChem*, 2015, **8**, 3746.
11. Z. Yu, F. Li and L. Sun, *Energy Environ. Sci.*, 2015, **8**(3), 760.
12. E. A. Gibson, *Chem. Soc. Rev.*, 2017, **46**(20), 6194.
13. P. B. Pati, L. Zhang, B. Philippe, R. Fernández-Terán, S. Ahmadi, L. Tian, H. Rensmo, L. Hammarström and H. Tian, *ChemSusChem*, 2017, **10**(11), 2480.
14. M. E. Michel-Beyerle, F. Rebentrost and H. Tributsch, *Electrochim. Acta*, 1968, **13**(6), 1509.
15. H. Gerischer and H. Tributsch, *Ber. Bunsenges. Phys. Chem.*, 1969, **73**, 251.
16. H. Gerischer and H. Tributsch, *Ber. Bunsenges. Phys. Chem.*, 1969, **73**, 850.
17. S. K. Deb, Dye-titanium dioxide photogalvanic cell, *US Pat.*, US4080488A, 1978.
18. N. Minami, T. Watanabe, A. Fujishima and K. Honda, *Ber. Bunsenges. Phys. Chem.*, 1979, **83**, 476.
19. N. Alonso-Vante, V. M. Beley, P. Chartier and V. Ern, *Rev. Phys. Appl.*, 1981, **16**, 5.
20. H. Kokado, T. Nayama and E. Inoue, *J. Phys. Chem. Solids*, 1974, **35**, 1169.
21. M. Matsumura, S. Matsudaira and H. Tsubomura, *Ind. Eng. Chem. Prod. Res. Dev.*, 1980, **19**, 415.
22. N. Vlachopoulos, P. Liska, J. Augustynski and M. Grätzel, *J. Am. Chem. Soc.*, 1988, **110**, 1216.
23. M. K. Nazeeruddin, A. Kay, I. Rodicio, R. Humphry-Baker, E. Müller, P. Liska, N. Vlachopoulos and M. Grätzel, *J. Am. Chem. Soc.*, 1993, **115**, 6382.
24. NREL certified PV efficiency chart, https://www.nrel.gov/pv/assets/images/efficiency-chart.png, last accessed, September 2017.
25. A. Hagfeldt and M. Grätzel, *Chem. Rev.*, 1995, **95**, 49.
26. M. Quintana, T. Edvinsson, G. Boschloo and A. Hagfeldt, *J. Phys. Chem. C*, 2007, **111**, 1035.
27. D. T. Cromer and K. Herrington, *J. Am. Chem. Soc.*, 1955, **77**, 4708.
28. J. K. Burdett, T. Hughbanks, G. J. Miller, J. James, W. Richardson and J. V. Smith, *J. Am. Chem. Soc.*, 1987, **109**, 3639.
29. M. Batzill and U. Diebold, *Prog. Surf. Sci.*, 2005, **79**, 47.
30. B. Raton, *Handbook of Chemistry and Physics*, CRC Press, 2004.
31. S. Roberts, *Phys. Rev.*, 1949, **76**, 1215.
32. H. Yoshikawa and S. Adachi, *Jpn. J. Appl. Phys.*, 1997, **36**, 6237.
33. L. Kavan, M. Grätzel, S. E. Gilbert, C. Klemenz and H. J. Scheel, *J. Am. Chem. Soc.*, 1996, **118**, 6716.
34. M. A. Butler and D. S. Ginley, *J. Electrochem. Soc.*, 1978, **125**, 228.
35. J. M. Bolts and M. S. Wrighton, *J. Phys. Chem.*, 1976, **80**(24), 2641.
36. L. Brus, *J. Phys. Chem.*, 1984, **80**, 4403.

37. H. Tang, K. Prasad, R. Sanjinbs, P. E. Schmid and F. Lévy, *J. Appl. Phys.*, 1994, **75**, 2042.
38. N. N. Syrbu, I. M. Tiginyanu, V. V. Zalamai, V. V. Ursaki and E. V. Rusu, *Phys. B*, 2004, **353**, 111.
39. M. Nagasawa, S. Shionoya and S. Makishima, *J. Phys. Soc. Jpn.*, 1965, **20**, 1093.
40. D. C. Look, D. C. Reynolds, J. R. Sizelove, R. L. Jones, C. W. Litton, G. Cantwe and W. C. Harsch, *Solid State Commun.*, 1998, **105**, 399.
41. C. H. Seagera and S. M. Myers, *J. Appl. Phys.*, 2003, **94**, 3675.
42. L. Forro, O. Chauvet, D. Emin, L. Zuppiroli, H. Berger and F. Lévy, *J. Appl. Phys.*, 1994, **75**, 633.
43. H. Mun, H. Yang, J. Park, C. Ju and K. Char, *APL Mater.*, 2015, **3**, 076107.
44. M. Kosmulski, *Adv. Colloid Interface Sci.*, 2002, **99**, 255.
45. N. Kallay, D. Babic and E. Matijevic, *Colloids Surf.*, 1986, **19**, 375.
46. M. Kosmulski, *J. Colloid Interface Sci.*, 2009, **337**, 439.
47. J. Wu, Z. Lan, J. Lin, M. Huang, Y. Huang, L. Fan, G. Luo, Y. Lin, Y. Xie and Y. Wei, *Chem. Soc. Rev.*, 2017, **46**, 5975.
48. S. Sumikura, S. Mori, S. Shimizu, H. Usami and E. Suzuki, *J. Photochem. Photobiol., A*, 2008, **194**, 143.
49. Z. Xu, D. Xiong, H. Wang, W. Zhang, X. Zeng, L. Ming, W. Chen, X. Xu, J. Cui, M. Wang, S. Powar, U. Bach and Y.-B. Cheng, *J. Mater. Chem. A*, 2014, **2**, 2968.
50. P. Ho, S. Thogiti, Y. H. Lee and J. H. Kim, *Sci. Rep.*, 2017, **7**, 2272.
51. J. Warnan, J. Gardner, L. Le Pleux, J. Petersson, Y. Pellegrin, E. Blart, L. Hammarström and F. Odobel, *J. Phys. Chem. C*, 2014, **118**, 103–113.
52. H. Tian, J. Oscarsson, E. Gabrielsson, S. K. Eriksson, R. Lindblad, B. Xu, Y. Hao, G. Boschloo, E. M. J. Johansson, J. M. Gardner, A. Hagfeldt, H. Rensmo and L. Sun, *Sci. Rep.*, 2014, **2**, 4282.
53. M. K. Nazeeruddin, P. Pechy, T. Renouard, S. M. Zakeeruddin, R. Humphry-Baker, P. Comte, P. Liska, L. Cevey, E. Costa, V. Shklover, L. Spiccia, G. B. Deacon, C. A. Bignozzi and M. Grätzel, *J. Am. Chem. Soc.*, 2001, **123**, 1613.
54. M. K. Nazeeruddin, S. M. Zakeeruddin, R. Humphry-Baker, M. Jirousek, P. Liska, N. Vlachopoulos, V. Shklover, C. H. Fischer and M. Grätzel, *Inorg. Chem.*, 1999, **38**, 6298.
55. G. Sauve, M. E. Cass, S. J. Doig, I. Lauermann, K. Pomykal and N. S. Lewis, *J. Phys. Chem. B*, 2000, **104**, 3488.
56. G. Sauve, M. E. Cass, G. Coia, S. J. Doig, I. Lauermann, K. E. Pomykal and N. S. Lewis, *J. Phys. Chem. B*, 2000, **104**, 6821.
57. G. M. Hasselmann and G. J. Meyer, *J. Phys. Chem. B*, 1999, **103**, 7671.
58. S. Ferrere and B. A. Gregg, *J. Am. Chem. Soc.*, 1998, **120**, 843.
59. A. Islam, H. Sugihara, K. Hara, L. P. Singh, R. Katoh, M. Yanagida, Y. Takahashi, S. Murata, H. Arakawa and G. Fujihashi, *Inorg. Chem.*, 2001, **40**, 5371.
60. E. A. M. Geary, L. J. Yellowlees, L. A. Jack, I. D. H. Oswald, S. Parsons, N. Hirata, J. R. Durrant and N. Robertson, *Inorg. Chem.*, 2005, **44**, 242.

61. N. Alonso-Vante, J.-F. Nierengarten and J.-P. Sauvage, *J. Chem. Soc., Dalton Trans.*, 1994, 1649.
62. T. L. Ma, K. Inoue, H. Noma, K. Yao and E. Abe, *J. Photochem. Photobiol., A*, 2002, **152**, 207.
63. C.-W. Lee, H.-P. Lu, C.-M. Lan, Y.-L. Huang, Y.-R. Liang, W.-N. Yen, Y.-C. Liu, Y.-S. Lin, R. W.-G. Diau and C.-Y. Yeh, *Chem. - Eur. J.*, 2009, **15**, 1403.
64. S. Eu, T. Katoh, T. Umeyama, Y. Matano and H. Imahori, *Dalton Trans.*, 2008, 5476.
65. A. Yella, H.-W. Lee, H. N. Tsao, C. Yi, A. K. Chandiran, M. K. Nazeeruddin, E. W.-G. Diau, C.-Y. Yeh, S. M. Zakeeruddin and M. Grätzel, *Science*, 2011, **334**, 629.
66. T. Kitamura, M. Ikeda, K. Shigaki, T. Inoue, N. A. Anderson, X. Ai, T. Q. Lian and S. Yanagida, *Chem. Mater.*, 2004, **16**, 1806.
67. J. H. Yum, D. P. Hagberg, S. J. Moon, K. M. Karlsson, T. Marinado, L. C. Sun, A. Hagfeldt, M. K. Nazeeruddin and M. Grätzel, *Angew. Chem., Int. Ed.*, 2009, **48**, 1576.
68. R. Li, X. Lv, D. Shi, D. Zhou, Y. Cheng, G. Zhang and P. Wang, *J. Phys. Chem. C*, 2009, **113**, 7469.
69. J. Wiberg, T. Marinado, D. P. Hagberg, L. C. Sun, A. Hagfeldt and B. Albinsson, *J. Phys. Chem. C*, 2009, **113**, 3881.
70. J. T. Lin, P.-C. Chen, Y.-S. Yen, Y.-C. Hsu, H.-H. Chou and M.-C. P. Yeh, *Org. Lett.*, 2009, **11**, 97.
71. D. P. Hagberg, T. Edvinsson, T. Marinado, G. Boschloo, A. Hagfeldt and L. Sun, *Chem. Commun.*, 2006, 2245.
72. D. P. Hagberg, J. H. Yum, H. Lee, F. De Angelis, T. Marinado, K. M. Karlsson, R. Humphry-Baker, L. Sun, A. Hagfeldt, M. Grätzel and M. K. Nazeeruddin, *J. Am. Chem. Soc.*, 2008, **130**, 6259.
73. S. Haid, M. Marszalek, A. Mishra, M. Wielopolski, J. Teuscher, J.-E. Moser, R. Humphry-Baker, M. S. Zakeeruddin, M. Grätzel and P. Baeuerle, *Adv. Funct. Mater.*, 2012, **22**, 1291.
74. P. B. Pati and S. S. Zade, *Tetrahedron*, 2013, **69**, 2167.
75. P. B. Pati, W. Yang and S. S. Zade, *Spectrochim. Acta, Part A*, 2017, **178**, 106.
76. K. Hara, Z. S. Wang, T. Sato, A. Furube, R. Katoh, H. Sugihara, Y. Dan-Oh, C. Kasada, A. Shinpo and S. Suga, *J. Phys. Chem. B*, 2005, **109**, 15476.
77. T. Horiuchi, H. Miura and S. Uchida, *Chem. Commun.*, 2003, 3036.
78. S. Ito, H. Miura, S. Uchida, M. Takata, K. Sumioka, P. Liska, P. Comte, P. Pechy and M. Grätzel, *Chem. Commun.*, 2008, 5194.
79. K. Hara, T. Sato, R. Katoh, A. Furube, T. Yoshihara, M. Murai, M. Kurashige, S. Ito, A. Shinpo, S. Suga and H. Arakawa, *Adv. Funct. Mater.*, 2005, **15**, 246.
80. H. Tian, X. Yang, R. Chen, Y. Pan, L. Li, A. Hagfeldt and L. Sun, *Chem. Commun.*, 2007, 3741.
81. Z. S. Wang, N. Koumura, Y. Cui, M. Takahashi, H. Sekiguchi, A. Mori, T. Kubo, A. Furube and K. Hara, *Chem. Mater.*, 2008, **20**, 3993.
82. Z. S. Wang, F. Y. Li and C. H. Huang, *Chem. Commun.*, 2000, 2063.

83. K. Sayama, S. Tsukagoshi, K. Hara, Y. Ohga, A. Shinpou, Y. Abe, S. Suga and H. Arakawa, *J. Phys. Chem. B*, 2002, **106**, 1363.
84. S. Alex, U. Santhosh and S. Das, *J. Photochem. Photobiol., A*, 2005, **172**, 63.
85. L. Li, E. A. Gibson, P. Qin, G. Boschloo, M. Gorlov, A. Hagfeldt and L. Sun, *Adv. Mater.*, 2010, **22**(15), 1759–1762.
86. P. Qin, J. Wiberg, E. A. Gibson, M. Linder, L. Li, T. Brinck, A. Hagfeldt, B. Albinsson and L. Sun, *J. Phys. Chem. C*, 2010, **114**(10), 4738–4748.
87. A. Nattestad, A. J. Mozer, M. K. R. Fischer, Y. B. Cheng, A. Mishra, P. Bauerle and U. Bach, *Nat. Mater.*, 2010, **9**(1), 31–35.
88. K. A. Click, D. R. Beauchamp, B. R. Garrett, Z. Huang, C. M. Hadad and Y. Wu, *Phys. Chem. Chem. Phys.*, 2014, **16**(47), 26103–26111.
89. C. J. Wood, G. H. Summers and E. A. Gibson, *Chem. Commun.*, 2015, **51**(18), 3915–3918.
90. M. Gennari, F. Légalité, L. Zhang, Y. Pellegrin, E. Blart, J. Fortage, A. M. Brown, A. Deronzier, M.-N. Collomb, M. Boujtita, D. Jacquemin, L. Hammarström and F. Odobel, *J. Phys. Chem. Lett.*, 2014, **5**(13), 2254–2258.
91. J. Cong, X. Yang, L. Kloo and L. Sun, *Energy Environ. Sci.*, 2012, **5**, 9180–9194.
92. G. Boschloo and A. Hagfeldt, *Acc. Chem. Res.*, 2009, **42**, 1819–1826.
93. R. Komiya, A. Fukui, N. Murofushi, N. Koide, R. Yamanaka, H. Katayama, in *Technical Digest, 21st International Photovoltaic Science and Engineering Conference*, Fukuoka, 2011, pp. 2C-5O-08.
94. H. Nusbaumer, J.-E. Moser, S. M. Zakeeruddin, M. K. Nazeeruddin and M. Grätzel, *J. Phys. Chem. B*, 2001, **105**, 10461.
95. S. A. Sapp, C. M. Elliott, C. Contado, S. Caramori and C. A. Bignozzi, *J. Am. Chem. Soc.*, 2002, **124**, 11215.
96. P. J. Cameron, L. M. Peter, S. M. Zakeeruddin and M. Grätzel, *Coord. Chem. Rev.*, 2004, **248**, 1447.
97. S. M. Feldt, E. A. Gibson, E. Gabrielsson, L. Sun, G. Boschloo and A. Hagfeldt, *J. Am. Chem. Soc.*, 2010, **132**, 16714.
98. S. Hattori, Y. Wada, S. Yanagida and S. Fukuzumi, *J. Am. Chem. Soc.*, 2005, **127**, 9648.
99. Y. Bai, Q. Yu, N. Cai, Y. Wang, M. Zhang and P. Wang, *Chem. Commun.*, 2011, **47**, 4376.
100. J. Cong, D. Kinschel, Q. Daniel, M. Safdari, E. Gabrielsson, H. Chen, P. H. Svensson, L. Sun and L. Kloo, *J. Mater. Chem. A*, 2016, **4**, 14550.
101. Y. Saygili, M. Söderberg, N. Pellet, F. Giordano, Y. Cao, A. B. Muñoz-García, S. M. Zakeeruddin, N. Vlachopoulos, M. Pavone, G. Boschloo, L. Kavan, J.-E. Moser, M. Grätzel, A. Hagfeldt and M. Freitag, *J. Am. Chem. Soc.*, 2016, **138**, 15087.
102. J. Li, X. Yang, Z. Yu, G. G. Gurzadyan, M. Cheng, F. Zhang, J. Cong, W. Wang, H. Wang, X. Li, L. Kloo, M. Wang and L. Sun, *RSC Adv.*, 2017, **7**, 4611.
103. S. Caramori, J. Husson, M. Beley, C. A. Bignozzi, R. Argazzi and P. C. Gros, *Chem. - Eur. J.*, 2010, **16**, 2611.

104. I. Rutkowska, A. Andrearczyk, S. Zoladek, M. Goral, K. Darowicki and P. Kulesza, *J. Solid State Electrochem.*, 2011, **15**, 2545.
105. T. Daeneke, T.-H. Kwon, A. B. Holmes, N. W. Duffy, U. Bach and L. Spiccia, *Nat. Chem.*, 2011, **3**, 211.
106. T. Daeneke, Y. Uemura, N. W. Duffy, A. J. Mozer, N. Koumura, U. Bach and L. Spiccia, *Adv. Mater.*, 2012, **24**, 1222.
107. T. C. Li, A. M. Spokoyny, C. She, O. K. Farha, C. A. Mirkin, T. J. Marks and J. T. Hupp, *J. Am. Chem. Soc.*, 2010, **132**, 4580.
108. A. M. Spokoyny, T. C. Li, O. K. Farha, C. W. Machan, C. She, C. L. Stern, T. J. Marks, J. T. Hupp and C. A. Mirkin, *Angew. Chem., Int. Ed.*, 2010, **49**, 5339.
109. Z. Zhang, P. Chen, T. N. Murakami, S. M. Zakeeruddin and M. Grätzel, *Adv. Funct. Mater.*, 2008, **18**, 341.
110. F. Kato, N. Hayashi, T. Murakami, C. Okumura, K. Oyaizu and H. Nishide, *Chem. Lett.*, 2010, **39**, 464.
111. J. Min, J. Won, Y. S. Kang and S. Nagase, *J. Photochem. Photobiol., A*, 2011, **219**, 148.
112. D. Li, H. Li, Y. Luo, K. Li, Q. Meng, M. Armand and L. Chen, *Adv. Funct. Mater.*, 2010, **20**, 3358.
113. Y. Liu, J. R. Jennings, M. Parameswaran and Q. Wang, *Energy Environ. Sci.*, 2011, **4**, 564.
114. W. Mingkui, N. Chamberland, L. Breau, J.-E. Moser, R. Humphry-Baker, B. Marsan, S. M. Zakeeruddin and M. Grätzel, *Nat. Chem.*, 2010, **2**, 385.
115. H. Wu, Z. Lv, Z. Chu, D. Wang, S. Hou and D. Zou, *J. Mater. Chem.*, 2011, **21**, 14815.
116. H. Tian, X. Jiang, Z. Yu, L. Kloo, A. Hagfeldt and L. Sun, *Angew. Chem., Int. Ed.*, 2010, **49**, 7328.
117. M. Cheng, X. Yang, S. Li, X. Wang and L. Sun, *Energy Environ. Sci.*, 2012, **5**, 6290.
118. J. Burschka, V. Brault, S. Ahmad, L. Breau, M. K. Nazeeruddin, B. Marsan, S. M. Zakeeruddin and M. Grätzel, *Energy Environ. Sci.*, 2012, **5**, 6089.
119. Z. Zhang, *Enhancing the Open-Circuit Voltage of Dye-Sensitized Solar Cells: Coadsorbents and Alternative Redox Couples*, PhD thesis, EPFL, 2008.
120. Z. Yu, H. Tian, E. Gabrielsson, G. Boschloo, M. Gorlov, L. Sun and L. Kloo, *RSC Adv.*, 2012, **2**, 1083–1087.
121. R. Argazzi, C. A. Bignozzi, T. A. Heimer, F. N. Castellano and G. J. Meyer, *J. Phys. Chem. B*, 1997, **101**, 2591.
122. J. Desilvestro, M. Graetzel, L. Kavan, J. Moser and J. Augustynski, *J. Am. Chem. Soc.*, 1985, **107**, 2988.
123. P. Bonhôte, J.-E. Moser, R. Humphry-Baker, N. Vlachopoulos, S. M. Zakeeruddin, L. Walder and M. Grätzel, *J. Am. Chem. Soc.*, 1999, **121**, 1324.
124. Z.-S. Wang, K. Sayama and H. Sugihara, *J. Phys. Chem. B*, 2005, **109**, 22449.
125. C. Teng, X. Yang, C. Yuan, C. Li, R. Chen, H. Tian, S. Li, A. Hagfeldt and L. Sun, *Org. Lett.*, 2009, **11**, 5542.

126. C. Teng, X. Yang, S. Li, M. Cheng, A. Hagfeldt, L.-Z. Wu and L. Sun, *Chem. - Eur. J.*, 2010, **16**, 13127.
127. M. Gorlov, H. Pettersson, A. Hagfeldt and L. Kloo, *Inorg. Chem.*, 2007, **46**, 3566.
128. G. Oskam, B. V. Bergeron, G. J. Meyer and P. C. Searson, *J. Phys. Chem. B*, 2001, **105**, 6867.
129. P. Wang, S. M. Zakeeruddin, J.-E. Moser, R. Humphry-Baker and M. Grätzel, *J. Am. Chem. Soc.*, 2004, **126**, 7164.
130. B. V. Bergeron, A. Marton, G. Oskam and G. J. Meyer, *J. Phys. Chem. B*, 2004, **109**, 937.
131. S. Cazzanti, S. Caramori, R. Argazzi, C. M. Elliott and C. A. Bignozzi, *J. Am. Chem. Soc.*, 2006, **128**, 9996.
132. J. Cong, X. Yang, Y. Hao, L. Kloo and L. Sun, *RSC Adv.*, 2012, **2**, 3625.
133. J. Cong, Y. Hao, L. Sun and L. Kloo, *Adv. Energy Mater.*, 2014, **4**, 1301273.
134. J. Cong, Y. Hao, G. Boschloo and L. Kloo, *ChemSusChem*, 2015, **8**, 264.
135. Y. Hao, W. Yang, L. Zhang, R. Jiang, E. Mijangos, Y. Saygili, L. Hammarström, A. Hagfeldt and G. Boschloo, *Nat. Commun.*, 2016, **7**, 13934.
136. B. I. Lemon and J. T. Hupp, *J. Phys. Chem. B*, 1997, **101**, 2426.
137. Y. Liu, A. Hagfeldt, X.-R. Xiao and S.-E. Lindquist, *Sol. Energy Mater. Sol. Cells*, 1998, **55**, 267.
138. N. Kopidakis, K. D. Benkstein, J. van de Lagemaat and A. J. Frank, *J. Phys. Chem. B*, 2003, **107**, 11307–11315.
139. C. Shi, S. Dai, K. Wang, X. Pan, F. Kong and L. Hu, *Vib. Spectrosc.*, 2005, **39**, 99.
140. G. Boschloo, L. Häggman and A. Hagfeldt, *J. Phys. Chem. B*, 2006, **110**, 13144.
141. S. Kim, J. K. Lee, S. O. Kang, J. Ko, J. H. Yum, S. Fantacci, F. De Angelis, D. Di Censo, M. K. Nazeeruddin and M. Grätzel, *J. Am. Chem. Soc.*, 2006, **128**, 16701–16707.
142. H. Tian, X. Yang, J. Cong, R. Chen, J. Liu, Y. Hao, A. Hagfeldt and L. Sun, *Chem. Commun.*, 2009, 6288.
143. P. Wang, S. M. Zakeeruddin, J. E. Moser, R. Humphry-Baker, P. Comte, V. Aranyos, A. Hagfeldt, M. K. Nazeeruddin and M. Grätzel, *Adv. Mater.*, 2004, **16**, 1806.
144. T. Horiuchi, H. Miura, K. Sumioka and S. Uchida, *J. Am. Chem. Soc.*, 2004, **126**, 12218.
145. J. Wu, Z. Lan, J. Lin, M. Huang, Y. Huang, L. Fan and G. Luo, *Chem. Rev.*, 2015, **115**, 2136.
146. A. F. Nogueira, C. Longo and M. A. De Paoli, *Coord. Chem. Rev.*, 2004, **248**, 1455.
147. R.-X. Dong, S.-Y. Shen, H.-W. Chen, C.-C. Wang, P.-T. Shih, C.-T. Liu, R. Vittal, J.-J. Lin and K.-C. Ho, *J. Mater. Chem. A*, 2013, **1**, 8471.
148. C. Wang, L. Wang, Y. Shi, H. Zhang and T. Ma, *Electrochim. Acta*, 2013, **91**, 302.
149. P. Wang, S. M. Zakeeruddin, P. Comte, I. Exnar and M. Grätzel, *J. Am. Chem. Soc.*, 2003, **125**, 1166.

150. W. S. Chi, D. K. Roh, S. J. Kim, S. Y. Heo and J. H. Kim, *Nanoscale*, 2013, **5**, 5341.
151. X. Zhang, H. Yang, H.-M. Xiong, F.-Y. Li and Y.-Y. Xia, *J. Power Sources*, 2006, **160**, 1451.
152. E. Stathatos, P. Lianos and C. Krontiras, *J. Phys. Chem. B*, 2001, **105**, 3486.
153. Z. Yu, N. Vlachopoulos, M. Gorlov and L. Kloo, *Dalton Trans.*, 2011, **40**, 10289.
154. K. Kakiage, Y. Aoyama, T. Yano, K. Oya, J.-i. Fujisawa and M. Hanaya, *Chem. Commun.*, 2015, **51**, 15894.
155. M. K. Kashif, M. Nippe, N. W. Duffy, C. M. Forsyth, C. J. Chang, J. R. Long, L. Spiccia and U. Bach, *Angew. Chem., Int. Ed.*, 2013, **52**, 5527.
156. S. Ito, M. K. Nazeeruddin, P. Liska, P. Comte, R. Charvet, P. Péchy, M. Jirousek, A. Kay, S. M. Zakeeruddin and M. Grätzel, *Prog. Photovoltaics Res. Appl.*, 2006, **14**, 589.
157. Q. Yu, D. Zhou, Y. Shi, X. Si, Y. Wang and P. Wang, *Energy Environ. Sci.*, 2010, **3**, 1722.
158. C. Law, S. C. Pathirana, X. Li, A. Y. Anderson, P. R. F. Barnes, A. Listorti, T. H. Ghaddar and B. C. O'Regan, *Adv. Mater.*, 2010, **22**, 4505.
159. H. Ellis, R. Jiang, S. Ye, A. Hagfeldt and G. Boschloo, *Phys. Chem. Chem. Phys.*, 2016, **18**, 8419.
160. M. Gorlov and L. Kloo, *Dalton Trans.*, 2008, 2655.
161. P. Wang, S. M. Zakeeruddin, I. Exnar and M. Gratzel, *Chem. Commun.*, 2002, 2972.
162. D. Shi, N. Pootrakulchote, R. Li, J. Guo, Y. Wang, S. M. Zakeeruddin, M. Grätzel and P. Wang, *J. Phys. Chem. C*, 2008, **112**, 17046.
163. Q. Li, Q. Tang, N. Du, Y. Qin, J. Xiao, B. He, H. Chen and L. Chu, *J. Power Sources*, 2014, **248**, 816.
164. Y. Bai, Y. Cao, J. Zhang, M. Wang, R. Li, P. Wang, S. M. Zakeeruddin and M. Gratzel, *Nat. Mater.*, 2008, **7**, 626.
165. B. Xu, H. Tian, L. Lin, D. Qian, H. Chen, J. Zhang, N. Vlachopoulos, G. Boschloo, Y. Luo, F. Zhang, A. Hagfeldt and L. Sun, *Adv. Energy Mater.*, 2015, **5**(3), 1401185.
166. U. Bach, D. Lupo, P. Comte, J. E. Moser, F. Weissortel, J. Salbeck, H. Spreitzer and M. Grätzel, *Nature*, 1998, **395**(6702), 583.
167. B. Xu, *Advanced Organic Hole Transport Materials for Solution-Processed Photovoltaic Devices*, PhD thesis, KTH Royal Institute of Technology, 2015.
168. L. Zhang, G. Boschloo, L. Hammarström and H. Tian, *Phys. Chem. Chem. Phys.*, 2016, **18**(7), 5080.
169. X. Li, B. Xu, P. Liu, Y. Hu, L. Kloo, J. Hua, L. Sun and H. Tian, *J. Mater. Chem. A*, 2017, **5**(7), 3157.
170. Z. Shen, B. Xu, P. Liu, Y. Hu, Y. Yu, H. Ding, L. Kloo, J. Hua, L. Sun and H. Tian, *J. Mater. Chem. A*, 2017, **5**(3), 1242.
171. H. Tian, A. Soto, B. Xu, L. Sun, A. Hagfeldt, F. Fabregat-Santiago, I. Mora-Sero, Y. S. Kang, J. Bisquert and E. M. Barea, *Nano*, 2014, **09**(05), 1440005.

172. P. Liu, W. Sharmoukh, B. Xu, Y. Y. Li, G. Boschloo, L. Sun and L. Kloo, *ACS Omega*, 2017, **2**(5), 1812.
173. M. Shinya, T. Akihiro, S. Juntaro, K. Naohiko, H. Kazuo, U. Koji, M. Katsuyoshi and T. Tatsuo, *Appl. Phys. Express*, 2012, **5**(11), 112302.
174. Q. B. Meng, K. Takahashi, X. T. Zhang, I. Sutanto, T. N. Rao, O. Sato, A. Fujishima, H. Watanabe, T. Nakamori and M. Uragami, *Langmuir*, 2003, **19**(9), 3572.
175. K. Tennakone, G. K. R. Senadeera, D. B. R. A. D. Silva and I. R. M. Kottegoda, *Appl. Phys. Lett.*, 2000, **77**(15), 2367.
176. G. R. R. A. Kumara, A. Konno, G. K. R. Senadeera, P. V. V. Jayaweera, D. B. R. A. De Silva and K. Tennakone, *Sol. Energy Mater. Sol. Cells*, 2001, **69**(2), 195.
177. J. Bandara and H. Weerasinghe, *Sol. Energy Mater. Sol. Cells*, 2005, **85**(3), 385.
178. I. Chung, B. Lee, J. He, R. P. H. Chang and M. G. Kanatzidis, *Nature*, 2012, **485**(7399), 486.
179. B. Xu, H. Tian, D. Bi, E. Gabrielsson, E. M. J. Johansson, G. Boschloo, A. Hagfeldt and L. Sun, *J. Mater. Chem. A*, 2013, **1**(46), 14467.
180. B. Xu, J. Zhang, Y. Hua, P. Liu, L. Wang, C. Ruan, Y. Li, G. Boschloo, E. M. J. Johansson, L. Kloo, A. Hagfeldt, A. K. Y. Jen and L. Sun, *Chem*, 2017, **2**(5), 676.
181. J. E. Kroeze, N. Hirata, L. Schmidt-Mende, C. Orizu, S. D. Ogier, K. Carr, M. Grätzel and J. R. Durrant, *Adv. Funct. Mater.*, 2006, **16**(14), 1832.
182. I. K. Ding, N. Tétreault, J. Brillet, B. E. Hardin, E. H. Smith, S. J. Rosenthal, F. Sauvage, M. Grätzel and M. D. McGehee, *Adv. Funct. Mater.*, 2009, **19**(15), 2431.
183. L. Yang, B. Xu, D. Bi, H. Tian, G. Boschloo, L. Sun, A. Hagfeldt and E. M. J. Johansson, *J. Am. Chem. Soc.*, 2013, **135**(19), 7378.
184. Y. Hua, J. Zhang, B. Xu, P. Liu, M. Cheng, L. Kloo, E. M. J. Johansson, K. Sveinbjörnsson, K. Aitola, G. Boschloo and L. Sun, *Nano Energy*, 2016, **26**, 108.
185. B. Xu, E. Sheibani, P. Liu, J. Zhang, H. Tian, N. Vlachopoulos, G. Boschloo, L. Kloo, A. Hagfeldt and L. Sun, *Adv. Mater.*, 2014, **26**(38), 6629.
186. B. Xu, D. Bi, Y. Hua, P. Liu, M. Cheng, M. Gratzel, L. Kloo, A. Hagfeldt and L. Sun, *Energy Environ. Sci.*, 2016, **9**(3), 873.
187. D. Bi, B. Xu, P. Gao, L. Sun, M. Grätzel and A. Hagfeldt, *Nano Energy*, 2016, **23**, 138.
188. T. Leijtens, I. K. Ding, T. Giovenzana, J. T. Bloking, M. D. McGehee and A. Sellinger, *ACS Nano*, 2012, **6**(2), 1455.
189. K. Walzer, B. Maennig, M. Pfeiffer and K. Leo, *Chem. Rev.*, 2007, **107**(4), 1233.
190. J. Burschka, A. Dualeh, F. Kessler, E. Baranoff, N.-L. Cevey-Ha, C. Yi, M. K. Nazeeruddin and M. Grätzel, *J. Am. Chem. Soc.*, 2011, **133**(45), 18042.
191. J. Burschka, F. Kessler, M. K. Nazeeruddin and M. Grätzel, *Chem. Mater.*, 2013, **25**(15), 2986.

192. A. Abate, T. Leijtens, S. Pathak, J. Teuscher, R. Avolio, M. E. Errico, J. Kirkpatrik, J. M. Ball, P. Docampo, I. McPherson and H. J. Snaith, *Phys. Chem. Chem. Phys.*, 2013, **15**(7), 2572.
193. U. B. Cappel, T. Daeneke and U. Bach, *Nano Lett.*, 2012, **12**(9), 4925.
194. D.-Y. Chen, W.-H. Tseng, S.-P. Liang, C.-I. Wu, C.-W. Hsu, Y. Chi, W.-Y. Hung and P.-T. Chou, *Phys. Chem. Chem. Phys.*, 2012, **14**(33), 11689.
195. B. Xu, J. Huang, H. Ågren, L. Kloo, A. Hagfeldt and L. Sun, *ChemSusChem*, 2014, **7**(12), 3252.
196. B. Xu, E. Gabrielsson, M. Safdari, M. Cheng, Y. Hua, H. Tian, J. M. Gardner, L. Kloo and L. Sun, *Adv. Energy Mater.*, 2015, **5**(10), 1402340.
197. S. Ardo and G. J. Meyer, *Chem. Soc. Rev.*, 2009, **38**, 115.
198. C. S. Ponseca Jr, P. Chábera, J. Uhlig, P. Persson and V. Sundström, *Chem. Rev.*, 2017, **117**, 10940.
199. A. Listort, B. O'Regan and J. R. Durrant, *Chem. Mater.*, 2011, **23**, 3381.
200. (a) G. Benkö, J. Kallioinen, J. E. L. Korppi-Tommola, A. P. Yartsev and V. Sundström, *J. Am. Chem. Soc.*, 2002, **124**, 489; (b) G. Benkö, J. Kallioinen, P. Myllyperkiö, F. Trif, J. E. L. Korppi-Tommola, A. P. Yartsev and V. Sundström, *J. Phys. Chem. B*, 2004, **108**, 2862.
201. P. G. Johansson, Y. Zhang, M. Abrahamsson, G. J. Meyer and E. Galoppini, *Chem. Commun.*, 2011, **47**, 6410.
202. (a) A. Y. Anderson, P. R. F. Barnes, J. R. Durrant and B. C. O'Regan, *J. Phys. Chem. C*, 2010, **114**, 1953; (b) J. N. Clifford, E. Palomares, M. K. Nazeeruddin, M. Gratzel and J. R. Durrant, *J. Phys. Chem. C*, 2007, **111**, 6561; (c) I. Montanari, J. Nelson and J. R. Durrant, *J. Phys. Chem. B*, 2002, **106**, 12203.
203. J. E. Moser and M. Grätzel, *Chem. Phys.*, 1993, **176**, 493.
204. (a) C. A. Kelly, F. Farzad, D. W. Thompson, J. M. Stipkala and G. J. Meyer, *Langmuir*, 1999, **15**, 7047; (b) B. O'Regan, J. Moser, M. Anderson and M. Gratzel, *J. Phys. Chem.*, 1990, **94**, 8720; (c) S. A. Haque, Y. Tachibana, R. L. Willis, J. E. Moser, M. Gratzel, D. R. Klug and J. R. Durrant, *J. Phys. Chem. B*, 2000, **104**, 538; (d) S. A. Haque, Y. Tachibana, D. R. Klug and J. R. Durrant, *J. Phys. Chem. B*, 1998, **102**, 1745; (e) I. Martini, J. H. Hodak and G. V. Hartland, *J. Phys. Chem. B*, 1999, **102**, 1745.
205. M. Abrahamsson, P. G. Johansson, S. Ardo, A. Kopecky, E. Galoppini and G. J. Meyer, *J. Phys. Chem. Lett.*, 2010, **1**(11), 1725.
206. (a) V. Saavedra Becerril, D. Franchi and M. Abrahamsson, *J. Phys. Chem. C*, 2016, **120**(36), 20016; (b) H. Paulsson, L. Kloo, A. Hagfeldt and G. Boschloo, *J. Electroanal. Chem.*, 2006, **586**, 56; (c) S. Mahanta, A. Furube, H. Matsuzaki, T. Murakami and H. Matsumoto, *J. Phys. Chem. C*, 2012, **116**, 20213; (d) F. Li, J. R. Jennings, X. Wang, L. Fan, Z. Y. Koh, H. Yu, L. Yan and Q. Wang, *J. Phys. Chem. C*, 2014, **118**, 17153; (e) H. M. Upadhyaya, N. Hirata, S. A. Haque, M. A. de Paoli and J. R. Durrant, *Chem. Commun.*, 2006, 877; (f) S. A. Haque, E. Palomares, H. M. Upadhyaya, L. Otley, R. J. Potter, A. B. Holmes and J. R. Durrant, *Chem. Commun.*, 2003, 3008; (g) J. N. de Freitas, A. F. Nogueira and M. A. De Paoli, *J. Mater. Chem.*, 2009, **19**, 5279.

207. D. Kuciauskas, M. S. Freund, H. B. Gray, J. R. Winkler and N. S. Lewis, *J. Phys. Chem. B*, 2001, **105**, 392.
208. (a) S. A. Haque, S. Handa, K. Peter, E. Palomares, M. Thelakkat and J. R. Durrant, *Angew. Chem., Int. Ed.*, 2005, **44**, 5740; (b) J. N. Clifford, E. Palomares, M. K. Nazeeruddin, M. Grätzel, J. Nelson, X. Li, N. J. Long and J. R. Durrant, *J. Am. Chem. Soc.*, 2004, **126**, 5225; (c) C. Kleverlaan, M. Alebbi, R. Argazzi, C. A. Bignozzi, G. M. Hasselmann and G. J. Meyer, *Inorg. Chem.*, 2000, **39**, 1342.
209. (a) S. Ardo, Y. Sun, A. Staniszewski, F. N. Castellano and G. J. Meyer, *J. Am. Chem. Soc.*, 2010, **132**, 6696; (b) U. B. Cappel, S. M. Feldt, J. Schöneboom, A. Hagfeldt and G. Boschloo, *J. Am. Chem. Soc.*, 2010, **132**, 9096.
210. E. M. J. Johansson, R. Lindblad, H. Siegbahn, A. Hagfeldt and H. Rensmo, *ChemPhysChem*, 2014, **15**(6), 1006.
211. U. B. Cappel, V. Lanzilotto, E. M. J. Johansson, T. Edvinsson and H. Rensmo, in *Molecular Devices for Solar Energy Conversion and Storage*, ed. H. Tian, G. Boschloo and A. Hagfeldt, Springer, Singapore, 2018, p. 433.
212. S. Tanuma, *Surf. Interface Anal.*, 2011, **43**(3), 689.
213. L. Patthey, H. Rensmo, P. Persson, K. Westermark, L. Vayssieres, A. Stashans, A. Petersson, P. A. Bruhwiler, H. Siegbahn, S. Lunell and N. Martensson, *J. Chem. Phys.*, 1999, **110**(12), 5913.
214. J. Schnadt, A. Henningsson, M. P. Andersson, P. G. Karlsson, P. Uvdal, H. Siegbahn and P. A. Bru, *J. Phys. Chem. B*, 2004, **2**(110), 3114.
215. E. M. J. Johansson, S. Plogmaker, L. E. Walle, R. Scholin, A. Borg, A. Sandell and H. Rensmo, *J. Phys. Chem. C*, 2010, **114**(35), 15015.
216. L. C. Mayor, J. B. Taylor, G. Magnano, A. Rienzo, C. J. Satterley, J. N. O'Shea and J. Schnadt, *J. Chem. Phys.*, 2008, **129**(11), 114701.
217. E. M. J. Johansson, T. Edvinsson, M. Odelius, D. P. Hasberg, L. Sun, A. Hagfeldt, H. Siegbahn and H. Rensmo, *J. Phys. Chem. C*, 2007, **111**(24), 8580.
218. M. Hahlin, E. M. J. Johansson, S. Plogmaker, M. Odelius, D. P. Hagberg, L. Sun, H. Siegbahn and H. Rensmo, *Phys. Chem. Chem. Phys.*, 2010, **12**(7), 1507.
219. H. Ellis, S. K. Eriksson, S. M. Feldt, E. Gabrielsson, P. W. Lohse, R. Lindblad, L. Sun, H. Rensmo, G. Boschloo and A. Hagfeldt, *J. Phys. Chem. C*, 2013, **117**(41), 21029.
220. T. Marinado, M. Hahlin, X. Jiang, M. Quintana, E. M. J. Johansson, E. Gabrielsson, S. Plogmaker, D. P. Hagberg, G. Boschloo, S. M. Zakeeruddin, M. Grätzel, H. Siegbahn, L. Sun, A. Hagfeldt and H. Rensmo, *J. Phys. Chem. C*, 2010, **114**(27), 11903.
221. J. Oscarsson, M. Hahlin, E. M. J. Johansson, S. K. Eriksson, R. Lindblad, A. I. K. Eriksson, A. Zia, H. Siegbahn and H. Rensmo, *J. Phys. Chem. C*, 2016, **120**(23), 12484.
222. M. Hahlin, E. M. J. Johansson, R. Schölin, H. Siegbahn and H. Rensmo, *J. Phys. Chem. C*, 2011, **115**(24), 11996.
223. E. M. J. Johansson, P. G. Karlsson, M. Hedlund, D. Ryan, H. Siegbahn and H. Rensmo, *Chem. Mater.*, 2007, **19**(8), 2071.

224. U. B. Cappel, A. L. Smeigh, S. Plogmaker, E. M. J. Johansson, H. Rensmo, L. Hammarström, A. Hagfeldt and G. Boschloo, *J. Phys. Chem. C*, 2011, **115**(10), 4345.
225. R. Schölin, M. H. Karlsson, S. K. Eriksson, H. Siegbahn, E. M. J. Johansson and H. Rensmo, *J. Phys. Chem. C*, 2012, **116**(50), 26300.
226. R. Lindblad, J. Oscarsson, K. Fredin, S. K. Eriksson, H. Siegbahn, E. M. J. Johansson and H. Rensmo, *J. Electron Spectrosc. Relat. Phenom.*, 2017, **224**, 100.
227. D. E. Starr, Z. Liu, M. Hävecker, A. Knop-Gericke and H. Bluhm, *Chem. Soc. Rev.*, 2013, **42**(13), 5833.
228. S. K. Eriksson, M. Hahlin, S. Axnanda, E. Crumlin, R. Wilks, M. Odelius, A. I. K. Eriksson, Z. Liu, J. Åhlund, A. Hagfeldt, D. E. Starr, M. Bär, H. Rensmo and H. Siegbahn, *Top. Catal.*, 2016, **59**(5–7), 583.
229. F. Fabregat-Santiago, G. Garcia-Belmonte, I. Mora-Sero and J. Bisquert, *Phys. Chem. Chem. Phys.*, 2011, **13**, 9083.
230. F. Fabregat-Santiago, J. Bisquert, G. Garcia-Belmonte, G. Boschloo and A. Hagfeldt, *Sol. Energy Mater. Sol. Cells*, 2005, **87**, 117.
231. Q. Wang, S. Ito, M. Grätzel, F. Fabregat-Santiago, I. Mora-Sero, J. Bisquert, T. Bessho and H. Imai, *J. Phys. Chem. B*, 2006, **110**, 19406.
232. F. Fabregat-Santiago, J. Bisquert, E. Palomares, L. Otero, D. Kuang, S. M. Zakeeruddin and M. Grätzel, *J. Phys. Chem. C*, 2007, **111**, 6550.
233. E. M. Barea, C. Zafer, B. Gultekin, B. Aydin, S. Koyuncu, S. Icli, F. Fabregat-Santiago and J. Bisquert, *J. Phys. Chem. C*, 2010, **114**, 19840.
234. A. Reynal, A. Forneli, E. Martinez-Ferrero, A. Sanchez-Diaz, A. Vidal-Ferran, B. C. O'Regan and E. Palomares, *J. Am. Chem. Soc.*, 2008, **130**, 13558.
235. E. Barea, R. Caballerro, F. Fabregat-Santiago, P. De la Cruz, F. Langa and J. Bisquert, *ChemPhysChem*, 2009, **11**, 245.
236. M. Miyashita, K. Sunahara, T. Nishikawa, Y. Uemura, N. Koumura, K. Hara, A. Mori, T. Abe, E. Suzuki and S. Mori, *J. Am. Chem. Soc.*, 2008, **130**, 17874.
237. J. Bisquert, *J. Phys. Chem. B*, 2002, **106**, 325.
238. F. Fabregat-Santiago, H. Randriamahazaka, A. Zaban, J. Garcia-Cañadas, G. Garcıa-Belmonte and J. Bisquert, *Phys. Chem. Chem. Phys.*, 2006, **8**, 1827.
239. J. Bisquert, F. Fabregat-Santiago, I. Mora-Sero, G. Garcia-Belmonte and S. Gimenez, *J. Phys. Chem. C*, 2009, **113**, 17278.

CHAPTER 4

Solution-processed Solar Cells: Perovskite Solar Cells

EVA L. UNGER*[a,b], OLEKSANDRA SHARGAIEVA[a], STEFFEN BRAUNGER[a] AND PABLO DOCAMPO[c]

[a]Helmholtz-Zentrum Berlin für Materialien und Energie GmbH, Young Investigator Group Hybrid Materials Formation and Scaling, Albert-Einstein Strasse 16, 12489, Berlin, Germany; [b]Lund University, Department of Chemistry, Naturvetarvägen 14, Lund, 223 62, Sweden; [c]Newcastle University, School of Engineering, Merz Court, NE1 7RU Newcastle upon Tyne, UK
*E-mail: eva.unger@helmholtz-berlin.de, eva.unger@chemphys.lu.se

4.1 (R)evolutionary Solar Cells

With a continuous reduction in cost, solar cells are becoming an increasingly viable economic option for sustainable energy conversion. Figure 4.1, adapted from the frequently updated plot published by the National Renewable Energy Laboratory (NREL)[1] illustrates that for many of the established solar cell technologies no significant increase of device performance has been reported during the past couple of years, indicating that they are approaching their theoretical limits. Exceptions are cadmium telluride (CdTe), quantum dot (QD) and alkali/organic metal halide perovskite (ABX_3)-based solar cells, with substantial relative improvements of device performance being achieved.

Inorganic Materials Series No. 3
Solar Energy Capture Materials
Edited by Elizabeth A. Gibson

Published by the Royal Society of Chemistry, www.rsc.org

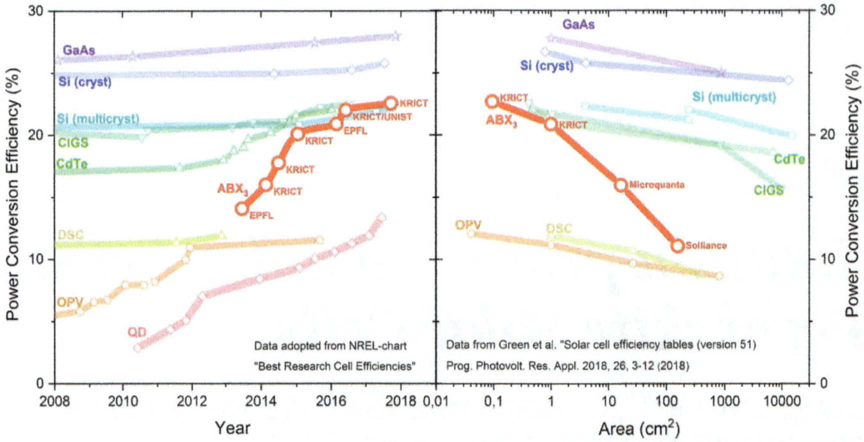

Figure 4.1 (Left) Best certified laboratory (small area) solar cell device efficiencies as a function of publication year.[1] (Right) Solar cell device performance as a function of area plotted from data collected by Green and co-workers[229] for various different solar cell device technologies.

The ABX_3 solar cell technology evolved from solid-state dye-sensitised solar cells (DSSCs) and is a solution-processable technology with many deposition strategies directly adapted from printable organic (OPV) and quantum dot (QD) photovoltaics. ABX_3-based devices have surpassed DSSCs, OPVs and QD-based devices for small-area devices, as illustrated in Figure 4.1.[2,3] This can be attributed to their excellent optoelectronic and structural properties and defect tolerance, enabling growth of high-quality ABX_3 layers from solutions, which we will discuss in Section 4.2.

Reaching power conversion efficiencies comparable to mature thin-film photovoltaic technologies, such as silicon, copper-indium-gallium-sulfide/selenide (CIGS), and cadmium telluride (CdTe), perovskite solar cells (PSCs) now have to measure up to these technologies in terms of their stability, scale and environmental compatibility as well. The right panel of Figure 4.1 compares the power conversion efficiency of different types of solar cell technologies as a function of the active device area, using datasets summarised by Green *et al.*[4] Evidently, mature solar cell technologies exhibit drops in power conversion efficiency (PCE) as a function of area of the order of −0.5% per decade for silicon, −0.75% for CdTe and −0.82% for OPV, due to inherent performance losses from increases in series resistance and shading losses of busbars in larger area devices.[5] In comparison, larger area alkali/organic metal halide perovskite (ABX_3)-based solar cells exhibit a performance drop of −3.7% per decade, which only reflects that ABX_3 solar cells are technologically still at an early stage of development.

The solution- and low-temperature processability of ABX_3 solar cells enable the realisation of low-cost, stable, single- and multi-junction solar energy conversion devices.[6,7] Solution- or wet-chemical processing allows the deposition of materials on substrates of almost any shape, and control over the resulting material layer thickness, homogeneity and morphology by adjusting the processing method and conditions. ABX_3-based solar cells already outperform OPVs and DSSCs for small-area devices and exhibit comparable performance for module-sized devices. For other inorganic photovoltaic (PV) materials, such as CdTe, CIGS and amorphous silicon (a-Si), fabrication by solution-processing methods has been reported.[8–10] However, in all three cases conventional high temperature and vacuum-based deposition techniques (*e.g.*, co-evaporation, sputtering) outperform solution-processed ones, as crystalline absorber materials with low defect densities and higher charge carrier mobility are a prerequisite for making efficient solar cells with these materials.

This chapter focuses on illuminating the material properties of ABX_3 perovskites that enable solution-processing at moderate temperatures (Section 4.2), recent insights into the solution chemistry of ABX_3 inks (Section 4.3) and how the optoelectronic quality of ABX_3 layers has been optimised through process and composition engineering (Section 4.4). Technological challenges when scaling metal halide perovskite-based device technology with solution-based methods include the adaptation of scalable deposition methods, such as inkjet printing and slot-die coating, to achieve controlled growth of high-quality ABX_3 layers, which includes the optimisation of ABX_3 precursor inks (Section 4.5). For large-scale deployment, stability and potential ecotoxicity concerns of lead-containing ABX_3 materials must be addressed (Section 4.6). Inspired by the evolution of ABX_3 materials, new materials are being developed that might provide environmentally benign alternatives to lead-containing perovskites in the future.

4.2 A Versatile Semiconductor

4.2.1 Crystal Structure and Tolerance Factor

The term *perovskite* refers to the ABX_3 crystal structure that these absorber materials, consisting of an organic or alkaline monovalent A-cation, a divalent B-cation and monovalent halide or pseudo-halide anion X^-, adopt. B-cation and X-anion form a network of corner-connected $[BX_6]$ octahedra, whereas the A-cation stabilises the crystal structure in the resulting 12-fold cuboctahedral coordinated site.

Whether a stable ideal cubic perovskite crystal structure can form is determined by the ratio of cations (r_A and r_B) and anion (r_X) radii according to the Goldschmidt tolerance factor, T_f, which was derived for oxide perovskites but has been shown to be applicable to halide perovskites.[11,12]

$$T_f = \frac{r_{A^+} + r_{X^-}}{\sqrt{2}\left(r_{B^{2+}} + r_{X^-}\right)} \tag{4.1}$$

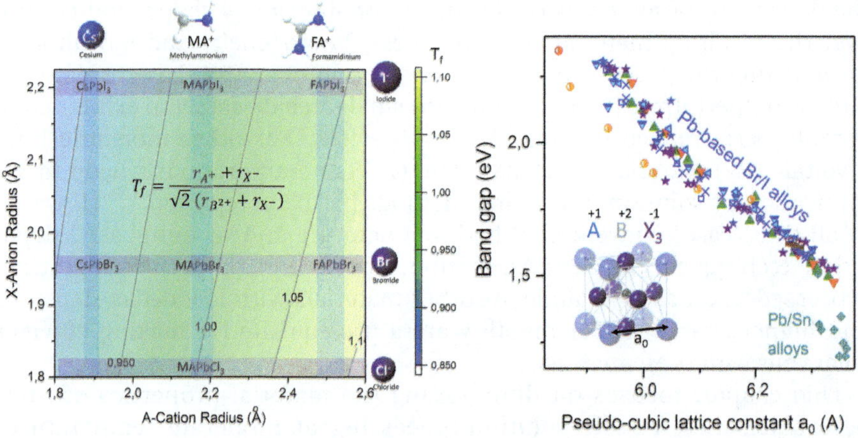

Figure 4.2 (Left) Representation of the generic cubic perovskite crystal structure composed of monovalent (Rb, Cs, MA, FA) and divalent (Pb, Sn) cations and halide anions (Cl, Br, I). The tolerance factor T_f (graph on right) can be estimated from the ratio of ion radii. Stable perovskite materials are predicted to exist for $0.8 < T_f < 1.05$.[11,12] (Right) The band gap of ABX_3 perovskites can be tuned by cation and anion substitution and scales with the pseudo-cubic lattice constant a_0 of the perovskite unit cell for a variety of different cation/anion alloys. Adapted from ref. 23 with permission from the Royal Society of Chemistry.

In Figure 4.2, the resulting tolerance factors, T_f, of lead-based ABX_3 materials are estimated as a function of anion radii ranging from chloride (Cl^-) to bromide (Br^-) and iodide (I^-) indicated on the *y*-axis, and cation radii ranging from caesium (Cs^+) to methylammonium (MA^+) and formamidinium (FA^+) plotted on the *x*-axis. For the halide ion radii, tabulated values reported by Shannon *et al.*[13] are used. A cation radius of 1.0 Å was used for the Pb^{2+} cation to account for the decreased effective cation radius when forming bonds with larger halide anions, due to a reduced ionic character of the bond according to Travis *et al.*[12] ABX_3 materials are predicted to form orthorhombic or tetragonal crystal structures due to distortions of the $[PbX_6]$ octahedral network when the cation radii give rise to a $T_f \ll 1$, cubic structures for T_f close to 1, and trigonal structures when $T_f > 1$.

Tolerance factors are useful to rationalise trends in ABX_3 perovskite materials. As an example, the first and to date most investigated metal halide perovskite material, methylammonium lead iodide, $MAPbI_3$ or $CH_3NH_3PbI_3$, exhibits a tetragonal crystal structure at room temperature with slightly tilted octahedra as the T_f is slightly too small, but the crystal structure becomes cubic when adding a fraction (>15%) of the smaller bromide anion.[14] Another example is that both caesium lead iodide, $CsPbI_3$, and formamidinium lead iodide, $FAPbI_3$, form non-perovskite *yellow* phases. The former has a T_f too small and forms an orthorhombic phase, whereas the latter has a T_f too large and preferentially forms trigonal or hexagonal non-perovskite crystal structures.[15,16]

Mixing formamidinium and caesium cations tunes the average A-cation radius towards a more favourable T_f leading to cubic ABX_3 compounds.[15–20] The mixed Cs/FA cation ABX_3 material exhibits higher temperature stability, which is of technological importance, as discussed in Section 4.6.

4.2.2 Electronic Structure and Defect Tolerance

An important motivation for cation and anion mixing in ABX_3 perovskites is to tune the electronic band gap to match the ABX_3 absorption with other absorber materials to realise multi-junction solar cells based on, for example, silicon as a bottom cell.[7,20,21] Figure 4.2 (right) shows the correlation between the absorption onset with the halide–lead–halide bond distances in ABX_3 perovskites expressed as a pseudo-cubic perovskite lattice parameter, a_0, as discussed elsewhere.[22,23] Noticeably, the absorption onset is, to a first approximation, linearly dependent on the halide–lead spacing in the crystal lattice. Due to their low T_f, pure $CsPbI_3$ materials are only metastable but can be stabilised when synthesised as quantum dots.[24] Note that, because of a non-negligible contribution of bound excitons to the absorption spectra of ABX_3 perovskites, the absorption onset is not equivalent to the material's band gap and the latter cannot be derived from Tauc plots as discussed by Green and co-workers.[25]

The A-cation only influences the absorption onset by affecting the halide–lead bond distance and bonding angle, which is consistent with the fact that bands from the A-site cation are far from the valence and conduction band in the electronic band structure. The electronic properties of ABX_3 perovskites are determined by the lead and halide orbitals. The A-cation interacts with the PbX_6 cage only *via* van der Waals forces and creates bands far from the band gap, shown in Figure 4.3 (left). The valence band is dominated by the mixing of Pb(6p) and I(5p) atomic orbitals, forming a deep valence band with dominant I(5p) and slight anti-bonding character. The minimum of the conduction band has dominant Pb(6p) character. The bonding interaction has significant charge transfer character from the Pb(6p) to the I(5p) orbital.[26–28]

Due to this particular band structure, the most prominent vacancies and interstitials are resonant with either the valence or conduction band, or are energetically very shallow.[29,30] This is consistent with often-observed steep absorption onsets and low Urbach tails of methylammonium lead iodide thin films.[31] Lead halide perovskite semiconductors are often referred to as *defect-tolerant* materials.[26,32,33] Metal halide perovskites typically exhibit very low non-radiative recombination rates and high photoluminescence quantum yields, which in turn enable high open circuit voltages.[34,35]

Defect tolerance does not presume the absence of defects, but, as the term implies, that the optoelectronic properties of the material are not negatively affected by the dominant defects in the material. A prerequisite for defect tolerance is that the defects with the lowest formation energy create shallow

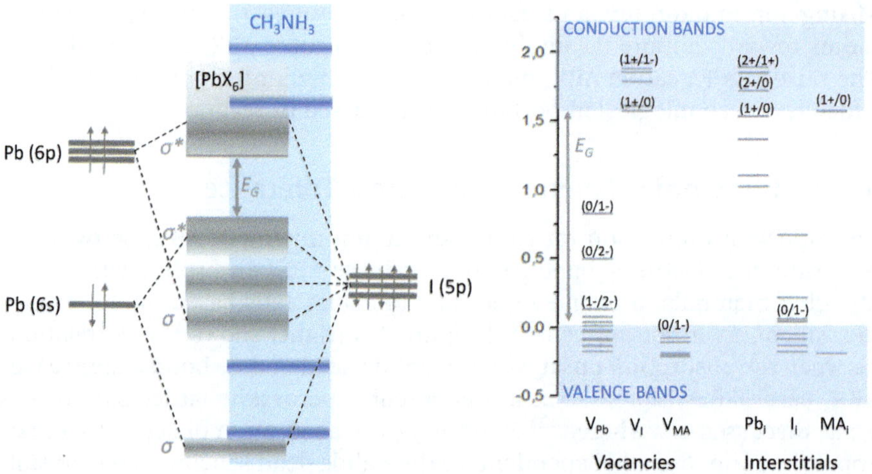

Figure 4.3 (Left) Band diagram derived from theoretical calculated band structures adapted from ref. 26 and 198. (Right) Calculated defect levels for ion vacancies and interstitials in $MAPbI_3$ derived from digitised data reported in ref. 38–40 and shown here in one plot for comparison.

levels that do not cause non-radiative recombination. Point defects with energy levels deep in the band gap have conversely high formation energies, but are unlikely to form in highly ionic ABX_3 perovskites. Likewise, surfaces and grain boundaries are predicted to be electronically benign.[36] The defect tolerance of ABX_3 perovskites is attributed to the high ionicity of the lead–halide bonds, which cleave heterolytically, preventing dangling bonds and forming only shallow defect levels.[37] Effective masses of electrons and holes are low and the static dielectric constant is high, screening free charge carriers from charged point defects.[26]

In Figure 4.3, theoretically calculated defect levels for prominent vacancies and interstitials from different literature sources[38–40] are summarised. According to calculations, some of the most likely defects to form are halide vacancies.[29,38] The redistribution of halide and other vacancies is interpreted to be the origin of the self-healing phenomena of defective ABX_3 perovskite under illumination or thermal heating, leading to the annihilation of defects.[41–45] Mosconi *et al.*[46] proposed that the photoinduced enhancement of the luminescence is due to the compensation of iodide vacancies (V^+_I) by migrating (interstitial) iodide species (I^-_i). These *Frenkel defect pairs* should be abundant in metal halide perovskites as both point defects have low formation energies and are predicted to form only shallow defect levels. The depth of halide interstitial I^-_i defect levels is currently under debate and dynamic phenomena in spectroscopic investigations have been attributed to iodine photochemical oxidation/reduction reactions.[40]

Halide migration has also been attributed with causing the emergence of low energy emission sites in mixed Br/I perovskites.[47] Ion migration has also been hypothesised to cause slow transient phenomena in current–voltage measurements, resulting in a scan rate-dependent discrepancy between forward and reverse scans, often referred to as hysteresis.[48–51] This phenomenon gives rise to an uncertainty in the reported power conversion efficiencies of ABX_3-based solar cells.

While shallow defects might not negatively affect device performance, they may still affect the charge carrier mobility. The mobility of charge carriers in ABX_3 semiconductors is sometimes quoted as being "large", which is only true when compared with molecular semiconductors, as conventional semiconductors such as GaAs or silicon exhibit substantially higher charge carrier mobilities.[52] Regardless of the depth of the defects, they could very well influence charge carrier transport phenomena, as charge carriers interact with shallow or "in-band" defects. Charge carrier diffusion lengths of the order of several micrometres have been reported[53,54] and are limited only by grain size or film thickness.

4.2.3 Enthalpy of Formation

The ability to form highly crystalline metal halide perovskite semiconductors from ionic salts dissolved in solution, but also by co-evaporation of precursors, is due to the low enthalpy of formation. Thermodynamically, the standard enthalpy of formation, ΔH_f, is the change in enthalpy when a material is formed with respect to its precursors. For the methylammonium lead halides, $MAPbI_3$, $MAPbBr_3$ and $MAPbCl_3$, the enthalpy of formation with respect to their lead and methylammonium halide precursors, notably the solid and not solvated species, was experimentally determined to be 8.25, 1.60 and −2.16 kcal mol^{-1} [55] and theoretically estimated to be −2.30, −5.77 and −16.14 kcal mol^{-1} by density functional theory (DFT) calculations.[56] Hence, $MAPbI_3$ has an enthalpy of formation (ΔH_f) of close to or even above 0 kcal mol^{-1} relative to the precursors lead iodide (PbI_2) and methylammonium iodide (MAI).[55,56] This explains why crystalline metal halide perovskite thin-film layers can be obtained by processing at moderately low temperatures compatible with printing and coating technologies. In comparison, the higher band gap analogues, from substituting iodide with the more electronegative bromide and chloride ($MAPbBr_3$ and $MAPbCl_3$), exhibit a more negative ΔH_f and higher stability.[56]

The low enthalpy of formation of some of the ABX_3 perovskites implies that they are as easy to break as they are to make.[55–57] Understanding the degradation mechanism of ABX_3 perovskite is of paramount importance to identify materials with increased stability and implement technical measures to prevent certain degradation pathways. Intrinsic materials and device stability are paramount for the technological use of ABX_3-based optoelectronic devices. Aspects of stability will be addressed in Section 4.6.

4.3 ABX_3 Solution Chemistry

When growing ABX_3 perovskite materials from solution, the microstructure and optoelectronic quality of the ABX_3 semiconductor critically depends on the crystallisation behaviour during processing. Layer uniformity, coverage and morphology are predetermined by the complex solution chemistry of ABX_3 precursor inks. Further insight into the solution chemistry and role of complexes and colloids in film formation will enable optimisation of ABX_3 inks with favourable process windows to form ABX_3 layers reproducibly using scalable solution-based methods, which will be discussed in Section 4.5.1.

Precursor solutions or *inks* for ABX_3 perovskite deposition typically contain the alkylammonium halide or alkaline halide (AX) salts and divalent group 14 metal halide (BX_2) salts in polar aprotic solvent or mixtures of solvents.[58,59] The choice of solvent or solvent mixture determines the maximum concentration of ABX_3 precursor solutions due to the specific solubility of precursor salts as indicated in Scheme 4.1. Experimental values for the solubility of PbX_2 salts are widely available for aqueous solutions, with

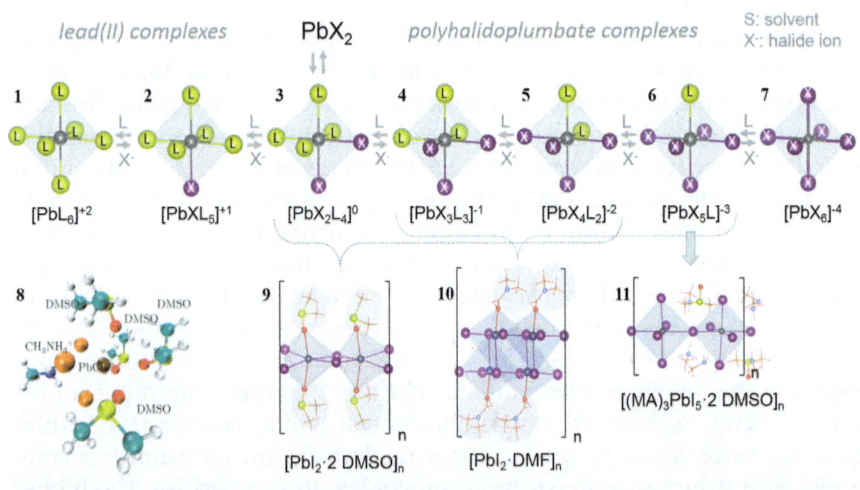

Scheme 4.1 Equilibrium between lead halide complexes with Lewis base solvent (S) forming various different Pb(II) complexes of $[PbS_{6-x}X_x]^{2-x}$, according to Rahimnejad *et al.*[63] An octahedral representation of solvent–lead–halide complexes was chosen to relate solution complexes to features expected in crystalline lead halide materials. This is an oversimplification as steric effects of the stereoactive Pb(II) lone pair are ignored and a more realistic solution complex structure is one calculated by Stevenson *et al.*[64] for the $[MA^+][PbCl_3]^-$ 4DMSO complex (**8**). Reproduced from ref. 64 with permission from American Chemical Society, Copyright 2017. Structures **9**, **10** and **11** are crystal structures of PbI_2–solvent complexes forming polymeric lead–halide chains reproduced from crystallographic information files (CIF) files reported by Cao *et al.*[72] using the open crystallographic visualisation program VESTA.[237]

the solubility decreasing from $PbCl_2$ and $PbBr_2$ to PbI_2, and the overall solubility being low.[60,61] In Lewis base solvents, such as dimethylformamide (DMF), dimethylsulfoxide (DMSO), gamma-butyrolactone (GBL), *N*-methyl-2-pyrrolidone (NMP), and acetonitrile (MeCN), the solubility of lead halide salts is typically of the order of 1 M at ambient temperature and the solvent competes with halides for coordination of lead.[62] The stronger the Lewis base, the greater the enthalpic advantage for the formation of the solvent adduct over a lead–halide bond.

In solution, various different lead–halide–solvent complexes and colloids will form in chemical equilibrium with each other.[63] In Scheme 4.1, the equilibrium between different lead–halide–solvent complexes with presumed octahedral coordination of $[PbX_x(S)_{6-x}]^{(x-2)-}$ (**1–7**) with increasing X at higher halide concentration is shown. Note, that the octahedral complexes chosen to represent the complexes in Scheme 4.1 are not representative of the real geometry of complexes in solution. As Stevenson *et al.*[64] point out, the real geometry of lead complexes in solution would be expected to adopt different geometries due to the stereoactive lone pair on lead and are more likely to form eight- or nine- coordinate complexes. To illustrate this, we have included the calculated structure of a $[MA^+][PbCl_3^-]$ monomer coordinated by four DMSO solvent molecules in Scheme 4.1 (**8**).[64]

The structure and equilibrium of $[PbX_x(S)_{6-x}]^{(x-2)-}$ complexes has been investigated by UV–visible spectroscopy and X-ray absorption spectroscopy, confirming the dynamic equilibrium between the various lead(Pb)–halide(X)–solvent(S) adducts illustrated in Scheme 4.1.[63,65,66] Complexes **4–7** are called halidoplumbate complexes as they are negatively charged, while the neutral or positively charged complexes **1–3** are referred to as lead(II) complexes. Mononuclear halidoplumbates, such as $[PbX_3(S)_3]^{1-}$ and $[PbX_4(S)_2]^{2-}$, have been identified in solutions of PbX_2 + MAX (with X = I, Br) in DMF or DMSO solvents by UV–visible spectroscopy[63,65,67,68] due to their characteristic ligand-to-metal charge transfer bands.[69] The addition of halide anions, for example in the co-dissolution of methylammonium salts, shifts the equilibrium shown in Scheme 4.1 towards the right.[63]

In the discussion of the different processing strategies in Section 4.4, the equilibrium between the various lead(II) species and their dependence on the relative concentration of lead and halide ions, the bonding strength of a specific halide anion as well as the bonding and/or coordination strength of the solvent are important factors to take into account.

With increasing halide ion coordination, the enthalpy of solvation of the Pb–X complexes decreases, but negatively charged complexes can be stabilised by hydrogen bond formation with methylammonium cations[64] and the formation of colloidal or chain-like polynuclear species.[70–72] These might be reminiscent of crystalline solvate phases, such as those identified by Cao *et al.*[72] from experimental XRD patterns. For this reason, we decided to depict the $[PbX_x(S)_{6-x}]^{(x-2)-}$ complexes in Scheme 4.1 in pseudo-octahedral geometry, as this allows an intuitive comparison of solution complexes with crystalline lead–halide–solvent intermediate phases.[72–74] As an example, the crystal structures of $[PbI_2{\cdot}2\ DMSO]_n$ (**9**) and $[PbI_2{\cdot}DMF]_n$ (**10**) are included in Scheme 4.1, as the polymer-like chains of PbX_n octahedra stabilised by solvent molecules can be derived from $[PbX_x(S)_{6-x}]^{(x-2)-}$ building blocks. In lead halide solutions, which also contain AX salts such as methylammonium iodide (MAI), polynuclear complexes of corner-sharing pentahalidoplumbates $[(MA)_3PbI_5{\cdot}2\ DMSO]_n$ (**11**) can form, as shown in Scheme 4.1. These poly-complexes are stabilised and are charge-neutral through coordination with three methylammonium cations (MA^+) and a DMSO solvent molecule. Dynamic light scattering results have revealed that lead halide solutions may contain colloids up to 1 µm in size.[75,76] Therefore, Yan *et al.*[75] postulated that perovskite inks with concentrations below 1 mM are solutions, but should be considered as colloidal dispersions at higher concentrations. If excess halide is present in solution, colloids are "consumed" and decrease in size and quantity in favour of an increasing concentration of higher order halidoplumbates.[75]

The complex and colloid chemistry defines the composition, defect density and morphology of the final ABX_3 films, as these *building blocks* define intermediate phases. Further studies on the coordination chemistry of lead(ɪɪ) halides in Lewis base solvents are required to better understand nucleation and crystal growth on a molecular level. Here, the nature of the solution species critically affects the crystallisation behaviour of perovskites, as we will discuss in Section 4.4. During the process of film formation, colloids and halidoplumbate oligomers probably serve as nucleation centres, affecting crystallisation dynamics, film morphology and grain size as will be further discussed in Section 4.4.1.[71,75,76]

4.4 Process Engineering

The evolution of metal halide perovskite deposition methods has been extensively reviewed.[3,77–79] We therefore focus here on discussing the current understanding of how precursor chemistry and intermediate phases during film formation determine the ABX_3 layer morphology, which strongly depends on preparation, as illustrated in Figure 4.4. Due to the low enthalpy of formation with respect to PbX_2 and MAX precursors,[57] ABX_3 semiconductors form readily from mixed precursor salt solutions (Figure 4.4a–c) or by co-evaporating solid reactants (Figure 4.4f).[80–82] ABX_3 perovskites can also readily form when reactants are deposited sequentially, *e.g.*, converting a PbI_2 to a $MAPbI_3$ thin film by dipping the former into a solution, or exposing to vapour of MAI (Figure 4.4d).[83–86] The discussion here will be limited to solution-based processing routes as we focus on aspects of solution-processed solar cells. Readers interested in vapour-phase deposition strategies are referred to reviews dedicated to this topic.[87]

The coordination chemistry and species formed in solution depend critically on the chemical equilibrium of precursor and solvent molecules, as discussed in Section 4.3. The preferred molecular interaction and coordination is maintained, and the equilibrium shifted during drying and annealing

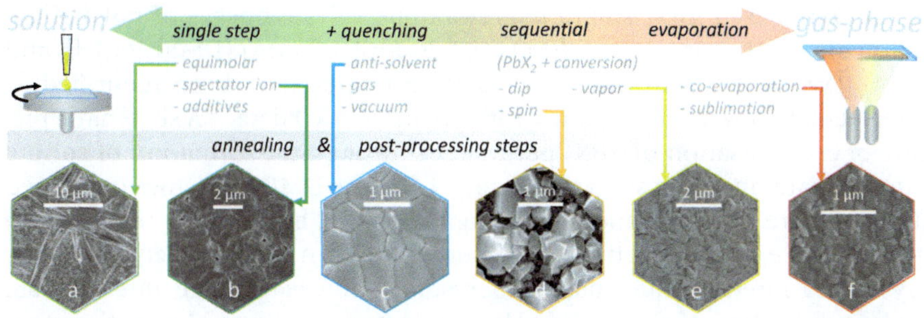

Figure 4.4 Comparison of different processing strategies resulting in ABX_3 layers with substantially different microstructure and optoelectronic properties. The SEM images of ABX_3 layers shown here were adapted from ref. 77 with permission from John Wiley and Sons, © 2017 WILEY-VCH Verlag GmbH & Co. KGaA, Weinheim. Original SEM images can be found in ref. 80, 230–233.

steps in ABX_3 processing. Recent work has highlighted the importance of defined crystalline precursor phases in determining the microstructural and morphological properties of ABX_3 perovskite semiconductor layers.[86,88–92] Understanding the solution chemistry and intermediates formed during film growth is of key importance in gaining control over nucleation and growth in order to manufacture high-quality thin films reproducibly. Here, substrate coverage, pinhole formation,[88,93] and grain size[94,95] critically influence the performance of ABX_3-based devices.

4.4.1 Weakly Coordinating Solvents

The first literature recipes on ABX_3 deposition utilised weakly coordinating solvents such as dimethylformamide (DMF) and gamma-butyrolactone (GBL).[96,97] In solutions of weakly coordinating solvents, the equilibrium of lead–halide–solvent complexes is shifted from lead(II) complexes towards halidoplumbate complexes with increasing halide concentration. Equally, the addition of methylammonium halide salts[75] or addition of hydrohalic acids (HI, HBr and HCl),[76,98–103] for example, increases the solubility of PbX_2 in Lewis base solvents, as the equilibrium is shifted towards polyhalide complexes.

Kamat and coworkers[65] investigated the solution equilibrium between different halidoplumbate complexes in diluted solutions from their UV–visible spectra. They determined the equilibrium complex formation constant between $[PbI_2(DMF)_4]$ (**3** in Scheme 4.1) and $[PbI_3(DMF)_3]^{1-}$ (**4** in Scheme 4.1) to be $K_{4/3}$ = 54 M^{-1}, while for the tetraiodoplumbate, $[PbI_4(DMF)_2]^{2-}$ (**5** in Scheme 4.1), the complex formation constant was determined to be $K_{5/4}$ = 6 M^{-1}. In more concentrated solutions, absorption bands of higher order halidoplumbates, such as $[PbI_5]^{3-}$, $[PbI_6]^{4-}$, and polymeric halidoplumbates $(PbI_x)_n$ are expected.[63,66]

For DMF-based ABX_3 solutions, crystalline solvate phases of polyhalido-plumbate–DMF complexes, $[(MA)_2(DMF)_2Pb_2I_6]$, $[(MA)_2(DMF)_2Pb_3I_8]$, and $[(MA)_3(DMF)PbI_5]$, with distinct crystal structures have been identified as intermediate phases by structure refinement of XRD data.[70] A schematic picture of crystallisation of ABX_3 perovskites through the formation of solvate intermediate phases is shown in Figure 4.5a. The DMF molecules contained in the solvate phases evaporate during annealing, but the film morphology and coverage of the resulting ABX_3 perovskite thin films is dictated by the crystalline intermediate phases.[90] Crystallisation *via* solvate intermediate phases consisting of chain-like polycomplexes as building blocks, therefore leads to dendritic ABX_3 layers with low coverage. A typical dendritic ABX_3 layer grown from a DMF solution is shown in the scanning electron microscopy (SEM) image in Figure 4.4a.

4.4.2 Spectator Ions

When ABX_3 perovskites are grown from chloride-containing precursors and an excess of methylammonium, the crystallisation kinetics are dramatically reduced and delayed, as illustrated in Figure 4.5b. Negligible amounts of chloride are found in annealed samples, suggesting that chloride only plays an important role in intermediate stages of film formation.[88,89,104–106] Chloride and bromide ions are expected to exhibit stronger coordination to lead(II) than iodide ions.[68] Chloride- or bromide-rich crystallites might, therefore, form intermittently and act as seed crystals.[57,107] In the presence of chloride or bromide, the direct formation of an ABX_3 perovskite phase and supression of solvate intermediates was demonstrated by *in situ* grazing incidence wide angle X-ray scattering (GIWAXS) experiments during spin-coating.[90] The preferred coordination with a chloride or bromide anion will suppresses the formation of higher iodoplumbate complexes such as $[PbI_4]^{2-}$ that have been correlated with an increased defect density and lower charge carrier lifetime in ABX_3 perovskite films.[107]

The crystal structure of the chloride-containing intermediate phase was recently solved by Stone *et al.*,[108] who identified the intermediate phase as MA_2PbI_3Cl. In this intermediate structure, four corner-sharing lead halide octahedra stack into columns, and chloride preferentially occupies apical octahedral positions along the chain. The fact that only one equivalent of Cl is included in the crystal structure rationalises the delayed crystallisation onset of ABX_3, as the second equivalent of chloride is probably evaporating first from excess methylammonium chloride (MACl) in the film. This is consistent with empirical findings that the best device performance is obtained for solutions containing $MACl:MAI:PbI_2$ in a 1 : 1 : 1 ratio.[75] The crystallisation kinetics of $MAPbI_3$ from Cl-containing precursors has been successfully described by an Avrami-type growth according to:

$$g(t) = 1 - e^{-\left(k_0 e^{-\frac{E_a}{RT}} t\right)^n} \tag{4.2}$$

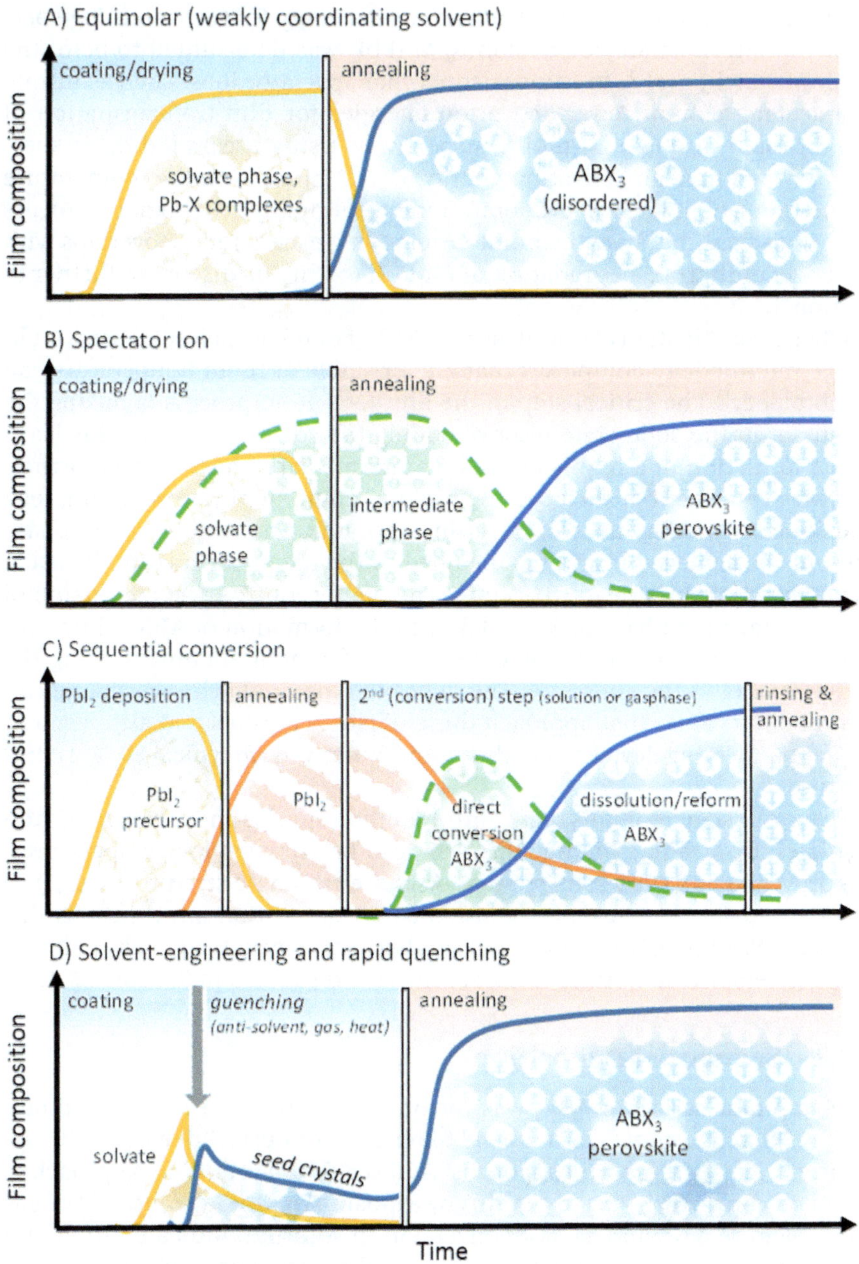

Figure 4.5 Schematic illustration of different solution-based strategies to deposit ABX_3 perovskite semiconductor layers. These schematic reaction mechanisms capture various experimental findings on the specifics and are deliberately generic to allow conceptual categorisation of different processing approaches: A) equimolar, B) spectator ion, C) sequential deposition, and D) solvent engineering and rapid quenching.

as discussed by Moore *et al.*[91] The activation energy for Cl-containing spectator ion intermediate conversion to $MAPbI_3$ was determined to be of the order of 86.6 kJ mol^{-1}. In comparison, other spectator ions, such as nitrate and acetate, exhibit lower activation energies for film transformation of 56.6 kJ mol^{-1} and 67.5 kJ mol^{-1}, respectively. Using excess iodide as spectator ion requires an activation energy of 97.3 kJ mol^{-1} and a temperature of the order of 150 °C to evaporate iodide thermally. This makes process control difficult because the point of removing excess iodide overlaps with the onset of thermal degradation of $MAPbI_3$ to PbI_2, as discussed further in Section 4.6.2.

When a combination of lead(II) acetate and a Lewis base solvent is used, alternative intermediate complexes that are air-stable at room temperature can be obtained.[95] The critical step in the film formation process is mixing the precursor in the solid state prior to adding the solvent mixture. This leads to the formation of a highly crystalline intermediate phase with an orthorhombic structure that includes the acetate group, methylammonium, lead and iodine. Notably, no solvent is incorporated. Interestingly, this phase can only be achieved if water is included in the precursor mixture by using the hydrated form of lead(II) acetate. The highly controllable conversion of this intermediate phase at 80 °C leads to the formation of ABX_3 films with grains sizes beyond 30 micrometres and preferential alignment along the *z*-axis parallel to the substrate. The deposited material achieves charge carrier mobility values that approach those of single crystals and also result in higher performing devices than those simply derived *via* typical DMF : DMSO mixtures.[95]

The use of spectator ions that control solution species and crystalline intermediates, from which high-quality ABX_3 can be grown at low temperatures, has proven to be a successful strategy. As these are single-step processes, they can be translated to scalable solution-based deposition strategies, as discussed in Section 4.5. Spectator ions and other molecules as likely additives in the development of ABX_3 precursor inks are discussed in Section 4.5.1.

4.4.3 Sequential Deposition

Two-step fabrication of ABX_3 was introduced by Liang *et al.*[238] and modified by Burschka *et al.*[83] as a means to reach high optical density ABX_3 layers in devices based on mesoporous TiO_2 scaffold layers. This approach is based on converting PbI_2 to $MAPbI_3$ by exposure to MAI in solution[83] or gas phase (Figure 4.4d and e). For conversion in solution, indicated in Figure 4.5c, two mechanisms have been identified as occurring, which critically affect the ABX_3 layer morphology. The *direct conversion* or *intercalation* process leads to ABX_3 layers maintaining high coverage of the PbI_2 layer as methylammonium cations simply diffuse through the structure. Growth is constrained in the lateral direction, leading to predominant growth perpendicular to the substrate.[85] Alternatively, in the so-called *dissolution-reformation* or *recrystallisation* process,[109] PbI_2 or $MAPbI_3$ crystals dissolve and the perovskite recrystallises in a thermodynamically favoured shape of larger and isolated cuboids, shown in Figure 4.4d. These two processes compete during crystallisation. Dominance of the *direct conversion* product is observed at high MAI concentrations. Interestingly, the *dissolution-reformation* results in crystals being tilted at an angle of about 55°, which corresponds to (202) planes being parallel to the substrate.[110]

Lewis adduct formation of PbI_2 crystals enables better control over the crystallisation process, leading to large, uniform crystals.[72] Retarding the crystallisation of PbI_2 by solvent adduct formation with strongly coordinating Lewis bases such as DMSO and NMP has been shown to enable efficient intramolecular exchange in sequential conversion.[111–113]

4.4.4 Solvent Engineering

Most commonly employed ABX_3 inks use mixtures of solvents such as DMF : DMSO in a 4 : 1 volume ratio or NMP : DMF in a 9 : 8 volume ratio. Mixing solvents can be expected to play a role in the specific coordination complexes formed and broaden the processing window, as solvents evaporate at different rates.[114,115] The affinity of solvents to coordinate to Pb^{2+} has been proposed to decrease as DMSO > DMF > GBL, based on UV–visible spectroscopy of 0.5 mM PbI_2 solutions.[63] These findings agree with solvent acceptor numbers (A_N) and donor numbers (D_N) according to Gutmann.[116] The coordination strength of a solvent should roughly scale with the sum of the acceptor and donor numbers. For some common solvents, these are: DMSO (A_N = 19.3 kcal mol^{-1}, D_N = 29.8 kcal mol^{-1}, Σ = 50.1 kcal mol^{-1}), DMF (A_N = 16 kcal mol^{-1}, D_N = 26.6 kcal mol^{-1}, Σ = 42.6 kcal mol^{-1}) and GBL (A_N = 18 kcal mol^{-1}, D_N = 17.3 kcal mol^{-1}, Σ = 35.3 kcal mol^{-1}).[116,117]

Regarding the equilibrium of halide–lead–solvent complexes shown in Scheme 4.1, more strongly coordinating Lewis bases such as DMSO inhibit halide coordination to lead, as this process becomes endergonic.[63] The formation of higher polyhalidoplumbate complexes (**5–7**, Scheme 4.1) becomes supressed and ABX_3 films are grown from more favourable $[PbI_3S_3]^-$ (**4**) and $[PbI_2S_4]^0$ (**3**) complexes, which are more ideal building blocks. In the case of more strongly coordinating Lewis bases, the formation of solvate intermediate phases may critically determine the ABX_3 film morphology. Jeon *et al.*[58] observed distinct crystalline phases of a MAI–PbI_2–DMSO intermediate phase. Cao *et al.*[72] identified these intermediate phases as $(MA)_2(DMSO)_2Pb_3I_8$ and $(MA)_3(DMSO)PbI_5$ (**10** in Scheme 4.1), analogous to the DMF solvate phases discussed in Section 4.4.1.[70] When crystalline intermediate phases adopt strong preferred morphology, such as the dendritic structure often observed for DMF solvate phases, the morphology of ABX_3 thin films is negatively affected. In solvent engineering a balance is reached between preventing higher halidoplumbate complexes detrimental for device performance[107] through high D_N-coordinating Lewis bases, and prevention of stable solvate intermediate phases with detrimental preferred film morphology, which negatively affect the film coverage and homogeneity.

Strong donor and acceptor solvents include pyridine (A_N = 14.2 kcal mol^{-1}, D_N = 33.1 kcal mol^{-1}, Σ = 47.3 kcal mol^{-1})[117] and hexamethylphosphoramide (HMPA) (A_N = 10.6 kcal mol^{-1}, D_N = 38.8 kcal mol^{-1}, Σ = 49.4 kcal mol^{-1}),[117] which have both been employed in perovskite deposition procedures,[118,119] but would not be suitable for large-scale ink deployment due to their carcinogenic properties. Hamill *et al.*[120] demonstrated that the high D_N solvent dimethylpropyleneurea (DMPU) is beneficial for obtaining high-quality ABX_3 layers. Categorisation of solvents according to Gutmann acceptor and donor numbers seems a suitable first approximation, but more advanced approaches to estimate the complexation effectiveness reliably, such as Mayer bond order, should be considered.[64]

Recently, tetrahydrothiophene-1-oxide (THTO) was highlighted as an interesting potential solvent by Foley *et al.*[121] *via* computational screening, as it has a stronger solvation efficacy than commonly used solvents. This solvent, in fact, suppresses the formation of intermediate structures, significantly retarding the crystallisation process and, thus, greatly reducing the density of nucleation sites. This results in a film of highly oriented perovskite crystals, as proven by *in situ* GIWAXS.[95]

Molecules with strong donor character, such as amines, have been successfully employed by Chen *et al.*[122] to synthesise methylamine adducts of lead(II) halides, $PbX_2{\cdot}(NH_2CH_3)_n$ and to obtain high-quality ABX_3 films.

In the choice of solvents for ABX_3 film processing it seems that the solvent's coordination strength and ability to form beneficial or detrimental crystalline intermediate solvate phases has to be balanced with the solvent evaporation rate(s) determining the processing window of film deposition. Insight into specific molecular interactions in solution and during film formation will provide the means to further optimise the ink composition.

4.4.5 Inducing Crystallisation

Beyond solvent engineering, controlling the nucleation and crystallisation of ABX_3 has become of crucial importance to obtain high-quality layers, indicated in Figure 4.5d.[58,59] In this context, the *anti-solvent* drip during the spin-coating process has emerged as one of the most used strategies. During an instantaneous drip of a large amount of a solvent, which dissolves the constituent ink solvents well but does not dissolve the ABX_3 perovskites (such as chlorobenzene, aniline or toluene), a substantial fraction of ABX_3 solvent(s) is removed. This drives the wet film into oversaturation. As crystallisation occurs preferentially on the top of the spinning wet film, seed crystals and, hence, crystallisation are induced, presumably from the top of the ABX_3 wet film rather than from the substrate. Homogeneous over heterogeneous grain growth was facilitated when processing ABX_3 layers on top of substrates where poor wetting of the ABX_3 solution occurs.[123] Unfortunately, dripping large volumes of *anti-solvents* on top of a wet film is incompatible with most of the scalable deposition methods discussed in Section 4.1. Alternative strategies to achieve controlled supersaturation during perovskite film deposition include processing at higher temperature,[124–126] gas quenching[127,128] and vacuum-assisted[129,130] solvent removal.

4.4.6 Compositional Engineering

Compositional engineering of ABX_3 materials is motivated by obtaining compounds with higher stability and to tune the band gap,[14,16,131–137] as discussed in Section 4.2.2. The latter is of particular importance for the realisation of high efficiency multi-junction devices based on ABX_3 perovskites on, for example, silicon or CIGS solar cells as bottom cells.[6,7] The ideal band gap for a top cell absorber would have to be close to 1.75 eV.[138] Figure 4.6 compares the device performance metrics of short circuit current density, J_{SC}, open circuit voltage, V_{OC}, and power conversion efficiency, PCE, for different cation/anion mixed ABX_3 perovskite compounds, as a function of their absorption onset compiled from literature surveys.[23,139,140]

ABX_3 compounds with absorption onsets larger than 1.75 eV exhibit a drop in expected power conversion efficiencies as the open circuit voltage becomes a limiting factor. This has been attributed to photoinduced halide migration, as mentioned in Section 4.2.2, leading to segregation of the absorber material into bromide-rich and iodide-rich domains, as first reported by Hoke

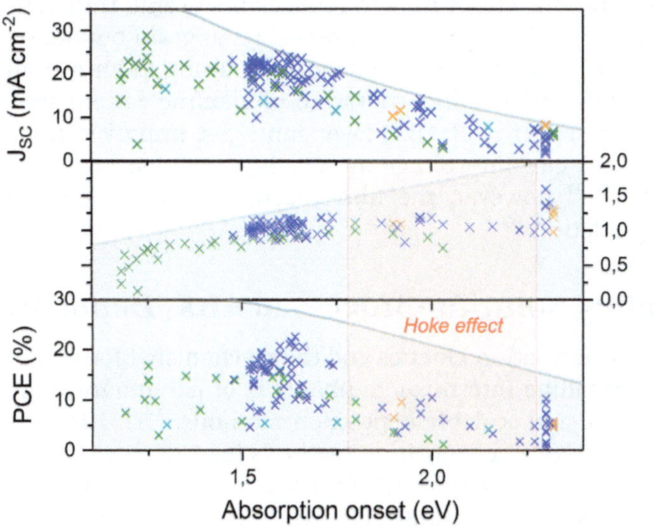

Figure 4.6 Comparison of device performance metrics of short circuit current density, J_{SC}, open circuit voltage, V_{OC}, and power conversion efficiency, PCE, plotted as a function of the absorption onset of the absorber material. Data shown here have been summarised in ref. 23 and 140 and similar plots can be found in ref. 23. Adapted from ref. 23 with permission from the Royal Society of Chemistry.

et al.[14,16,131] Potentially, this effect can be suppressed by preventing ion migration across grain boundaries by, for example, passivating grain boundaries with larger organic cations.[141]

Apart from tuning the ionic radii to achieve more favourable tolerance factors (T_f), growth from precursors containing multiple cations and anions affects the crystallisation kinetics and process window.[21,115] Kamat and coworkers[57,68] highlighted that the complex chemistry dictates the composition and phase homogeneity of mixed halide ABX_3 solid solutions. The equilibrium constants of the bromide analogues between complexes **3**/**4** and **4**/**5** (Scheme 4.1) are about four times larger compared with those of the iodide ones. Bromide will therefore coordinate lead preferentially and prevent the formation of higher iodoplumbates, as discussed for the formation from mixed chloride/iodide precursors in Section 4.4.2. The stronger coordination and lower enthalpy of formation of bromide ABX_3 perovskites induces preferred crystallisation of bromide-rich perovskite phases as demonstrated by Munir *et al.* by *in situ* GIWAXS experiments.[90] Consequently, the synthesis of mixed halide Br/I perovskites with statistical distribution of halide ions is far from trivial and lower energy subphases originate from the synthesis.[142,143]

Cation alloys and, particularly, the incorporation of small amounts of alkaline cations such as caesium (Cs) and rubidium (Rb) were shown to result in single-junction devices with performances higher than 20% and increased stability.[15,115] These two additives play different roles: caesium is incorporated into the structure and reduces the overall trap density of the material, while rubidium predominantly stays at grain boundaries, passivating defects that affect charge transport, resulting in an overall increase in charge mobility.[144] The incorporation of alkaline cations was found to improve the crystallinity of ABX_3 layers and give improved photostability and also reduced current–voltage hysteresis, which has been linked to ion migration;[17,145–147] however, the films become more prone to moisture-induced degradation.[148]

4.5 Scaling Solution-processed ABX_3 Deposition

Insight into the formation kinetics and the mechanism of ABX_3 growth from potentially crystalline intermediate phases is of paramount importance to engineer reliable and scalable deposition methods.[57] This is a prerequisite for developing large-area deposition methods for ABX_3-based optoelectronic devices. As illustrated in Figure 4.1, for larger area devices the huge discrepancy in PCE between perovskite solar cells and other technologies becomes evident. The reason, as described in Section 4.4, is that the layer quality in ABX_3 processing strongly depends on processing conditions and the precursor chemistry. This section compares and discusses literature examples of scalable deposition techniques utilised for the preparation of larger area perovskite devices and modules. Focusing on solution-based deposition methods, we will contrast non-scalable spin-coating with scalable deposition

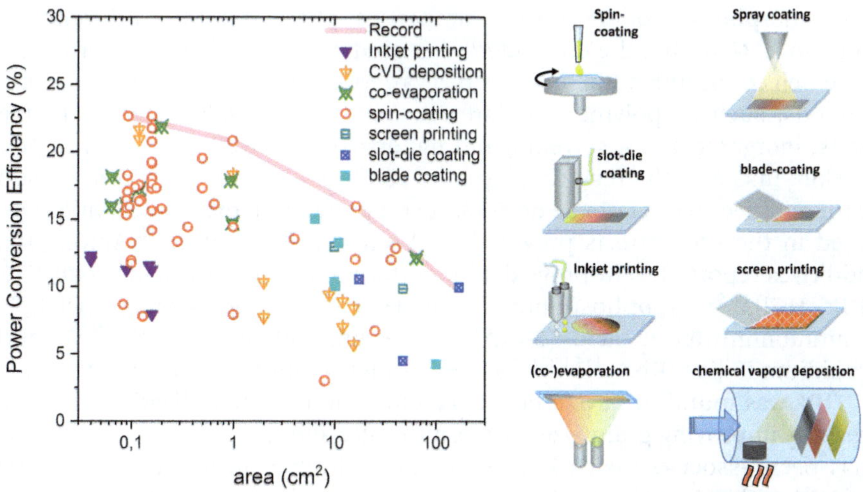

Figure 4.7 Power conversion efficiency of ABX_3-based devices as a function of area, for distinguishing different solution-based deposition methods. The data plotted here were taken from tabulated values in recent review papers.[5,87,149,150,171] References to original articles can be found therein.

methods such as spray-coating, slot-die coating, inkjet printing, screen printing and blade coating. Representative results for larger area devices are summarised in Figure 4.7 as a function of deposition method and reported device area from recent review papers[149–151] and book chapters.[152,153] Gas-phase deposition methods, co-evaporation and chemical vapour deposition, are included for comparison, but these will not be explicitly discussed in this chapter to maintain our focus on solution processing. Progress in this field of research can be found in recent comprehensive review papers.[5,82]

4.5.1 Perovskite Inks

As discussed in Section 4.3, ABX_3 solutions typically contain a mixture of dissolved salts that form ionic complexes and colloids. As further discussed in Section 4.4, certain complexes and colloids in solutions may determine the thermodynamically preferred crystalline intermediate phases that have a strong influence on the perovskite crystallisation dynamics and affect the film morphology. When moving towards large-scale solution deposition of ABX_3 perovskites, ink engineering by varying precursors, stoichiometry, solvents and additives is needed, to enable control over nucleation and crystal growth kinetics and enable reproducible manufacturing of high-quality thin films. Inks of frequently employed standard ABX_3 recipes, such as PbI_2 : MAI (1 : 1) in DMF (*e.g.*, TCI P2415)[154] and $PbCl_2$: MAI (1 : 3) in DMF (*e.g.*, Ossila I101),[155] are commercially available.

Striving to improve material properties, additives in perovskite inks have been employed to modulate the perovskite nucleation and growth dynamics, layer morphology, optoelectronic properties, phase stability, and moisture tolerance. Note that the term *additive* is used ambiguously in the perovskite research community. In recent years a variety of organic and inorganic compounds, such as polymers, fullerenes, metal halide salts, organic halide salts, inorganic acids, solvents, and nanoparticles have been added to perovskite inks and the resulting layers tested in PSCs, usually in comparison with additive-free reference devices. A comprehensive overview of additives used in the literature is provided in the review by Li *et al.*[156] Among the additives reported to improve device performance are hydrohalic acids (HI, HBr, HCl),[76,98] hypophosphorous acid (HPA),[157–159] diiodo-octane (DIO),[160] 5-ammoniumvaleric acid (5-AVA),[161] caesium halide (CsX)[15,17,145,162] and rubidium iodide (RbX),[115,146,163] as well as methyl amine (MAH).[122,164]

HPA was found to have a beneficial effect on the crystallisation kinetics, thereby improving grain size, layer uniformity and defect density.[157–159] This has been associated with $PbHPO_3$ colloids, formed in a reaction cascade of HPA with HI, MAH and PbX_2, which act as nucleation centres in the perovskite ink.[158] In contrast, hydrohalic acids such as HI, HBr and HCl increase the solubility of dissolved lead(II) species,[98] thereby triggering the dissolution of plumbate colloids.[76] Hydrohalic acid additives have, however, been shown to be detrimental to the shelf-life of ABX_3 inks. Due to the increased concentration of supersaturation and lower colloid concentration, nucleation and crystal growth decelerate, resulting in smooth perovskite layers with large grains.[102,103] This shows how two sorts of inorganic acid additives improve crystallisation dynamics, although based on completely different mechanisms.

4.5.2 Spin-coating

As illustrated in Figure 4.7, high efficiency ABX_3 perovskite solar cells are currently being predominantly fabricated by spin-coating. Spin-coating is among the most widely used methods to deposit materials from solutions. A small amount of precursor solution is deposited onto the substrate mounted in the spin-coater, often by vacuum, and rotated at high speed. This causes the solution to spread and solvents to evaporate. For a detailed discussion of the physical principles of this deposition method the interested reader is referred to publications elsewhere.[152,153] The resulting film thickness is determined by the concentration of the precursors, the viscosity of the solution/ink, temperature and spin speed. When spin-coating ABX_3 perovskites, the crystallisation behaviour of material during spin-coating needs to be considered.

Major drawbacks of spin-coating are that up to 90% of the material is spun off the substrate during deposition and that deposition methods and process parameters are very difficult to translate to larger area manufacturing. Radial differences in the film crystallisation kinetics give rise to inhomogeneity in spin-coated ABX_3 perovskite thin films. This can be suppressed partially by optimised solvent or gas-quenching,[165] but, ultimately, spin-coating cannot be scaled indefinitely.

It is not trivial to translate the processing strategies optimised for small-area spin-coated devices, summarised in Figure 4.4, to larger areas. Understanding the intricacies of ABX_3 crystallisation and morphology-determining intermediate phases, as discussed in Section 4.3, is of paramount importance to adopt deposition strategies for high-quality ABX_3 layers using scalable deposition methods that allow the production of modules.[57] In the next few paragraphs, scalable deposition methods that have been utilised for the deposition of ABX_3 perovskite layers will be discussed.

4.5.3 Blade Coating

Blade coating has been among the most successfully adapted methods for the realisation of larger area perovskite solar cells.[166] This deposition method relies on spreading the perovskite precursor ink using a sharp blade placed at a defined position over the substrate. Layer morphology and thickness depends on the composition and concentration of the precursor ink as well as on the distance between the blade and substrate and relative speed of motion. Ink utilisation is generally high, with little wastage. Power conversion efficiencies up to 19% for small-area devices,[114,167] 15% for a square-inch (6.25 cm^2) device, and 15.6% for a 10.36 cm^2 device have been demonstrated, and a blade-coated mini-module with power conversion efficiencies of about 5% for 100 cm^2 modules has been achieved.[151,168,169] A major advantage of scalable deposition methods such as blade coating and slot-die coating is that the substrate temperature can be controlled. Zhong *et al.*[126] demonstrated that favourable large grained ABX_3 layers can be obtained by blade coating at higher temperature (130 °C), as crystalline solvate intermediate phases that template dendritic film morphologies are circumvented, as illustrated in Figure 4.8.

4.5.4 Slot-die Coating

Slot-die coating is another scalable deposition technique and, in its deposition principles, is comparable to blade coating. As the ink is always fed directly into the slot-die the requirements for inks in terms of their processing window are quite flexible.[152,170] Width and thickness of the deposited film are controlled by slot-die dimensions, ink concentration, feed rate and relative speed of motion with the slot-die. As the ink meniscus between slot-die and substrate has to stabilise, the first few centimetres of the coated layer are often more inhomogeneous. As larger volumes of ink are needed, slot-die coating is less suitable for the development of inks but, once inks have been optimised, slot-die coating is probably one of the most viable wet-chemical methods for larger area module fabrication.[150] Using slot-die coating for the deposition of the ABX_3 layer, a mini-module with efficiencies up to 10.6% for an active area of 17.3 cm^2 has been demonstrated.[171] Most notably, a 10% device on 168 cm^2 has been announced by Solliance.[172]

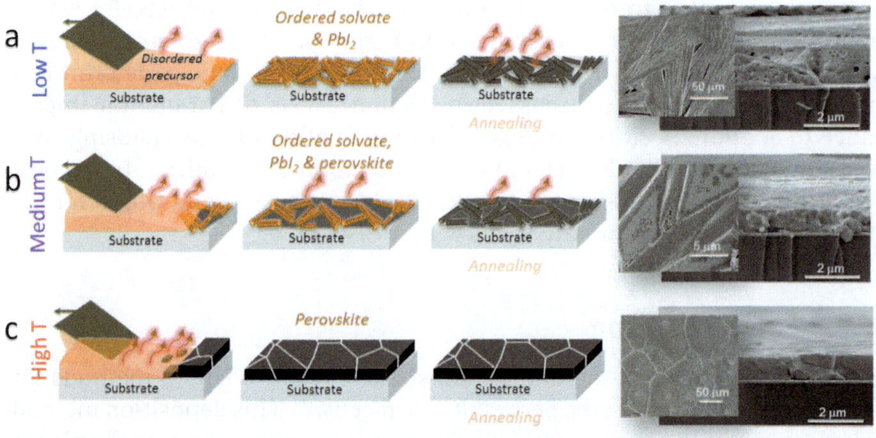

Figure 4.8 Schematic illustration of the film transformation behaviour of $MAPbI_3$ perovskite thin films upon blade coating at different temperatures (a) low *ca.* 25–50 °C, (b) *ca.* 80 °C, and (c) >100 °C, with resulting film morphology depicted in SEM top-view and cross-section images. Reproduced from ref. 126 with permission from American Chemical Society, Copyright 2018.

4.5.5 Spray-coating

Spray-coating relies on the deposition of a material onto a substrate by spraying small droplets of ink generated by leading a carrier gas through precursor solution *via* a nozzle. This method can be easily scaled. Film homogeneity and the quality of spray-coated functional materials depends on the size and local overlap of ink droplets propelled to the substrate surface. As a constant flux of ink droplets reach the surface, deposited material may redissolve. This can be avoided by keeping the substrate surface at higher temperature to promote quick evaporation of the solvent.[173] Layer homogeneity also depends on droplet size determined by the geometry of the nozzle and mode of droplet generation, by gas pressure or ultrasound.[5] ABX_3 perovskite thin films with high layer homogeneity have been obtained by ultrasonic spray-coating.[174]

4.5.6 Screen Printing

Screen printing relies on material transfer through a patterned mesh screen that is placed over a substrate. Inks used for screen printing have to be fairly viscous and are spread across the screen using a squeegee. Screen printing has been commonly used to deposit thicker layers of mesoporous TiO_2 and carbon back-contacting materials in dye-sensitised solar cell manufacturing, that have been adapted for perovskite devices. Mei *et al.*[175] used a mesoporous TiO_2/ZrO_2 scaffold and infiltrated the ABX_3 from solution containing the additive 5-ammoniumvaleric acid (5-AVA) and reported a 12.8% efficient device with a demonstrated stability of > 1000 h. Hu *et al.*[161] used a similar architecture and demonstrated a 10.4% efficient device with an active area of 49 cm^2 and their team demonstrated a fully solution-processed ABX_3 device of 7 m^2.

4.5.7 Inkjet Printing

Inkjet printing enables material deposition in defined patterns by droplet deposition from arrays of nozzles. Inkjet printing can be scaled to large areas when high printing speeds can be used and industrial-scale nozzle arrays are used. The first demonstration of inkjet-printed perovskite devices on a small area (0.16 cm^2) yielded up to 12.3% power conversion efficiency.[176] Meanwhile, 12.9% PCE on 0.09 cm^2 for triple cation inkjet-printed devices in a regular architecture was achieved.[177] Liang *et al.* demonstrated 17.04% PCE on 0.04 cm^2 and 13.27% PCE on 4 cm^2 by optimising the process window through solvent engineering and a post-printing vacuum-assisted thermal annealing step, to improve the film homogeneity of printed perovskite layers.[178] Compared with other scalable deposition methods, inkjet printing has clear advantages in the optimisation of ink formulations, as small ink volumes are consumed and it can even be used for rapid combinatorial screening by multi-component printing.[179]

4.6 Modules, Stability and Alternative Materials

4.6.1 Perovskite Modules

ABX_3 devices have been realised in many different architecture types. Apart from the challenges in obtaining homogenous ABX_3 layers on larger areas, the device performance decreases due to geometrical and electrical losses in the transparent conducting contacts. As for other thin-film devices, resistance losses can be reduced by patterning the large-area device into series-interconnected sub-cells. This will decrease the active area by 3–10% due to area lost to interconnections, and optimisation of the module requires the reduction of geometric losses *vs.* series resistance losses.[5] The geometric fill factor (GFF) is defined as the ratio between the active area and total area of the module. When reporting device performance, a distinction between active and total area needs to be made and the GFF should be stated. In solution-processed devices, interconnection schemes can either be defined by printing or be established by laser patterning, as for other types of thin-film devices. Laser scribing for series interconnection in ABX_3 modules with GFFs of *ca.* 95% have been demonstrated.[180,181] When laser patterning perovskite materials, potential material degradation at trenches needs to be considered. The degradation to PbI_2 by the thermal removal of MAI in areas adjacent to laser scribes might be beneficial, as this higher band gap material might passivate interfaces, while reduction to metallic lead would introduce non-radiative recombination sites at interfaces.

4.6.2 Stability

As mentioned in Section 4.2.3, the low enthalpy of formation of ABX_3 perovskites implies easy formation of the materials as well as easy decomposition.[57] We will, here, comment on the various degradation pathways induced by heat, light and ambient humidity, indicated in Figure 4.9 (left) for the most thoroughly investigated methylammonium lead iodide perovskite. Within this section, we deliberately refer to $MAPbI_3$ as $CH_3NH_3PbI_3$ to highlight degradation due to the fragmentation of the methylammonium cation. More in-depth reviews of ABX_3 degradation mechanisms and device stability can be found in recent review papers.[182,183]

Understanding the decomposition mechanisms of ABX_3 perovskites is an important step towards identification of necessary material design requirements and technical measures to improve the intrinsic material and extrinsic device stability. During the operation of a photovoltaic device, chemical and compositional changes in the absorber, contact layers and at interfaces may be induced by temperature, light and bias, as illustrated in Figure 4.9. Atmospheric species such as oxygen and water may induce specific chemical degradation reactions. A comparison of experimental lifetimes of ABX_3 solar cells in comparison with organic (OPV) and dye-sensitised solar cells is shown in Figure 4.9 (right). Compared with mature technologies, such as silicon or CIGS thin-film devices, ABX_3-based devices require improvements to become technologically relevant. It is crucial to understand key degradation pathways and investigate means to suppress them, by improving either the intrinsic stability of ABX_3 materials or the extrinsic stability of ABX_3 devices.

Detrimental effects on device performance due to reaction with atmospheric species, such as oxygen and water, can be reduced through device

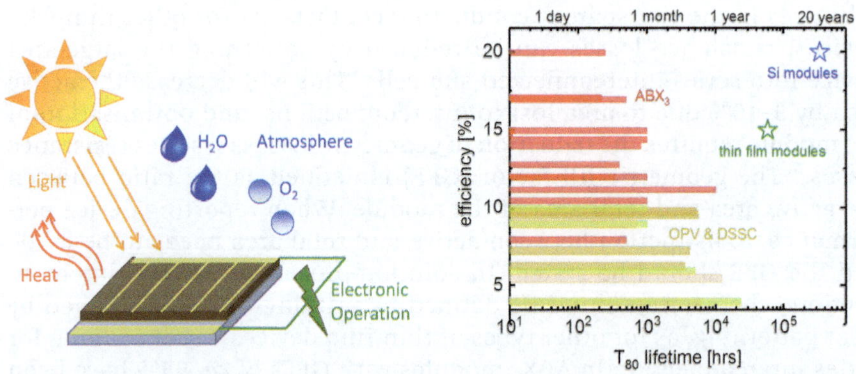

Figure 4.9 Illustration of stimuli potentially inducing degradation of ABX_3 solar cells (left) and reported operational lifetimes of ABX_3 devices in comparison with OPVs and DSSCs compared with guaranteed operational lifetimes of commercial silicon and thin-film solar cell modules plotted from data discussed in ref. 194 and 234. T_{80} lifetime refers to the time over which the efficiency of the device drops to 80% of its initial value.

encapsulation. However, in the case of a mechanical and encapsulation failure, these degradation pathways become inevitable.

Upon exposure to ambient humidity, two types of hydrates (complexes with water molecules) have been reported.[184] The formation of hydrates can be reversible (eqn (4.3)) but will determine the film morphology negatively, as di-hydrate phases tend to form needle-shaped crystals. At higher ambient humidity, ABX_3 may dissolve and decompose according to (eqn (4.4)).

$$4(CH_3NH_3)PbI_3 + 4\,H_2O \rightleftarrows 4[CH_3NH_3PbI_3{\cdot}H_2O] \rightleftarrows (CH_3NH_3)_4PbI_6{\cdot}2H_2O + 3PbI_2 + 2H_2O \quad (4.3)$$

$$(CH_3NH_3)_4PbI_6{\cdot}2H_2O_{(s)} \rightarrow 4CH_3NH_3I_{(l)} + PbI_{2(s)} + 2H_2O \quad (4.4)$$

To avoid decomposition and non-semiconductor hydrate phase formation, numerous approaches involving the usage of hydrophobic capping layers,[185–189] and even the chemical modification of perovskite, have been employed.[190–194]

Another degradation pathway is thermal decomposition. $MAPbI_3$ thermally decomposes at temperatures above 148 °C. This fact is related to the volatility of methylammonium iodide, which undergoes a phase transition at 150 °C and then evaporates at $T = 190$ °C.[195] To improve the thermal stability, methylammonium or organic cations in general can be substituted with inorganic cations, such as Cs and Rb. In fact, even partial substitution of the MA cation with about 0.09 of Cs ($Cs_{0.09}(CH_3NH_3)_{0.91}PbI_3$) leads to drastic improvement of device stability when tested at 85 °C.[196] Fully inorganic ABX_3 materials resist temperatures above 300 °C without decomposition.[197] Compositional engineering, as discussed in Section 4.4.6, is, hence, a viable strategy to increase the intrinsic stability of ABX_3 perovskites.

Light may degrade hybrid perovskites through different mechanisms. Light-induced ion migration, discussed in Sections 4.2.2 and 4.4.6, can lead to compositional changes and decomposition of the material. More severely, the reported photodissociation of the organic cation, reported by Nickel *et al.*[198] at photon energies of $E_{ph} \geq 2.7$ eV, leads to the structural collapse to PbI_2 (eqn (4.5)) through loss of highly volatile species such as methylamine, hydrogen, and iodine gas. PbI_2 photodegrades under prolonged illumination to metallic lead and iodide gas (eqn (4.6)).[199–203] The photodissociation reactions of $CH_3NH_3PbI_3$ perovskites can be summarised according to the equations:

$$2CH_3NH_3PbI_{3(s)} + 2e^- \rightarrow 2CH_3NH_{2(g)}\uparrow + H_{2(g)}\uparrow + I_{2(g)}\uparrow + 2PbI_{2(s)} \quad (4.5)$$

$$PbI_{2(s)} \rightarrow Pb + I_{2(g)} \quad (4.6)$$

In summary, moisture-, light-, heat- and even bias-induced degradation mechanisms need to be considered in the operation of ABX_3 devices and when screening for alternative perovskite-inspired materials with superior stability and comparable optoelectronic properties.

4.6.3 ABX_3-inspired Materials

The ecotoxicology of lead is a presumed obstacle for commercialisation of the current generation of ABX_3 perovskite photovoltaics. Efforts to replace lead with other elements are ongoing but this is proving to be non-trivial. There are only a few experimental datasets available for pure Sn-based ABX_3 perovskites. Unlike Pb^{2+}, which is stabilised by the inert pair effect, Sn^{2+} quickly oxidises to Sn^{4+}, resulting in metallic rather than semiconducting compounds.[204] A strategy to increase the material stability and to obtain materials with lower band gaps, down to 1.2 eV, has involved mixing Sn^{2+} and Pb^{2+}.[205] These solid solutions exhibit a pronounced non-linear band gap behaviour, shown in Figure 4.2, that was rationalised by an energetic mismatch between Sn and Pb atomic orbitals. This mismatch leads to inversion of the band edges, with the valence band being formed by Sn 5s and I 6p orbitals, while the conduction band minimum is formed by Pb 6p and I 6p orbitals, resulting in smaller values of the absorption onset than for pure Sn- and Pb-based perovskites.[205,206]

As shown in Section 4.6.2, hybrid perovskites are prone to numerous decomposition pathways, such as through the influence of oxygen, water, temperature, and light. There are several approaches to improve the stability of the material. Since photoinstability is closely related to the presence of weakly bonded N–H groups, substitution of the organic cations with inorganic cations has been shown to improve stability.[198] All-inorganic perovskites exhibit substantially higher thermal and photostability.[196,207,208] However, such materials often demonstrate unfavourable T_f and tend not to form stable perovskite phases at room temperature, as discussed in Section 4.2.1.[208]

Inspired by research on ABX_3 perovskite materials, computational screening efforts for novel materials are being carried out. Haupt and Wells have demonstrated a diversity of stoichiometry of inorganic perovskites.[209,210] Cs- and Rb-based lead halides adopt structures of A_2BX_4, inspiring the design of new perovskite-like materials.

Varying the valency of the B-site cation enables exploration of a broad range of stoichiometries of $A_xB_yX_z$ compounds, as summarised in Table 4.1.[26,211] Hybrid perovskites with 3+ B-site cations can form layered structures with general formulae ABX_4 or $A_3B_2X_9$. This concept was first reported by Lindqvist in 1968[212] and recently $A_3B_2X_9$ materials, where A = Cs, Rb, or MA, B = Bi, Ga, or Sb and X = Cl, Br or I, have been explored in solar cells. So far $Cs_3Bi_2I_9$-, $MA_3Bi_2I_9$- and $Cs_3Sb_2I_9$-based solar cells have exhibited up to 2% PCE, which is a promising result for a comparatively unexplored material class.[213–215]

In a similar manner, substitution of the B-site with a 4+ cation results in the formation of metal-deficient perovskites with general formula A_2BX_6, such as Cs_2SnX_6.[216–218] Table 4.1 shows structures of typical representatives of B^{3+} and B^{4+} perovskites, where the green colour corresponds to caesium-based coordination centres and purple corresponds to bismuth-based coordination centres.

Table 4.1 Multi-valent metal halide materials inspired by ABX_3 metal halide perovskite semiconductor research. Structures of $CsPbBr_3$, $Cs_3Bi_2Br_9$, and Cs_2SnI_6 were plotted from CIF files published by Jain *et al.*[235] The structure of Cs_2AgBiI_6 was plotted from a CIF file reported by Savory *et al.*[236] using a three-dimensional visualisation system for crystallographic studies and electronic state calculations VESTA.[237]

3D and 2D perovskites	Layered/dimer Perovskites	Metal deficient Perovskites	Double Perovskites
Divalent	Trivalent	Tetravalent	Double valency
B^{2+}	B^{3+}	B^{4+}	B^{1+}, B^{3+}
ABX_3	ABX_4	A_2BX_6	$A_2B^{1+}B^{3+}X_6$
A_2BX_4	$A_3B_2X_9$		
Pb^{2+}, Sn^{2+}, Ge^{2+}	Sb^{3+}, Bi^{3+}	Sn^{4+}, Ge^{4+}	K^+, Cu^+, Ag^+, Au^+ Bi^{3+}, In^{3+}, Tl^{3+}
$CsPbBr_3$	$Cs_3Bi_2Br_9$	Cs_2SnI_6	Cs_2AgBiI_6
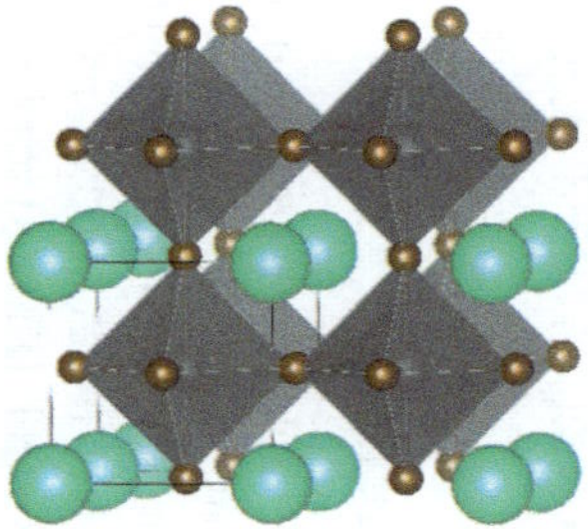	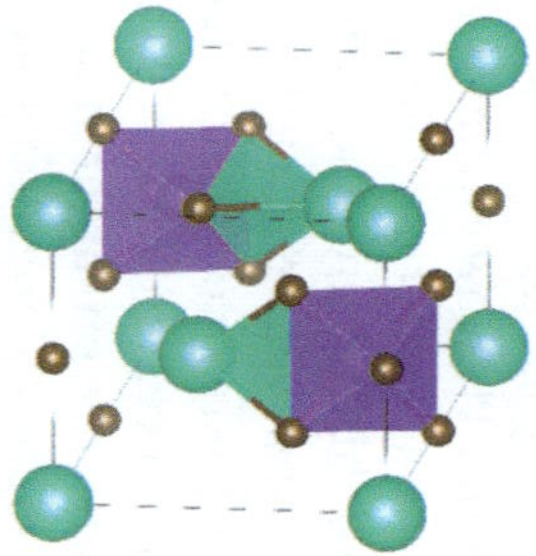	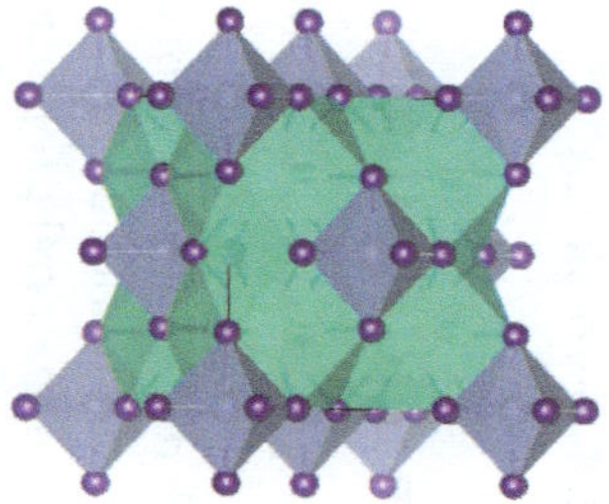	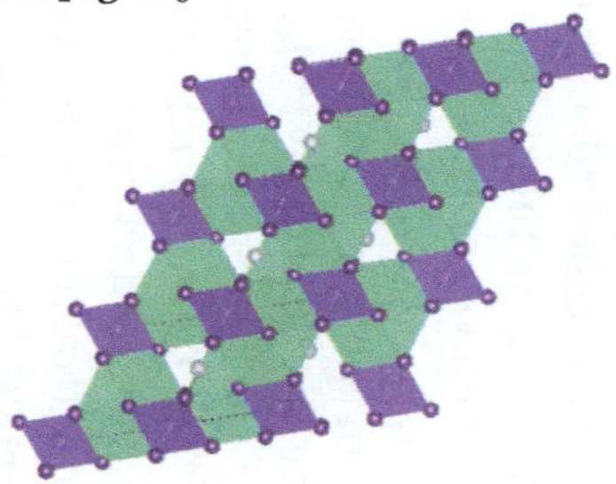

Another approach is to mix two B-site cations with different valences. This leads to the formation of so-called “double perovskites” or *elpasolites*, which adapt structures of $A_2B'B''X_6$. In the simplest version of a double perovskite, B′- and B″-site cations with valences of +1 (*e.g.*, Li, Na, K, Rb, Cs, and Tl) and +3 (*e.g.*, Al, Ga, In, V, Cr, Fe, Co, Ni, and Bi) result in an average valence of +2. $A_2B'B''X_6$ materials consisting of +1 cations as A and/or B′ as B″ were extensively studied over the last 50 years[219–223] but are only recently being considered for solar energy conversion, in contrast to the progress in lead-based ABX_3 photovoltaics.[224–228] Currently reported double perovskites often consist of Ag and Bi (A_2AgBiX_6) and reports using other monovalent cations, such as Cu and Au, and trivalent cations, such as In, Ga and Sb, are emerging. Interestingly, these materials demonstrate rather remarkable optoelectronic properties, such as a band gap in the visible range and long charge carrier lifetime and mobility, while exhibiting higher stability and reduced toxicity in comparison with ABX_3 perovskites.

4.7 Conclusion and Outlook

The rather serendipitous evolution of ABX_3 perovskites as solar energy conversion materials is likely to be just the beginning of a renaissance of known-but-not-considered semiconductors, and screening for, as well as the development of, novel semiconductors for solar energy conversion applications. A rationalisation of material properties responsible for the unforeseen success of metal halide perovskites, such as their intrinsic defect tolerance discussed in Section 4.2.2, helps to define parameters for experimental and computational screening for novel materials with beneficial properties for solar energy conversion. The low enthalpy of formation with respect to reactants of lead-based ABX_3 materials allows low-temperature solution-based fabrication. ABX_3 instabilities and the various degradation pathways of current ABX_3 materials, however, require the design of more stable derivatives for long-term use in energy conversion technology. While ABX_3 materials are the best solution-processed solar cell technology for small-area devices, reaching high performance on larger areas will require optimisation of scalable deposition techniques to manufacture larger area ABX_3 layers reproducibly. Scaling, stability and ecologically benign chemical components are the three most important challenges to enable large-scale technological use of ABX_3 or, in the future, ABX_3-inspired materials.

Acknowledgement

E. U., S. B. and O. S. acknowledge financial support from the HySPRINT Helmholtz Innovation Laboratory. E. U. acknowledges funding from the Swedish Research Council (Project 2015-00163) and Marie Skłodowska Curie Actions, Cofund, Project INCA 600398, as well as from the German Ministry of Education and Research (BMBF) for the Young Investigator Group Hybrid

Materials Formation and Scaling (HyPerFORME) within the programme 'NanoMatFutur' (grant no. 03XP0091). O. S. acknowledges a PhD fellowship from the graduate school Hybrid4Energy. S. B. acknowledges the Alexander von Humboldt Foundation for a Feodor Lynen Return Fellowship.

References

1. K. Emery, Best Research-Cell Efficiencies, https://www.nrel.gov/pv/cell-efficiency.html.
2. T. M. Brenner, D. A. Egger, L. Kronik, G. Hodes and D. Cahen, *Nat. Rev. Mater.*, 2016, **1**, 15007.
3. M. A. Green, A. Ho-Baillie and H. J. Snaith, *Nat. Photonics*, 2014, **8**, 506–514.
4. M. A. Green, Y. Hishikawa, E. D. Dunlop, D. H. Levi, J. Hohl-Ebinger and A. W. Y. Ho-Baillie, *Prog. Photovoltaics*, 2018, **26**, 3–12.
5. Z. Li, T. R. Klein, D. H. Kim, M. Yang, J. J. Berry, M. F. A. M. Van Hest and K. Zhu, *Nat. Rev. Mater.*, 2018, **3**, 1–20.
6. S. Albrecht and B. Rech, *Nat. Energy*, 2017, **2**, 16196.
7. C. D. Bailie, M. G. Christoforo, J. P. Mailoa, A. R. Bowring, E. L. Unger, W. H. Nguyen, J. Burschka, N. Pellet, J. Z. Lee, M. Grätzel, R. Noufi, T. Buonassisi, A. Salleo and M. D. McGehee, *Energy Environ. Sci.*, 2015, **8**, 956.
8. H. Xue, R. Wu, Y. Xie, Q. Tan, D. Qin, H. Wu and W. Huang, *Appl. Sci.*, 2016, **6**, 197.
9. T. Zhang, Y. Yang, D. Liu, S. C. Tse, W. Cao, Z. Feng, S. Chen and L. Qian, *Energy Environ. Sci.*, 2016, **9**, 3674–3681.
10. T. Sontheimer, D. Amkreutz, K. Schulz, P. H. Wöbkenberg, C. Guenther, V. Bakumov, J. Erz, C. Mader, S. Traut, F. Ruske, M. Weizman, A. Schnegg, M. Patz, M. Trocha, O. Wunnicke and B. Rech, *Adv. Mater. Interfaces*, 2014, **1**, 1–6.
11. G. Kieslich, S. Sun and A. K. Cheetham, *Chem. Sci.*, 2015, **6**, 3430–3433.
12. W. Travis, E. N. K. Glover, H. Bronstein, D. O. Scanlon and R. G. Palgrave, *Chem. Sci.*, 2016, **7**, 4548–4556.
13. R. D. Shannon and C. T. Prewitt, *Acta Crystallogr., Sect. B: Struct. Crystallogr. Cryst. Chem.*, 1969, **25**, 925–946.
14. J. H. Noh, S. H. Im, J. H. Heo, T. N. Mandal and S. Il Seok, *Nano Lett.*, 2013, **13**, 1764–1769.
15. M. Saliba, T. Matsui, J. Y. Seo, K. Domanski, J. P. Correa-Baena, M. K. Nazeeruddin, S. M. Zakeeruddin, W. Tress, A. Abate, A. Hagfeldt and M. Grätzel, *Energy Environ. Sci.*, 2016, **9**, 1989–1997.
16. N. J. Jeon, J. H. Noh, W. S. Yang, Y. C. Kim, S. Ryu, J. Seo and S. Il Seok, *Nature*, 2015, **517**, 476–480.
17. J. W. Lee, D. H. Kim, H. S. Kim, S. W. Seo, S. M. Cho and N. G. Park, *Adv. Energy Mater.*, 2015, **5**, 1501310.
18. C. Yi, J. Luo, S. Meloni, A. Boziki, N. Ashari-Astani, C. Grätzel, S. M. Zakeeruddin, U. Röthlisberger and M. Grätzel, *Energy Environ. Sci.*, 2016, **9**, 656–662.

19. Z. Li, M. Yang, J.-S. Park, S.-H. Wei, J. Berry and K. Zhu, *Chem. Mater.*, 2016, **28**, 284–292.
20. S. Albrecht, M. Saliba, J. P. Correa Baena, F. Lang, L. Kegelmann, M. Mews, L. Steier, A. Abate, J. Rappich, L. Korte, R. Schlatmann, M. K. Nazeeruddin, A. Hagfeldt, M. Grätzel and B. Rech, *Energy Environ. Sci.*, 2016, **9**, 81–88.
21. D. P. McMeekin, G. Sadoughi, W. Rehman, G. E. Eperon, M. Saliba, M. T. Hörantner, A. Haghighirad, N. Sakai, L. Korte, B. Rech, M. B. Johnston, L. M. Herz and H. J. Snaith, *Science*, 2016, **351**, 151–155.
22. J. H. Noh, S. H. Im, J. H. Heo, T. N. Mandal and S. Il Seok, *Nano Lett.*, 2013, **13**, 1764–1769.
23. E. L. Unger, L. Kegelmann, K. Suchan, D. Sörell, L. Korte and S. Albrecht, *J. Mater. Chem. A*, 2017, **5**, 11401–11409.
24. A. Swarnkar, A. R. Marshall, E. M. Sanehira, B. D. Chernomordik, D. T. Moore, J. A. Christians, T. Chakrabarti and J. M. Luther, *Science*, 2016, **354**, 92–96.
25. M. A. Green, Y. Jiang, A. M. Soufiani and A. Ho-Baillie, *J. Phys. Chem. Lett.*, 2015, **6**, 4774–4785.
26. R. E. Brandt, V. Stevanovic, D. S. Ginley and T. Buonassisi, *MRS Commun.*, 2015, **5**, 265–275.
27. T. Umebayashi, K. Asai, T. Umebayashi, K. Asai, T. Kondo, T. Kondo and A. Nakao, *Phys. Rev. B: Condens. Matter Mater. Phys.*, 2003, **67**, 2–7.
28. J. S. Manser, J. A. Christians and P. V. Kamat, *Chem. Rev.*, 2016, **116**, 12956–13008.
29. W. J. Yin, T. Shi and Y. Yan, *Appl. Phys. Lett.*, 2014, **104**, 0–4.
30. W. J. Yin, T. Shi and Y. Yan, *J. Phys. Chem. C*, 2015, **119**, 5253–5264.
31. S. de Wolf, J. Holovsky, S.-J. Moon, P. Löper, B. Niesen, M. Ledinsky, F. Haug, J. Yum and C. Ballif, *J. Phys. Chem. C*, 2014, **5**, 1035–1139.
32. K. X. Steirer, P. Schulz, G. Teeter, V. Stevanovic, M. Yang, K. Zhu and J. J. Berry, *ACS Energy Lett.*, 2016, **1**, 360–366.
33. A. Walsh and A. Zunger, *Nat. Mater.*, 2017, **16**, 964–967.
34. W. Tress, N. Marinova, O. Inganäs, M. K. Nazeeruddin, S. M. Zakeeruddin and M. Graetzel, *Adv. Energy Mater.*, 2015, **5**, 1–6.
35. K. Tvingstedt, O. Malinkiewicz, A. Baumann, C. Deibel, H. J. Snaith, V. Dyakonov and H. J. Bolink, *Sci. Rep.*, 2014, **4**, 1–7.
36. J. Haruyama, K. Sodeyama, L. Han and Y. Tateyama, *Acc. Chem. Res.*, 2016, **49**, 554–561.
37. J. Kim, S.-H. Lee, J. H. Lee and K.-H. Hong, *J. Phys. Chem. Lett.*, 2014, **5**, 1312–1317.
38. W. J. Yin, T. Shi and Y. Yan, *Adv. Mater.*, 2014, **26**, 4653–4658.
39. A. Buin, P. Pietsch, J. Xu, O. Voznyy, A. H. Ip, R. Comin and E. H. Sargent, *Nano Lett.*, 2014, **14**, 6281–6286.
40. D. Meggiolaro, S. G. Motti, E. Mosconi, A. J. Barker, J. Ball, C. Andrea Riccardo Perini, F. Deschler, A. Petrozza and F. De Angelis, *Energy Environ. Sci.*, 2018, **11**, 702–713.

41. C. Eames, J. M. Frost, P. R. F. Barnes, B. C. O'Regan, A. Walsh and M. S. Islam, *Nat. Commun.*, 2015, **6**, 2–9.
42. N. Aristidou, C. Eames, I. Sanchez-Molina, X. Bu, J. Kosco, M. Saiful Islam and S. A. Haque, *Nat. Commun.*, 2017, **8**, 1–10.
43. Y. X. Tian, M. Peter, E. L. Unger, M. A. Abdellah, K. Zheng, T. Pullerits, A. P. Yartsev, V. Sundström and I. G. Scheblykin, *Phys. Chem. Chem. Phys.*, 2015, **17**, 24978–24987.
44. A. Walsh, D. O. Scanlon, S. Chen, X. G. Gong and S. H. Wei, *Angew. Chem., Int. Ed.*, 2015, **54**, 1791–1794.
45. W. Nie, J. C. Blancon, A. J. Neukirch, K. Appavoo, H. Tsai, M. Chhowalla, M. A. Alam, M. Y. Sfeir, C. Katan, J. Even, S. Tretiak, J. J. Crochet, G. Gupta and A. D. Mohite, *Nat. Commun.*, 2016, **7**, 1–9.
46. E. Mosconi, D. Meggiolaro, H. J. Snaith, S. D. Stranks and F. De Angelis, *Energy Environ. Sci.*, 2016, **9**, 3180–3187.
47. E. T. Hoke, D. J. Slotcavage, E. R. Dohner, A. R. Bowring, H. I. Karunadasa and M. D. McGehee, *Chem. Sci.*, 2015, **6**, 613–617.
48. E. L. Unger, E. T. Hoke, C. D. Bailie, W. H. Nguyen, A. R. Bowring, T. Heumüller, M. G. Christoforo and M. D. McGehee, *Energy Environ. Sci.*, 2014, 7, 3690–3698.
49. H. J. Snaith, A. Abate, J. M. Ball, G. E. Eperon, T. Leijtens, N. K. Noel, S. D. Stranks, J. T. W. Wang, K. Wojciechowski and W. Zhang, *J. Phys. Chem. Lett.*, 2014, **5**, 1511–1515.
50. W. Tress, N. Marinova, T. Moehl, S. M. Zakeeruddin, M. K. Nazeeruddin and M. Grätzel, *Energy Environ. Sci.*, 2015, **8**, 995–1004.
51. W. Tress, J. P. Correa Baena, M. Saliba, A. Abate and M. Graetzel, *Adv. Energy Mater.*, 2016, **6**, 1600396.
52. T. M. Brenner, D. A. Egger, A. M. Rappe, L. Kronik, G. Hodes and D. Cahen, *J. Phys. Chem. Lett.*, 2015, **6**, 4754–4757.
53. S. D. Stranks, S. D. Stranks, G. E. Eperon, G. Grancini, C. Menelaou, M. J. P. Alcocer, T. Leijtens, L. M. Herz, A. Petrozza and H. J. Snaith, *Science*, 2014, **342**, 341–344.
54. T. Dittrich, F. Lang, O. Shargaieva, J. Rappich, N. H. Nickel, E. Unger and B. Rech, *Appl. Phys. Lett.*, 2016, **109**, 1–5.
55. G. P. Nagabhushana, R. Shivaramaiah and A. Navrotsky, *Proc. Natl. Acad. Sci. U. S. A.*, 2016, **113**, 7717–7721.
56. A. Buin, R. Comin, J. Xu, A. H. Ip and E. H. Sargent, *Chem. Mater.*, 2015, **27**, 4405–4412.
57. J. S. Manser, M. I. Saidaminov, J. A. Christians, O. M. Bakr and P. V. Kamat, *Acc. Chem. Res.*, 2016, **49**, 330–338.
58. N. J. Jeon, J. H. Noh, Y. C. Kim, W. S. Yang, S. Ryu and S. Il Seok, *Nat. Mater.*, 2014, **13**, 897–903.
59. Y. Rong, Z. Tang, Y. Zhao, X. Zhong, S. Venkatesan, H. Graham, M. Patton, Y. Jing, A. M. Guloy and Y. Yao, *Nanoscale*, 2015, **7**, 10595–10599.
60. H. L. Clever and F. J. Johnston, *J. Phys. Chem. Ref. Data*, 1980, **9**, 751–784.
61. P. Patnaik, *Handbook of Inorganic Chemicals*, McGraw-Hill, 2002.

62. S. E. Denmark and G. L. Beutner, *Angew. Chem., Int. Ed.*, 2008, **47**, 1560–1638.
63. S. Rahimnejad, A. Kovalenko, S. M. Forés, C. Aranda and A. Guerrero, *ChemPhysChem*, 2016, 2795–2798.
64. J. Stevenson, B. Sorenson, V. H. Subramaniam, J. Raiford, P. P. Khlyabich, Y. L. Loo and P. Clancy, *Chem. Mater.*, 2017, **29**, 2435–2444.
65. K. G. Stamplecoskie, J. S. Manser and P. V. Kamat, *Energy Environ. Sci.*, 2015, **8**, 208–215.
66. A. Sharenko, C. Mackeen, L. Jewell, F. Bridges and M. F. Toney, *Chem. Mater.*, 2017, **29**, 1315–1320.
67. R. J. Stewart, C. Grieco, A. V. Larsen, G. S. Doucette and J. B. Asbury, *J. Phys. Chem. C*, 2016, **120**, 12392–12402.
68. S. J. Yoon, K. G. Stamplecoskie and P. V. Kamat, *J. Phys. Chem. Lett.*, 2016, **7**, 1368–1373.
69. K. Oldenburg, A. Vogler, I. Mikó and O. Horváth, *Inorg. Chim. Acta*, 1996, **248**, 107–110.
70. A. A. Petrov, I. P. Sokolova, N. A. Belich, G. S. Peters, P. V. Dorovatovskii, Y. V. Zubavichus, V. N. Khrustalev, A. V. Petrov, M. Grätzel, E. A. Goodilin and A. B. Tarasov, *J. Phys. Chem. C*, 2017, **121**, 20739–20743.
71. Y. Guo, K. Shoyama, W. Sato, Y. Matsuo, K. Inoue, K. Harano, C. Liu, H. Tanaka and E. Nakamura, *J. Am. Chem. Soc.*, 2015, **137**, 15907–15914.
72. J. Cao, X. Jing, J. Yan, C. Hu, R. Chen, J. Yin, J. Li and N. Zheng, *J. Am. Chem. Soc.*, 2016, **138**, 9919–9926.
73. P. P. Khlyabich and Y.-L. Loo, *Chem. Mater.*, 2016, **28**, 9041–9048.
74. A. A. Petrov, I. P. Sokolova, N. A. Belich, G. S. Peters, P. V. Dorovatovskii, Y. V. Zubavichus, V. N. Khrustalev, A. V. Petrov, M. Grätzel, E. A. Goodilin and A. B. Tarasov, *J. Phys. Chem. C*, 2017, **121**, 20739–20743.
75. K. Yan, M. Long, T. Zhang, Z. Wei, H. Chen, S. Yang and J. Xu, *J. Am. Chem. Soc.*, 2015, **137**, 4460–4468.
76. D. P. McMeekin, Z. Wang, W. Rehman, F. Pulvirenti, J. B. Patel, N. K. Noel, M. B. Johnston, S. R. Marder, L. M. Herz and H. J. Snaith, *Adv. Mater.*, 2017, **29**, 1607039.
77. M. L. Petrus, J. Schlipf, C. Li, T. P. Gujar, N. Giesbrecht, P. Müller-Buschbaum, M. Thelakkat, T. Bein, S. Hüttner and P. Docampo, *Adv. Energy Mater.*, 2017, **7**, 1–27.
78. J.-W. Lee, H.-S. Kim and N.-G. Park, in *Organic-Inorganic Halide Perovskite Photovoltaics*, ed. N.-G. Park, M. Grätzel and T. Miyasaka, Springer International Publishing Switzerland, 1st edn, 2016, pp. 223–254.
79. S. D. Stranks and H. J. Snaith, in *Photovoltaic Solar Energy: From Fundamentals to Applications*, ed. A. Reinders, P. Verlinden, W. van Sark and A. Freundlich, John Wiley & Sons, Ltd., 1st edn, 2017, pp. 277–291.
80. M. Liu, M. B. Johnston and H. J. Snaith, *Nature*, 2013, **501**, 395–398.
81. J. Ávila, C. Momblona, P. P. Boix, M. Sessolo and H. J. Bolink, *Joule*, 2017, **1**, 431–442.
82. L. K. Ono, M. R. Leyden, S. Wang and Y. Qi, *J. Mater. Chem. A*, 2016, **4**, 6693–6713.

83. J. Burschka, N. Pellet, S.-J. Moon, R. Humphry-Baker, P. Gao, M. K. Nazeeruddin and M. Grätzel, *Nature*, 2013, **499**, 316–320.
84. A. Ummadisingu and M. Grätzel, *Sci. Adv.*, 2018, **4**, 1–10.
85. J. Schlipf, P. Docampo, C. J. Schaffer, V. Körstgens, L. Bießmann, F. Hanusch, N. Giesbrecht, S. Bernstorff, T. Bein and P. Müller-Buschbaum, *J. Phys. Chem. Lett.*, 2015, **6**, 1265–1269.
86. I. Grill, K. Handloser, F. C. Hanusch, N. Giesbrecht, T. Bein, P. Docampo, M. Handloser and A. Hartschuh, *Sol. Energy Mater. Sol. Cells*, 2017, **166**, 269–275.
87. L. K. Ono, M. R. Leyden, S. Wang and Y. Qi, *J. Mater. Chem. A*, 2016, **4**, 6693–6713.
88. A. Binek, I. Grill, N. Huber, K. Peters, A. G. Hufnagel, D. M. Handloser, D. P. Docampo, P. A. Hartschuh and P. T. Bein, *Chem. - Asian J.*, 2016, **11**, 1199–1204.
89. E. L. Unger, A. R. Bowring, C. J. Tassone, V. L. Pool, A. Gold-Parker, R. Cheacharoen, K. H. Stone, E. T. Hoke, M. F. Toney and M. D. McGehee, *Chem. Mater.*, 2014, **26**, 7158–7165.
90. R. Munir, A. D. Sheikh, M. Abdelsamie, H. Hu, L. Yu, K. Zhao, T. Kim, O. El Tall, R. Li, D. M. Smilgies and A. Amassian, *Adv. Mater.*, 2017, **29**, 1604113.
91. D. T. Moore, H. Sai, K. W. Tan, D. M. Smilgies, W. Zhang, H. J. Snaith, U. Wiesner and L. A. Estroff, *J. Am. Chem. Soc.*, 2015, **137**, 2350–2358.
92. W. Zhang, M. Saliba, D. T. Moore, S. K. Pathak, M. T. Hörantner, T. Stergiopoulos, S. D. Stranks, G. E. Eperon, J. A. Alexander-Webber, A. Abate, A. Sadhanala, S. Yao, Y. Chen, R. H. Friend, L. A. Estroff, U. Wiesner and H. J. Snaith, *Nat. Commun.*, 2015, **6**, 6142.
93. G. E. Eperon, V. M. Burlakov, P. Docampo, A. Goriely and H. J. Snaith, *Adv. Funct. Mater.*, 2013, **24**, 151–157.
94. N. Giesbrecht, J. Schlipf, L. Oesinghaus, A. Binek, T. Bein, P. Müller-Buschbaum and P. Docampo, *ACS Energy Lett.*, 2016, **1**, 150–154.
95. N. Giesbrecht, J. Schlipf, I. Grill, P. Rieder, V. Dyakonov, T. Bein, A. Hartschuh, P. Müller-Buschbaum and P. Docampo, *J. Mater. Chem. A*, 2018, **6**, 4822–4828.
96. M. M. Lee, J. Teuscher, T. Miyasaka, T. N. Murakami and H. J. Snaith, *Science*, 2012, **338**, 643–647.
97. H. S. Kim, C. R. Lee, J. H. Im, K. B. Lee, T. Moehl, A. Marchioro, S. J. Moon, R. Humphry-Baker, J. H. Yum, J. E. Moser, M. Grätzel and N. G. Park, *Sci. Rep.*, 2012, **2**, 1–7.
98. J. H. Heo, D. H. Song and S. H. Im, *Adv. Mater.*, 2014, **26**, 8179–8183.
99. J. H. Heo, D. H. Song, H. J. Han, S. Y. Kim, J. H. Kim, D. Kim, H. W. Shin, T. K. Ahn, C. Wolf, T. W. Lee and S. H. Im, *Adv. Mater.*, 2015, **27**, 3424–3430.
100. J. Kim, J. S. Yun, X. Wen, A. M. Soufiani, C. F. J. Lau, B. Wilkinson, J. Seidel, M. A. Green, S. Huang and A. W. Y. Ho-Baillie, *J. Phys. Chem. C*, 2016, **120**, 11262–11267.
101. J. Huang, M. Wang, L. Ding, Z. Yang and K. Zhang, *RSC Adv.*, 2016, **6**, 55720–55725.

102. J. Pan, C. Mu, Q. Li, W. Li, D. Ma and D. Xu, *Adv. Mater.*, 2016, 8309–8314.
103. G. E. Eperon, S. D. Stranks, C. Menelaou, M. B. Johnston, L. M. Herz and H. J. Snaith, *Energy Environ. Sci.*, 2014, **7**, 982.
104. D. T. Moore, K. W. Tan, H. Sai, K. P. Barteau, U. Wiesner and L. A. Estroff, *Chem. Mater.*, 2015, **27**, 3197–3199.
105. C. G. Bischak, E. M. Sanehira, J. T. Precht, J. M. Luther and N. S. Ginsberg, *Nano Lett.*, 2015, **15**, 4799–4807.
106. D. P. Nenon, J. A. Christians, L. M. Wheeler, J. L. Blackburn, E. M. Sanehira, B. Dou, M. L. Olsen, K. Zhu, J. J. Berry and J. M. Luther, *Energy Environ. Sci.*, 2016, **9**, 2072–2082.
107. R. J. Stewart, C. Grieco, A. V. Larsen, G. S. Doucette and J. B. Asbury, *J. Phys. Chem. C*, 2016, **120**, 12392–12402.
108. K. H. Stone, A. Gold-Parker, V. L. Pool, E. L. Unger, A. R. Bowring, M. D. McGehee, M. F. Toney and C. J. Tassone, *Nat. Commun.*, 2018, **9**, 3458.
109. S. Yang, Y. C. Zheng, Y. Hou, X. Chen, Y. Chen, Y. Wang, H. Zhao and H. G. Yang, *Chem. Mater.*, 2014, **26**, 6705–6710.
110. L. Oesinghaus, J. Schlipf, N. Giesbrecht, L. Song, Y. Hu, T. Bein, P. Docampo and P. Müller-Buschbaum, *Adv. Mater. Interfaces*, 2016, **3**, 1600403.
111. Y. Wu, A. Islam, X. Yang, C. Qin, J. Liu, K. Zhang, W. Peng and L. Han, *Energy Environ. Sci.*, 2014, **7**, 2934–2938.
112. W. Fu, J. Yan, Z. Zhang, T. Ye, Y. Liu, J. Wu, J. Yao, C. Z. Li, H. Li and H. Chen, *Sol. Energy Mater. Sol. Cells*, 2016, **155**, 331–340.
113. Y. Jo, K. S. Oh, M. Kim, K. Kim, H. Lee, C. Lee and D. S. Kim, *Adv. Mater. Interfaces*, 2016, **3**, 1500768.
114. M. Yang, Z. Li, M. O. Reese, O. G. Reid, D. H. Kim, S. Siol, T. R. Klein, Y. Yan, J. J. Berry, M. F. A. M. Van Hest and K. Zhu, *Nat. Energy*, 2017, **2**, 1–9.
115. M. Saliba, T. Matsui, K. Domanski, J.-Y. Seo, A. Ummadisingu, S. M. Zakeeruddin, J.-P. Correa-Baena, W. R. Tress, A. Abate, A. Hagfeldt and M. Gratzel, *Science*, 2016, **354**, 206–209.
116. V. Gutmann, *The Donor-Acceptor Approach to Molecular Interactions*, Springer US, 1978.
117. C. Reichardt, *Solvents and Solvent Effects in Organic Chemistry*, 2003.
118. Y. Zhang, P. Gao, E. Oveisi, Y. Lee, Q. Jeangros, G. Grancini, S. Paek, Y. Feng and M. K. Nazeeruddin, *J. Am. Chem. Soc.*, 2016, **138**, 14380–14387.
119. H. Zhang, J. Cheng, D. Li, F. Lin, J. Mao, C. Liang, A. K. Y. Jen, M. Grätzel and W. C. H. Choy, *Adv. Mater.*, 2017, **29**, 1604695.
120. J. C. Hamill, J. Schwartz and Y. L. Loo, *ACS Energy Lett.*, 2018, **3**, 92–97.
121. B. J. Foley, J. Girard, B. A. Sorenson, A. Z. Chen, J. Scott Niezgoda, M. R. Alpert, A. F. Harper, D. M. Smilgies, P. Clancy, W. A. Saidi and J. J. Choi, *J. Mater. Chem. A*, 2017, **5**, 113–123.
122. H. Chen, F. Ye, W. Tang, J. He, M. Yin, Y. Wang, F. Xie, E. Bi, X. Yang, M. Grätzel and L. Han, *Nature*, 2017, **550**, 92–95.
123. C. Bi, Q. Wang, Y. Shao, Y. Yuan, Z. Xiao and J. Huang, *Nat. Commun.*, 2015, **6**, 1–7.
124. Y. Deng, E. Peng, Y. Shao, Z. Xiao, Q. Dong and J. Huang, *Energy Environ. Sci.*, 2015, **8**, 1544–1550.

125. W. Nie, H. Tsai, R. Asadpour, J.-C. Blancon, A. J. Neukirch, G. Gupta, J. J. Crochet, M. Chhowalla, S. Tretiak, M. A. Alam, H.-L. Wang and A. D. Mohite, *Science*, 2015, **347**, 522–525.
126. Y. Zhong, R. Munir, J. Li, M.-C. Tang, M. R. Niazi, K. Zhao and A. Amassian, *ACS Energy Lett.*, 2018, **3**, 1078–1085.
127. Y. Zhou and N. P. Padture, *ACS Energy Lett.*, 2017, **2**, 2166–2176.
128. G. Cotella, J. Baker, D. Worsley, F. De Rossi, C. Pleydell-Pearce, M. Carnie and T. Watson, *Sol. Energy Mater. Sol. Cells*, 2017, **159**, 362–369.
129. X. Li, D. Bi, C. Yi, J.-D. Décoppet, J. Luo, S. M. Zakeeruddin, A. Hagfeldt and M. Grätzel, *Science*, 2016, **8060**, 1–10.
130. F. Mathies, T. Abzieher, A. Hochstuhl, K. Glaser, A. Colsmann, U. W. Paetzold, G. Hernandez-Sosa, U. Lemmer and A. Quintilla, *J. Mater. Chem. A*, 2016, **4**, 19207–19213.
131. E. T. Hoke, D. J. Slotcavage, E. R. Dohner, A. R. Bowring, H. I. Karunadasa and M. D. McGehee, *Chem. Sci.*, 2014, **6**, 613–617.
132. G. E. Eperon, T. Leijtens, K. A. Bush, R. Prasanna, T. Green, J. T. W. Wang, D. P. McMeekin, G. Volonakis, R. L. Milot, R. May, A. Palmstrom, D. J. Slotcavage, R. A. Belisle, J. B. Patel, E. S. Parrott, R. J. Sutton, W. Ma, F. Moghadam, B. Conings, A. Babayigit, H. G. Boyen, S. Bent, F. Giustino, L. M. Herz, M. B. Johnston, M. D. McGehee and H. J. Snaith, *Science*, 2016, **354**, 861–865.
133. B. Zhao, M. Abdi-Jalebi, M. Tabachnyk, H. Glass, V. S. Kamboj, W. Nie, A. J. Pearson, Y. Puttisong, K. C. Gödel, H. E. Beere, D. A. Ritchie, A. D. Mohite, S. E. Dutton, R. H. Friend and A. Sadhanala, *Adv. Mater.*, 2017, **29**, 1604744.
134. Z. Yang, A. Rajagopal, S. B. Jo, C. C. Chueh, S. Williams, C. C. Huang, J. K. Katahara, H. W. Hillhouse and A. K. Y. Jen, *Nano Lett.*, 2016, **16**, 7739–7747.
135. D. Zhao, Y. Yu, C. Wang, W. Liao, N. Shrestha, C. R. Grice, A. J. Cimaroli, L. Guan, R. J. Ellingson, K. Zhu, X. Zhao, R.-G. Xiong and Y. Yan, *Nat. Energy*, 2017, **2**, 17018.
136. M. Kulbak, D. Cahen and G. Hodes, *J. Phys. Chem. Lett.*, 2015, **6**, 2452–2456.
137. S. N. Habisreutinger, D. P. McMeekin, H. J. Snaith and R. J. Nicholas, *APL Mater.*, 2016, **4**, 91503.
138. S. Albrecht, M. Saliba, J. P. Correa-Baena, K. Jäger, L. Korte, A. Hagfeldt, M. Grätzel and B. Rech, *J. Opt.*, 2016, **18**, 1–10.
139. E. L. Unger and M. Saliba, in *Annual Meeting of the German Physical Society*, 2017, pp. 1–10.
140. L. K. Ono, E. J. Juarez-Perez and Y. Qi, *ACS Appl. Mater. Interfaces*, 2017, **9**, 30197–30246.
141. A. J. Barker, A. Sadhanala, F. Deschler, M. Gandini, S. P. Senanayak, P. M. Pearce, E. Mosconi, A. J. Pearson, Y. Wu, A. R. Srimath Kandada, T. Leijtens, F. De Angelis, S. E. Dutton, A. Petrozza and R. H. Friend, *ACS Energy Lett.*, 2017, **2**, 1416–1424.
142. T. Jesper Jacobsson, J. P. Correa-Baena, M. Pazoki, M. Saliba, K. Schenk, M. Grätzel and A. Hagfeldt, *Energy Environ. Sci.*, 2016, **9**, 1706–1724.

143. A. Sadhanala, F. Deschler, T. H. Thomas, S. E. Dutton, K. C. Goedel, F. C. Hanusch, M. L. Lai, U. Steiner, T. Bein, P. Docampo, D. Cahen and R. H. Friend, *J. Phys. Chem. Lett.*, 2014, 140709102712008.
144. Y. Hu, E. M. Hutter, P. Rieder, I. Grill, J. Hanisch, M. F. Aygüler, A. G. Hufnagel, M. Handloser, T. Bein, A. Hartschuh, K. Tvingstedt, V. Dyakonov, A. Baumann, T. J. Savenije, M. L. Petrus and P. Docampo, *Adv. Energy Mater.*, 2018, **1703057**, 1–11.
145. C. Yi, J. Luo, S. Meloni, A. Boziki, N. Ashari-Astani, C. Grätzel, S. M. Zakeeruddin, U. Röthlisberger and M. Grätzel, *Energy Environ. Sci.*, 2016, **9**, 656–662.
146. M. Zhang, J. S. Yun, Q. Ma, J. Zheng, C. F. J. Lau, X. Deng, J. Kim, D. Kim, J. Seidel, M. A. Green, S. Huang and A. W. Y. Ho-Baillie, *ACS Energy Lett.*, 2017, **2**, 438–444.
147. W. Rehman, D. P. McMeekin, J. B. Patel, R. L. Milot, M. B. Johnston, H. J. Snaith and L. M. Herz, *Energy Environ. Sci.*, 2017, **10**, 361–369.
148. Y. Hu, M. F. Aygüler, M. L. Petrus, T. Bein and P. Docampo, *ACS Energy Lett.*, 2017, **2**, 2212–2218.
149. L. K. Ono and Y. Qi, *J. Phys. D: Appl. Phys.*, 2018, **51**, 93001.
150. Z. Yang, S. Zhang, L. Li and W. Chen, *J. Mater.*, 2017, **3**, 231–244.
151. S. Razza, S. Castro-Hermosa, A. Di Carlo and T. M. Brown, *APL Mater.*, 2016, **4**, 91508.
152. A. Kovalenko and M. Hrabal, in *Printable Solar Cells*, ed. N. D. Sankir andM. Sankir, Scrivener Publishing LLC, 2017, pp. 163–202.
153. J. Ciro, J. F. Montoya, R. Betancur and F. Jaramillo, in *Printable Solar Cells*, ed. N. D. Sankir and M. Sankir, Scrivener Publishing LLC, 2017, pp. 37–64.
154. TCI P2415, www.tcichemicals.com/eshop/de/de/commodity/P2415/.
155. Ossila I101, https://www.ossila.com/products/perovskite-ink-air.
156. T. Li, Y. Pan, Z. Wang, Y. Xia, Y. Chen and W. Huang, *J. Mater. Chem. A*, 2017, **5**, 12602–12652.
157. W. Zhang, S. Pathak, N. Sakai, T. Stergiopoulos, P. K. Nayak, N. K. Noel, A. A. Haghighirad, V. M. Burlakov, D. W. deQuilettes, A. Sadhanala, W. Li, L. Wang, D. S. Ginger, R. H. Friend and H. J. Snaith, *Nat. Commun.*, 2015, **6**, 10030.
158. I. Levchuk, Y. Hou, M. Gruber, M. Brandl, P. Herre, X. Tang, F. Hoegl, M. Batentschuk, A. Osvet, R. Hock, W. Peukert, R. R. Tykwinski and C. J. Brabec, *Adv. Mater. Interfaces*, 2016, **3**, 1–10.
159. Z. Xiao, D. Wang, Q. Dong, Q. Wang, W. Wei, J. Dai, X. Zeng and J. Huang, *Energy Environ. Sci.*, 2016, **9**, 867–872.
160. C.-C. Chueh, C.-Y. Liao, F. Zuo, S. T. Williams, P.-W. Lianga and A. K.-Y. Jen, *J. Mater. Chem. A*, 2015, **3**, 9058–9062.
161. Y. Hu, S. Si, A. Mei, Y. Rong, H. Liu, X. Li and H. Han, *Sol. RRL*, 2017, **1**, 160019.
162. M. Deepa, M. Salado, L. Calio, S. Kazim, S. M. Shivaprasad and S. Ahmad, *Phys. Chem. Chem. Phys.*, 2017, **19**, 4069–4077.
163. T. Duong, Y. L. Wu, H. Shen, J. Peng, X. Fu, D. Jacobs, E. C. Wang, T. C. Kho, K. C. Fong, M. Stocks, E. Franklin, A. Blakers, N. Zin, K. McIntosh,

W. Li, Y. B. Cheng, T. P. White, K. Weber and K. Catchpole, *Adv. Energy Mater.*, 2017, 7, 1–11.
164. Z. Liu, J. Hu, H. Jiao, L. Li, G. Zheng, Y. Chen, Y. Huang, Q. Zhang, C. Shen, Q. Chen and H. Zhou, *Adv. Mater.*, 2017, **29**, 1–8.
165. J. Kim, J. S. Yun, Y. Cho, S. L. Da, B. Wilkinson, A. M. Soufiani, X. Deng, J. Zheng, A. Shi, S. Lim, S. Chen, Z. Hameiri, M. Zhang, C. F. J. Lau, S. Huang, M. A. Green and A. W. Y. Ho-Baillie, *ACS Energy Lett.*, 2017, **2**, 1978–1984.
166. A. T. Mallajosyula, K. Fernando, S. Bhatt, A. Singh, B. W. Alphenaar, J.-C. Blancon, W. Nie, G. Gupta and A. D. Mohite, *Appl. Mater. Today*, 2016, **3**, 96–102.
167. S. Tang and E. Al, *Adv. Energy Mater.*, 2017, **7**, 17038.
168. S. Razza, F. Di, F. Matteocci, L. Cin, A. Lorenzo, S. Casaluci, P. Cameron, A. D. Epifanio, S. Licoccia, A. Reale, T. M. Brown and A. Di, *J. Power Sources*, 2015, **277**, 286–291.
169. M. Yang, D. H. Kim, T. R. Klein, Z. Li, M. O. Reese, B. J. Tremolet de Villers, J. J. Berry, M. F. A. M. van Hest and K. Zhu, *ACS Energy Lett.*, 2018, **3**, 322–328.
170. D. Angmo, M. Gao and D. Vak, in *Printable Solar Cells*, ed. N. D. Sankir and M. Sankir, Scrivener Publishing LLC, 2017, pp. 313–362.
171. Y. Rong, Y. Ming, W. Ji, D. Li, A. Mei, Y. Hu and H. Han, *J. Phys. Chem. Lett.*, 2018, **9**, 2707–2713.
172. Press release: Solliance 10% 168 cm^2 module, https://www.solliance.eu/2016/solliance-realizes-a-first-up-scaled-perovskite-based-pv-module-with-10-efficiency/.
173. J. H. Heo, M. H. Lee, M. H. Jang and S. H. Im, *J. Mater. Chem. A*, 2016, **4**, 17636–17642.
174. S. Das, B. Yang, G. Gu, P. C. Joshi, I. N. Ivanov, C. M. Rouleau, T. Aytug, D. B. Geohegan and K. Xiao, *ACS Photonics*, 2015, **2**, 680–686.
175. A. Mei, X. Li, L. Liu, Z. Ku, T. Liu, Y. Rong, M. Xu, M. Hu, J. Chen, Y. Yang, M. Grätzel and H. Han, *Science*, 2014, **345**, 295–298.
176. S.-G. Li, K.-J. Jiang, M.-J. Su, X.-P. Cui, J.-H. Huang, Q.-Q. Zhang, X.-Q. Zhou, L.-M. Yang and Y.-L. Song, *J. Mater. Chem. A*, 2015, **3**, 9092–9097.
177. F. Mathies, H. Eggers, B. S. Richards, G. Hernandez-Sosa, U. Lemmer and U. W. Paetzold, *ACS Appl. Energy Mater.*, 2018, **1**, 1834–1839.
178. C. Liang, P. Li, H. Gu, Y. Zhang, F. Li, Y. Song, G. Shao, N. Mathews and G. Xing, *Sol. RRL*, 2018, **2**, 1770150.
179. M. Bag, Z. Jiang, L. A. Renna, S. P. Jeong, V. M. Rotello and D. Venkataraman, *Mater. Lett.*, 2016, **164**, 472–475.
180. S.-J. Moon, J.-H. Yum, L. Löfgren, A. Walter, L. Sansonnes, M. Benkhaira, S. Nicolay, J. Bailat and C. Ballif, *IEEE J. Photovoltaics*, 2015, **5**, 1087–1092.
181. J. Dagar, S. Castro-Hermosa, M. Gasbarri, A. L. Palma, L. Cina, F. Matteocci, E. Calabrò, A. Di Carlo and T. M. Brown, *Nano Res.*, 2018, **11**, 2669–2681.
182. T. A. Berhe, W.-N. Su, C.-H. Chen, C.-J. Pan, J.-H. Cheng, H.-M. Chen, M.-C. Tsai, L.-Y. Chen, A. A. Dubale and B.-J. Hwang, *Energy Environ. Sci.*, 2016, **9**, 323–356.
183. G. Niu, X. Guo and L. Wang, *J. Mater. Chem. A*, 2015, **3**, 8970–8980.

184. A. M. A Leguy, Y. Hu, M. Campoy-Quiles, M. Isabel Alonso, O. J. Weber, P. Azarhoosh, M. van Schilfgaarde, M. T. Weller, T. Bein, J. Nelson, P. Docampo and P. R. F. Barnes, *Chem. Mater.*, 2015, **27**, 3397–3407.
185. G.-W. Kim, G. Kang, M. Malekshahi Byranvand, G.-Y. Lee and T. Park, *ACS Appl. Mater. Interfaces*, 2017, **9**, 27720–27726.
186. I. Hwang, I. Jeong, J. Lee, M. J. Ko and K. Yong, *ACS Appl. Mater. Interfaces*, 2015, **7**, 17330–17336.
187. S.-W. Lee, S. Kim, S. Bae, K. Cho, T. Chung, L. E. Mundt, S. Lee, S. Park, H. Park, M. C. Schubert, S. W. Glunz, Y. Ko, Y. Jun, Y. Kang, H.-S. Lee and D. Kim, *Sci. Rep.*, 2016, **6**, 38150.
188. G.-H. Kim, H. Jang, Y. J. Yoon, J. Jeong, S. Y. Park, B. Walker, I.-Y. Jeon, Y. Jo, H. Yoon, M. Kim, J.-B. Baek, D. S. Kim and J. Y. Kim, *Nano Lett.*, 2017, **17**, 6385–6390.
189. J. S. Yoo, G. S. Han, S. Lee, M. C. Kim, M. Choi, H. S. Jung and J.-K. Lee, *Nano Res.*, 2017, **10**, 3885–3895.
190. I. C. Smith, E. T. Hoke, D. Solis-Ibarra, M. D. McGehee and H. I. Karunadasa, *Angew. Chem., Int. Ed.*, 2014, **53**, 11232–11235.
191. Z. Wang, Q. Lin, F. P. Chmiel, N. Sakai, L. M. Herz and H. J. Snaith, *Nat. Energy*, 2017, **2**, 17135.
192. D. B. Mitzi, S. Wang, C. A. Feild, C. A. Chess and A. M. Guloy, *Science*, 1995, **267**, 1473–1476.
193. G. Jia, Z.-J. Shi, Y.-D. Xia, Q. Wei, Y.-H. Chen, G.-C. Xing and W. Huang, *Opt. Express*, 2018, **26**, 66–74.
194. G. Grancini, C. Roldán-Carmona, I. Zimmermann, E. Mosconi, X. Lee, D. Martineau, S. Narbey, F. Oswald, F. De Angelis, M. Graetzel and M. K. Nazeeruddin, *Nat. Commun.*, 2017, **8**, 15684.
195. Z. Song, S. C. Watthage, A. B. Phillips, B. L. Tompkins, R. J. Ellingson and M. J. Heben, *Chem. Mater.*, 2015, **27**, 4612–4619.
196. G. Niu, W. Li, J. Li, X. Liang and L. Wang, *RSC Adv.*, 2017, **7**, 17473–17479.
197. R. J. Sutton, G. E. Eperon, L. Miranda, E. S. Parrott, B. A. Kamino, J. B. Patel, M. T. Hörantner, M. B. Johnston, A. A. Haghighirad, D. T. Moore and H. J. Snaith, *Adv. Energy Mater.*, 2016, **6**, 1–6.
198. N. H. Nickel, F. Lang, V. V. Brus, O. Shargaieva and J. Rappich, *Adv. Electron. Mater.*, 2017, **3**, 1700158.
199. R. I. Dawood, A. J. Forty and M. R. Tubbs, *Proceedings of the Royal Society of London*, The Royal Society, 1965, vol. 284, pp. 1–21.
200. M. R. Tubbs and A. J. Forty, *Br. J. Appl. Phys.*, 1964, **15**, 1553–1558.
201. X. Tang, M. Brandl, B. May, I. Levchuk, Y. Hou, M. Richter, H. Chen, S. Chen, S. Kahmann, A. Osvet, F. Maier, H.-P. Steinrück, R. Hock, G. J. Matt and C. J. Brabec, *J. Mater. Chem. A*, 2016, **4**, 15896–15903.
202. F.-S. Zu, P. Amsalem, I. Salzmann, R.-B. Wang, M. Ralaiarisoa, S. Kowarik, S. Duhm and N. Koch, *Adv. Opt. Mater.*, 2017, **5**, 1700139.
203. J. Schoonman, *Chem. Phys. Lett.*, 2015, **619**, 193–195.
204. C. C. Stoumpos, C. D. Malliakas and M. G. Kanatzidis, *Inorg. Chem.*, 2013, **52**, 9019–9038.

205. F. Hao, C. C. Stoumpos, R. P. H. Chang and M. G. Kanatzidis, *J. Am. Chem. Soc.*, 2014, **136**, 8094–8099.
206. A. Goyal, S. McKechnie, D. Pashov, W. Tumas, M. Van Schilfgaarde and V. Stevanovic, *Chem. Mater.*, 2018, **30**, 3920–3928.
207. R. E. Beal, D. J. Slotcavage, T. Leijtens, A. R. Bowring, R. A. Belisle, W. H. Nguyen, G. F. Burkhard, E. T. Hoke and M. D. McGehee, *J. Phys. Chem. Lett.*, 2016, **7**, 746–751.
208. Z. Li, M. Yang, J.-S. Park, S.-H. Wei, J. J. Berry and K. Zhu, *Chem. Mater.*, 2015, **28**, 284–292.
209. H. Haupt, F. Huber and H. Preut, *Z. Anorg. Allg. Chem.*, 1974, **408**, 209–213.
210. H. L. Wells, *Z. Anorg. Allg. Chem.*, 1893, **3**, 195–210.
211. S. Chakraborty, W. Xie, N. Mathews, M. Sherburne, R. Ahuja, M. Asta and S. G. Mhaisalkar, *ACS Energy Lett.*, 2017, **2**, 837–845.
212. O. Lindqvist, *Acta Chem. Scand.*, 1968, **22**, 2943–2952.
213. B.-W. Park, B. Philippe, X. Zhang, H. Rensmo, G. Boschloo and E. M. J. Johansson, *Adv. Mater.*, 2015, **27**, 6806–6813.
214. B. Ghosh, B. Wu, H. K. Mulmudi, C. Guet, K. Weber, T. C. Sum, S. G. Mhaisalkar and N. Mathews, *ACS Appl. Mater. Interfaces*, 2018, **10**, 35000–35007.
215. A. Singh, K. M. Boopathi, A. Mohapatra, Y. F. Chen, G. Li and C. W. Chu, *ACS Appl. Mater. Interfaces*, 2018, **10**, 2566–2573.
216. X. Qiu, B. Cao, S. Yuan, X. Chen, Z. Qiu, Y. Jiang, Q. Ye, H. Wang, H. Zeng, J. Liu and M. G. Kanatzidis, *Sol. Energy Mater. Sol. Cells*, 2017, **159**, 227–234.
217. B. Lee, C. C. Stoumpos, N. Zhou, F. Hao, C. Malliakas, C.-Y. Yeh, T. J. Marks, M. G. Kanatzidis and R. P. H. Chang, *J. Am. Chem. Soc.*, 2014, **136**, 15379–15385.
218. Y. Cai, W. Xie, H. Ding, Y. Chen, K. Thirumal, L. H. Wong, N. Mathews, S. G. Mhaisalkar, M. Sherburne and M. Asta, *Chem. Mater.*, 2017, **29**, 7740–7749.
219. K. Knox and D. W. Mitchell, *J. Inorg. Nucl. Chem.*, 1961, **21**, 253–258.
220. L. R. Morrs and W. R. Robinson, *Acta Crystallogr., Sect. B*, 1972, **28**, 653–654.
221. D. Babel, R. Hägele, G. Pausewang and F. Wall, *Mater. Res. Bull.*, 1973, **8**, 1371–1382.
222. H. G. F. Winkler, *Heidelberger Beitr. Mineral. Petrogr.*, 1952, **3**, 297–306.
223. D. J. Kubicki, D. Prochowicz, A. Hofstetter, S. M. Zakeeruddin, M. Grätzel and L. Emsley, *J. Am. Chem. Soc.*, 2017, **139**, 14173–14180.
224. A. H. Slavney, T. Hu, A. M. Lindenberg and H. I. Karunadasa, *J. Am. Chem. Soc.*, 2016, **138**, 2138–2141.
225. G. Volonakis, M. R. Filip, A. A. Haghighirad, N. Sakai, B. Wenger, H. J. Snaith and F. Giustino, *J. Phys. Chem. Lett.*, 2016, **7**, 1254–1259.
226. E. T. Mcclure, M. R. Ball, W. Windl and P. M. Woodward, *Chem. Mater.*, 2016, **28**, 1348–1354.

227. M. R. Filip, S. Hillman, A. A. Haghighirad, H. J. Snaith and F. Giustino, *J. Phys. Chem. Lett.*, 2016, **7**, 2579–2585.
228. F. Wei, Z. Deng, S. Sun, F. Xie, G. Kieslich, D. M. Evans, M. A. Carpenter, P. D. Bristowe and A. K. Cheetham, *Mater. Horiz.*, 2016, **3**, 328–332.
229. M. A. Green, Y. Hishikawa, E. D. Dunlop, D. H. Levi, J. Hohl-Ebinger and A. W. Y. Ho-Baillie, *Prog. Photovoltaics*, 2018, **26**, 3–12.
230. M. Xiao, F. Huang, W. Huang, Y. Dkhissi, Y. Zhu, J. Etheridge, A. Gray-Weale, U. Bach, Y. B. Cheng and L. Spiccia, *Angew. Chem., Int. Ed.*, 2014, **53**, 9898–9903.
231. J.-H. Im, I.-H. Jang, N. Pellet, M. Grätzel and N.-G. Park, *Nat. Nanotechnol.*, 2014, **9**, 927–932.
232. H. Zhou, Q. Chen, G. Li, S. Luo, T. B. Song, H. S. Duan, Z. Hong, J. You, Y. Liu and Y. Yang, *Science*, 2014, **345**, 542–546.
233. T. P. Gujar and P. M. Thelakkat, *Energy Technol.*, 2016, **4**, 449–457.
234. F. Lang, O. Shargaieva, V. V. Brus, H. C. Neitzert, J. Rappich and N. H. Nickel, *Adv. Mater.*, 2018, **30**, 1–22.
235. A. Jain, S. P. Ong, G. Hautier, W. Chen, W. D. Richards, S. Dacek, S. Cholia, D. Gunter, D. Skinner, G. Ceder and K. A. Persson, *APL Mater.*, 2013, **1**, 11002.
236. C. N. Savory, A. Walsh and D. O. Scanlon, *ACS Energy Lett.*, 2016, **1**, 949–955.
237. K. Momma and F. Izumi, *J. Appl. Crystallogr.*, 2011, **44**, 1272–1276.
238. K. Liang, D. B. Mitzi and M. T. Prikas, *Chem. Mater.*, 1998, **10**, 403–411.

CHAPTER 5

Theoretical Analyses of Copper-based Solar Cell Materials for the Next Generation of Photovoltaics

R. CHEN[a,b], S. ZAMULKO[a], DAN HUANG[c] AND C. PERSSON*[a,d]

[a]University of Oslo, Department of Physics/Centre for Materials Science and Nanotechnology, P.O. Box 1048 Blindern, Oslo, 0316, Norway; [b]Uppsala University, Department of Engineering Sciences, P.O. Box 256, Uppsala, 75105, Sweden; [c]Guangxi University, Guangxi Key Laboratory of Processing for Non-Ferrous Metallic and Featured Materials/School of Physical Science and Technology, Nanning, 530004, China; [d]KTH Royal Institute of Technology, Department of Materials Science and Engineering, Stockholm, 10044, Sweden
*E-mail: Clas.Persson@fys.uio.no

5.1 Introduction

In this chapter, we demonstrate how theory can be used to investigate inorganic solar cell materials by applying first-principles atomistic methods within the framework of density functional theory (DFT). The aim is to apply computational understanding to accelerate progress in developing environmentally friendly p–n junction photovoltaics (PV), especially towards technologies with highly efficient and inexpensive ultrathin films. As discussed

Inorganic Materials Series No. 3
Solar Energy Capture Materials
Edited by Elizabeth A. Gibson

Published by the Royal Society of Chemistry, www.rsc.org

in Chapters 1–4, crystalline silicon (c-Si) is today by far the most commercialised material in PV modules, with over 90% of the market,[1] despite having an indirect gap with band gap energy $E_g \approx 1.2$ eV. For laboratory solar cells, the energy conversion efficiency for single crystal c-Si without concentrator is $\eta_{max} = 25.8\%$,[2] but the active absorber material is rather thick, often 100–200 μm due to insufficient light absorption. Thin-film PV technologies rely on 100 times thinner photoactive layers (typically around 1 μm), and the absorbers must therefore have a higher capacity to harness sunlight in the energy region from ~1 to ~3 eV. Most traditional inorganic solar cells[2] are based on either amorphous silicon (a-Si or its hydrogenated compound a-Si : H, with $E_g = 1.7$ eV and $\eta_{max} = 14.0\%$), cadmium telluride (CdTe; $E_g = 1.5$ eV and $\eta_{max} = 22.1\%$), thin-film gallium arsenide (GaAs; $E_g = 1.5$ eV and $\eta_{max} = 28.8\%$), or copper indium gallium selenide ($CuIn_{1-x}Ga_xSe_2$; $E_g \approx 1.2$ eV for $x = 0.3$ and $\eta_{max} = 22.6\%$), and these materials have their "pros and cons" in terms of prospective large-scale production.

Developments of even thinner absorber materials (*i.e.*, for ultrathin inorganic solar cells) is an emerging concept to produce competitive PV cells or to complement the c-Si technologies with a proper top-cell in tandem structures. Here, ultrathin implies absorber layers with a thickness less than 100 nm (perhaps as thin as 10–30 nm), which is then at least more than ten times thinner than in traditional thin-film technologies. Ultrathin layers will not only be lighter and save material, fabrication, and handling costs, but they also imply a shorter path for the photon-excited carriers to reach the electrodes. An ultrathin solar cell material is expected not only to have a very high ability to absorb sunlight, but also (as for the traditional PV technologies) to be environmentally friendly and of low cost. It requires not only that the cost of syntheses, manufacturing process, recycling, and handling of the modules should be sufficiently low, but also that the device should have a long operating lifetime. Moreover, the raw material elements should be Earth-abundant, non-toxic, and inexpensive. In addition to that, the materials should be thermodynamically stable even in operation, easily doped or be natively p- or n-type, the materials should contain no devastating defects, have high carrier mobility, and have good interface physics with optimum band alignment with other device layers.

From that perspective, copper-based chalcogenides can be regarded as attractive absorber materials for the development of ultrathin devices. Chalcopyrite copper indium gallium selenide ($CuIn_{1-x}Ga_xSe_2$, called CIGS) is today a commercially important thin-film PV material ($\eta_{max} \approx 22.6\%$[2]), and it is also a very intriguing material from a scientific point of view.[3–7] Kesterite, copper zinc tin sulphide/selenide (Cu_2ZnSnS_4, CZTS, and $Cu_2ZnSnSe_4$, CZTSe) is considered as an alternative absorber material to CIGS with the advantage that it does not involve indium, which is a costly element.[8–13] Its solar cell efficiency is, to date, some 10% lower than that of CIGS, that is $\eta_{max} = 12.6\%$.[2] Attempts to improve the device efficiency to optimise the band gap energy, as well as to form a built-in electric field arising from a compositional grading of the materials (*e.g.*, $Cu_2ZnSn(S_{1-x}Se_x)_4$, CZTSSe, and $Cu_2Zn(Sn_{1-x}Ge_x)S_4$, CZTGS), have shown some enhanced effects on optoelectronic performances.[14–21] One of the earliest recognised solar cell materials (already known in the 1920s, thus long before Si technologies) is cuprous oxide Cu_2O,[22] but the material quality and the PV performance were rather poor during attempts to develop low-cost devices in the 1970s.[23] Today, the compound is again relevant, especially as a top-cell in a device structure. The many good basic qualities of these copper-based chalcogenides, and the ability to fabricate rather inexpensive PV devices with polycrystalline, non-stoichiometric, as-grown p-type materials, is motivating researchers to explore copper-based chalcogenides further. For example, ternary Cu–group IV chalcogenides, like the Cu_2SnS_3 (CTS) compound[24–30] or the corresponding Sn/Ge-poor $Cu_4(Sn/Ge)S_4$ phase,[31–34] are candidates for solar cell absorber materials. In addition, their relevant isovalent cation alloys $Cu_2Sn_{1-x}Ge_xSe_3$ (CTGS) and also $Cu_2Sn_{1-x}Si_xSe_3$ (CTSS) can be attractive due to their tuneable band gap energies.[35,36] An alternative to cation alloying as in CTSS, anion alloying can be utilised to optimise both the band gap energy and the energy position of the valence band maximum (VBM). Furthermore, ternary compounds with heavy group V elements Sb or Bi, such as $CuSbS_2$ (CSS) or $CuBiS_2$ (CBS),[37–44] often have high absorption coefficients. Anion alloying $CuSb(Se_{1-x}Te_x)_2$ (CSST) and $CuBi(S_{1-x}Se_x)_2$ (CBSS) can tune the electronic band energies.[45] In addition, Sb/Bi-poor phases $Cu_3(Sb/Bi)S_3$ are interesting for wide-gap solar cell materials.[46–50] Both the VBM and the conduction band minimum (CBM) can be optimised for the overall band profile of the solar cell device, therefore, also including the buffer layers and window layer as well as the electrodes, by judiciously selecting the cation and anion alloying configurations. Simultaneously, the lattice mismatching between the absorber and buffer layers can also be minimised to a certain extent (this can be achieved by choosing an appropriate buffer layer material).

The theoretical studies of materials are based on many-particle quantum physics, where, in principle, all interactions between nuclei and electrons are considered. Based on DFT,[51] this complex many-particle problem can be transferred into single-electron equations, utilising an auxiliary interaction potential that artificially describes the complex interaction of each single electron with all other electrons. The single-electron equation is the so-called Kohn–Sham equation (KS-eq[52]), and the auxiliary interaction potential is called the exchange-correlation potential. This DFT/KS-eq is described in detail in Section 5.2. The exact formula for the exchange-correlation potential is not known, and there are numerous approximations to it. In this chapter, we theoretically model and explore several of these emerging Cu-based chalcogenides (*i.e.*, Cu–X–(O/S/Se/Te) and some of their isovalent alloys) based on first-principles atomistic calculations using the projector augmented wave (PAW) and the full-potential, augmented plane wave basis set formalisms, as implemented in the VASP[53,54] and WIEN2k[55] program packages, respectively. The calculations rely on the exchange-correlation potentials that are described by the generalised gradient approximations by Perdew, Burke, and Ernzerhof (PBE),[56] and the corresponding functional revised for solids

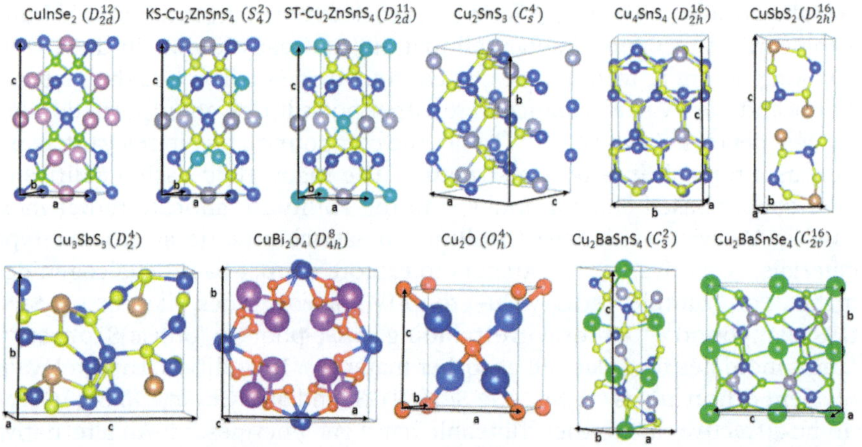

Figure 5.1 Crystalline structures of various Cu-based chalcogenides: Cu atoms in dark blue, Sn atoms in grey, S atoms in yellow and O atoms in red.

(PBEsol),[57] both with or without an on-site Coulomb interaction correction U_d[58] of the d-like orbitals for the transition metal atoms (*i.e.*, the PBE+U and PBEsol+U approaches), as well as the hybrid functional by Heyd, Scuseria, and Ernzerhof (HSE06).[59,60]

The crystalline structures of the compounds can be determined from total energy calculations; examples of the different crystalline structures are depicted in Figure 5.1. The electronic properties are analysed in terms of the electronic structures of the topmost valence bands (VBs) and the bottommost conduction bands (CBs), the fundamental band gap energy, as well as of the total density-of-states (DOS), with a detailed description from the atom and angular momentum-resolved DOS (*i.e.*, PDOS). The optical properties are analysed in terms of the complex dielectric function $\varepsilon(E) = \varepsilon_1(E) + i\varepsilon_2(E)$ and the absorption coefficient $\alpha(E)$. The very high absorption coefficients in the energy region $E_g \leq E < (E_g + 2\text{ eV})$ for certain Cu-based chalcogenides are directly explained by the electronic band structures, and these compounds are highly relevant for further investigation and developments of ultrathin solar cells. To calculate the optical properties one has to consider transitions from all valence bands and also from all possible **k**-points in the reciprocal space. Detailed analysis of the CTS compound, in particular, reveals that a very dense **k**-mesh of the irreducible Brillouin zone (IBZ) is needed, together with a proper **k**-space integration method to fully describe the optical properties of this material.

The performance of the materials as a solar cell absorber are analysed within the Shockley–Queisser (SQ) model for the theoretical limit of the maximum efficiency, η_{max}, considering a single p–n junction structure described by the Shockley ideal diode equation (see eqn (5.11)–(5.16) in Section 5.2.3). We include a thickness dependence of the absorptivity in the model, and we then describe how the maximum efficiency decreases when the absorber layer is thinner. From that analysis, we can tailor make emerging solar cell materials that have sufficiently high efficiency also for ultrathin layers. Our target materials should have direct band gaps and multi-valley band edges, and the energetically topmost VBs as well as the bottommost CBs should have rather flat energy dispersions. These requirements can be achieved by, for instance, considering compounds with heavy elements, since those elements have typically more energetically localised orbitals. Below, we compare the results for the emerging Cu-based chalcogenides with the corresponding data for traditional thin-film solar cell materials, to demonstrate the possibilities for developing highly efficient, ultrathin solar cell devices. We show that it is possible to design absorber materials with a theoretical maximum efficiency as high as $\eta_{max} \approx 30\%$ for film thickness of $d \approx 50$–100 nm. However, it is important that the flat band dispersions in these materials do not form an indirect gap, with an energy smaller than that of the fundamental direct gap, because the Auger effect can then lower the efficiency by some 3 to 5%.

5.2 Description of the First-principles Analysis Methods for the Materials

DFT in conjunction with the KS-eq, and with different types of approximations for the exchange-correlation potential, has, over the last 40 years, been extremely successful in describing systems with a large number of atoms, such as solids, nanostructures, and large molecules. Employing DFT, theoreticians in material physics and chemistry are supporting experimental studies, analysing the underlying physics of many-atom systems, as well as exploring new material and molecular structures. DFT is one of the most popular approaches to explore many-atom systems, and, in 1998, Prof. John A. Pople and Prof. Walter Kohn shared the Nobel Prize in Chemistry for their contributions to developing DFT and computational methods.

The computational methods are based on ground-state DFT[51] as realised with the KS-eq.[52] DFT is based on two main theorems; the Hohenberg–Kohn theorems. The first theorem states that the density of the system determines all ground-state properties of that system. The second theorem states that there exists a variational principle for the energy density functional, such that the ground-state total energy can be obtained by varying the density. Thus, there exists a universal function for the total energy $E_{tot}[n]$ of the electron density $n(\mathbf{r})$ for all electron systems, and it is obtained through $E_{tot}[n_0] < E_{tot}[n]$, where n_0 is the ground-state density. The idea behind DFT is to consider only the electron density in the material instead of using the complex many-electron wave function, and this would be a huge simplification of the problem. However, the Hohenberg–Kohn theorems do not tell us the form of $E_{tot}[n]$, and the expression of the functional is so far unknown.

The KS-eq is derived from DFT, and, in principle, it is a method to solve the DFT exactly. The approach relies, however, on auxiliary Kohn–Sham (KS) single-electron eigenfunctions $\psi_i^{KS}(\mathbf{r})$ to generate the exact total density $n(\mathbf{r})$ of the system. Assuming the corresponding auxiliary many-electron wave function is Hartree-like, then the KS-eq reads

$$\left\{-\frac{\hbar^2}{2m_e}\nabla^2 + V_{eff}(\mathbf{r})\right\}\psi_i^{KS}(\mathbf{r}) = \varepsilon_i^{KS}\psi_i^{KS}(\mathbf{r}) \quad (5.1)$$

$$V_{eff}(\mathbf{r}) = V_H(\mathbf{r}) + V_{en}(\mathbf{r}) + V_{xc}(\mathbf{r}) \quad (5.2)$$

$$E_{tot}[n] = \sum_i \varepsilon_i^{KS} - \frac{1}{2}\iint q^2 \frac{n(\mathbf{r})n(\mathbf{r}')}{|\mathbf{r}-\mathbf{r}'|}d\mathbf{r}\,d\mathbf{r}' + E_{xc}[n] - \int V_{xc}(\mathbf{r})n(\mathbf{r})d\mathbf{r} \quad (5.3)$$

$$n(\mathbf{r}) = \sum_i |\psi_i^{KS}(\mathbf{r})|^2 \; i=1,2\ldots Ne \text{ for the Ne occupied electronic states.} \quad (5.4)$$

Here, the effective potential $V_{eff}(\mathbf{r})$ comprises the well-defined Hartree potential $V_H(\mathbf{r})$, the system-dependent external electron–nuclei interaction potential $V_{en}(\mathbf{r})$, and the so-called exchange-correlation potential $V_{xc}(\mathbf{r})$. The KS-eq is an approach to map the many-electron problem into many single-electron equations with an effective potential $V_{eff}(\mathbf{r})$. The effective potential includes all the electron–electron interactions (but also some part of the kinetic energies of the electrons). The advantage of describing all the electron–electron interactions by an effective potential is immense. Importantly, the derivation proves that it is possible to generate the exact ground-state density and the exact ground-state total energy if one finds the universal expression of the effective potential that correctly includes the many-electron interactions. It is worth remarking here that the KS-eq is typically Fourier-transformed and solved in reciprocal space (**k**-space) in order to benefit from the translation symmetry of periodic systems, as considered in this chapter. The eigenfunctions and the corresponding eigenvalues are then represented for the jth energy band at the specific quantum state **k**, thus $\psi_{j,\mathbf{k}}^{KS}(\mathbf{r})$ and $\varepsilon_{j,\mathbf{k}}^{KS}(\mathbf{r})$, and the superscript KS is often neglected.

The KS-eq in DFT is, in principle, an exact method. However, to date we do not have the full explicit expression of the universal exchange-correlation potential $V_{xc}(\mathbf{r})$ that should include the often complex many-electron interactions. Here lies the main approximation within the Kohn–Sham approach: one has to rely on approximations of the exchange-correlation potential, and the local density approximation (LDA) is the first model of $V_{xc}(\mathbf{r})$. The approximation is based on the free electron gas, and it uses locally the exact description of the exchange-correlation of a homogeneous electron gas (free electron gas or "jellium"). Despite its simplicity, LDA has, however, been extremely successful in describing the many-electron interactions, although the approximation has deficiencies. Also, many improved approximations beyond the LDA are available; for example, the generalised gradient approximation (GGA) wherein the gradient of the electron density is also taken into account. The LDA and GGA are fairly accurate methods, but have limitations in terms of inaccurate band gap energies and localisation of d- and f-like states. Development of further improved exchange-correlation potential is still an ongoing and important research activity. Hybrid functional is an approach that mixes two or more types of exchange-correlation functionals, very often with a contribution from pure non-local Fock exchange potential to localise states, in order to optimise the computed material properties, especially the band gap energies as obtained from the single-electron energies. The LDA+U (or GGA+U) are methods to include a Hubbard-like on-site Coulomb correction potential to the DFT exchange-correlation potential in order to better describe strongly localised (strongly correlated) d- and f-systems.

Since the exchange-correlation potential $V_{xc}(\mathbf{r})$ is not exact, the total energy and the KS single-electron energies and eigenfunctions are not exact. However, the existing approximations to $V_{xc}(\mathbf{r})$ normally yield very good results. The drawback is the band gap energies, E_g, for semiconductors and insulators, which are underestimated by typically 50%. Approaches to more accurately determine the E_g (for example, the hybrid functionals with non-local Fock exchange) are often very time and computationally demanding, especially for systems with many atoms in the unit cells. Furthermore, even with an exact description of the exchange-correlation potential, the auxiliary KS single-electron and the corresponding total wave functions are not described exactly, since the KS-eq uses (in the approach above) a Hartree-like form of the total wave function, and thus its single-electron energies cannot be exact. Hence, if one determines the band gap energy from the KS eigenfunctions (instead of from total energies), one should not expect to obtain the exact gap energy even for the exact $V_{xc}(\mathbf{r})$. However, it is believed that the KS eigenfunctions can fairly accurately describe the true single-electron wave functions.

5.2.1 Basic Computational Details

The calculations of crystalline, electronic and optical properties are based on KS approach to solve the ground-state DFT. The approaches rely on either the PAW basis set in the VASP package[53,54] or the FPLAPW basis set in the WIEN2k package.[55] Unless specifically stated, the following described computational method and computational parameters are used.

The compounds and their crystalline structures are presented in Figure 5.1. The compounds are modelled with their primitive cells by the PAW potentials as a generalisation of the pseudopotential formalisms. The crystalline structures are fully relaxed until the total energy and the residual force on each atom from the quasi-Newton algorithm show convergence within 0.1 meV and 10 meV Å^{-1}, respectively, using the hybrid functional HSE06 with the standard parameters for mixing and screening,[59,60] *i.e.*, 25% non-local Fock exchange for the short-range interactions with the range separation of 0.2 Å^{-1}.

The total energy as well as the total DOS and the local, atomic- and angular-resolved DOS (*i.e.*, PDOS) were obtained from the modified tetrahedron integration method with a Γ-centred Monkhorst–Pack-like **k**-mesh. For cubic crystal structures with eight atoms per primitive cell, we typically use **k**-mesh grids of the size of $6 \times 6 \times 6$ in the IBZ, and the grid sizes change inversely with the cell sizes. We excluded the spin–orbit coupling (SOC), unless SOC causes a strong effect on the band dispersions, because typically there is no SOC-induced energy split in the degeneracies at the band edges that can make a strong impact on the dispersion. For some materials with heavy elements, SOC can narrow the band gap energy, but by only 0.1–0.2 eV, and such an effect does not change the material from being indirect to direct, and it has only a moderate effect on the DOS.

PBE+U with an on-site correction potential U_d is used to correct the d-like states.[58] This approach can better localise the d-state character of the semiconductor,[61–63] actually better than HSE06 in some cases, but the total energy is less accurately derived. Moreover, although the method can be applied to s- and p-orbitals to increase the gap energy, such an approach can easily generate unrealistic band curvatures,[64] and, therefore, we only use a correction potential to localise the d-like states. Herein, we choose $U_d(\text{Mn}) = 4.0$ eV, $U_d(\text{Fe}) = 4.6$ eV, $U_d(\text{Ni}) = 5.1$ eV, $U_d(\text{Cu}) = 4$ eV, and $U_d(\text{Zn}) = 7.5$ eV, according to Setyawan *et al.*[65] With the corrected potential on the d-like orbitals, the d-like energy states are corrected and the band gap is also opened moderately (typically by ~0.3 eV or less).

5.2.2 Calculations of the Optical Properties of Absorber Materials

The optical properties are evaluated based on the complex dielectric function $\varepsilon(E) = \varepsilon_1(E) + i\varepsilon_2(E)$. The dielectric function measures the electric displacement field due to the presence of an electric field in a dielectric material. The dielectric function is also an important property for describing the screening of the charges in the material near dopants, defects, and for other structural or electronic perturbations of the crystal, for example when an electromagnetic field (*i.e.*, light) is interacting with the material. The dielectric function is described as a three-dimensional rank 2 tensor (with components α and β), and the interband contribution to the imaginary part $\varepsilon_2(E)$ of the dielectric function can be calculated within linear response theory in the long wavelength limit when the change in wave number **q** goes to zero ($\lambda_{\mathbf{q}} = 2\pi/\mathbf{q} \to \infty$):[66]

$$\begin{aligned}\varepsilon_{2,\alpha\beta}(E) &= \lim_{\mathbf{q}\to 0} \frac{4\pi^2 e^2}{\Omega \mathbf{q}^2} \sum_{c,v,\mathbf{k}} 2w_{\mathbf{k}} \delta\left(E_c(\mathbf{k}) - E_v(\mathbf{k}) - E\right) \\ &\quad \times \langle u_c(\mathbf{k}+\mathbf{e}_\alpha \mathbf{q}) | u_v(\mathbf{k}) \rangle \langle u_c(\mathbf{k}+\mathbf{e}_\beta \mathbf{q}) | u_v(\mathbf{k}) \rangle^* .\end{aligned} \tag{5.5}$$

Here, $u_v(\mathbf{k})$ and $u_c(\mathbf{k})$ are the cell periodic parts of the eigenfunctions for the VB and CB states with energies $E_v(\mathbf{k})$ and $E_c(\mathbf{k})$, respectively. Ω is the volume of the primitive cell, $w_\mathbf{k}$ is the weight of the **k**-points in the IBZ, and $\mathbf{e}_\alpha$ is the unit vector in the Cartesian coordinates x, y, and z. A slightly different way to calculate the dielectric response utilises the optical matrix elements,[67] as in, for example, the WIEN2k package. Herein, indexing of the VBs is v_1, v_2, v_3, ..., where v_1 is the energetically topmost VB for each **k**-point, regardless of whether bands cross or not along the symmetry lines. Similarly, indexing of the CBs is c_1, c_2, c_3, ..., where c_1 is the bottommost CB for each **k**-point.

The summation over the **k**-states in eqn (5.5) is a computationally important aspect when analysing details in the dielectric spectra. In principle, a regular summation should be sufficient, but since there are as many **k**-states as there are primitive cells in the macroscopic crystal, it is not possible to perform such a summation. Instead, one has to rely on an approximation of the summation, and in this work we use the linear tetrahedron integration approach, with a finite-size **k**-mesh.[68] The dielectric response function is directly related to the joint DOS (*i.e.*, the Dirac-delta function $\delta(E)$ in the equation above), and the electronic band structure is therefore the main underlying property for analysing the polarisation response. This is exemplified for $CuIn_{0.5}Ga_{0.5}Se_2$ in Figure 5.2, where the total $\varepsilon_2(E)$ spectrum has been decomposed into the different contributions from each VB-to-CB transitions ($v_j \rightarrow c_{j'}$) in the summation of eqn (5.5).[69] By comparing the different VB-to-CB transitions with the electronic structure along the symmetry lines in the IBZ, one observes that the main E1 and E2 peaks at ~3 eV in the $\varepsilon_2(E)$ spectrum (thus, a strong contribution to absorption of sunlight) arise from transitions from the energetically topmost (v_1) and second topmost (v_2) VBs to the bottommost CB (c_1) at the edge of the IBZ. Transitions from the third topmost (v_3) VBs to the bottommost CB (c_1) show a clear double peak (E3 and E4) at around 3.2 and 3.7 eV. In the energy region 4–7 eV there are contributions from several VBs but also to several CBs.

The corresponding real part $\varepsilon_1(E)$ of the dielectric function is obtained *via* the Kramers–Kronig transformation relation

$$\varepsilon_{1,\alpha\beta}(E) = 1 + \frac{2}{\pi} P \int_0^\infty \frac{E' \varepsilon_{2,\alpha\beta}(E')}{E'^2 - E^2 + \mathrm{i}\eta} \mathrm{d}E', \tag{5.6}$$

where P is the principal value and η is a sufficiently small number, preferably an infinitesimally small number, but in practice it is larger than the step length of the energy mesh. The high-frequency dielectric constant ε_∞ is determined from the dielectric function at $\varepsilon_1(E = 0)$ when the electron–optical phonon coupling is excluded, and the static dielectric constant ε_0 is calculated from the Born effective charges, taking into account the ionic contribution. Experimentally, the high-frequency dielectric constant ε_∞ is determined typically in the mid-gap region, that is $\varepsilon_\infty \approx \varepsilon_1(0 \ll E \ll E_g)$, which can therefore be slightly different from the theoretically determined values, especially for small gap materials and when the frequencies of the optical phonon modes are high.

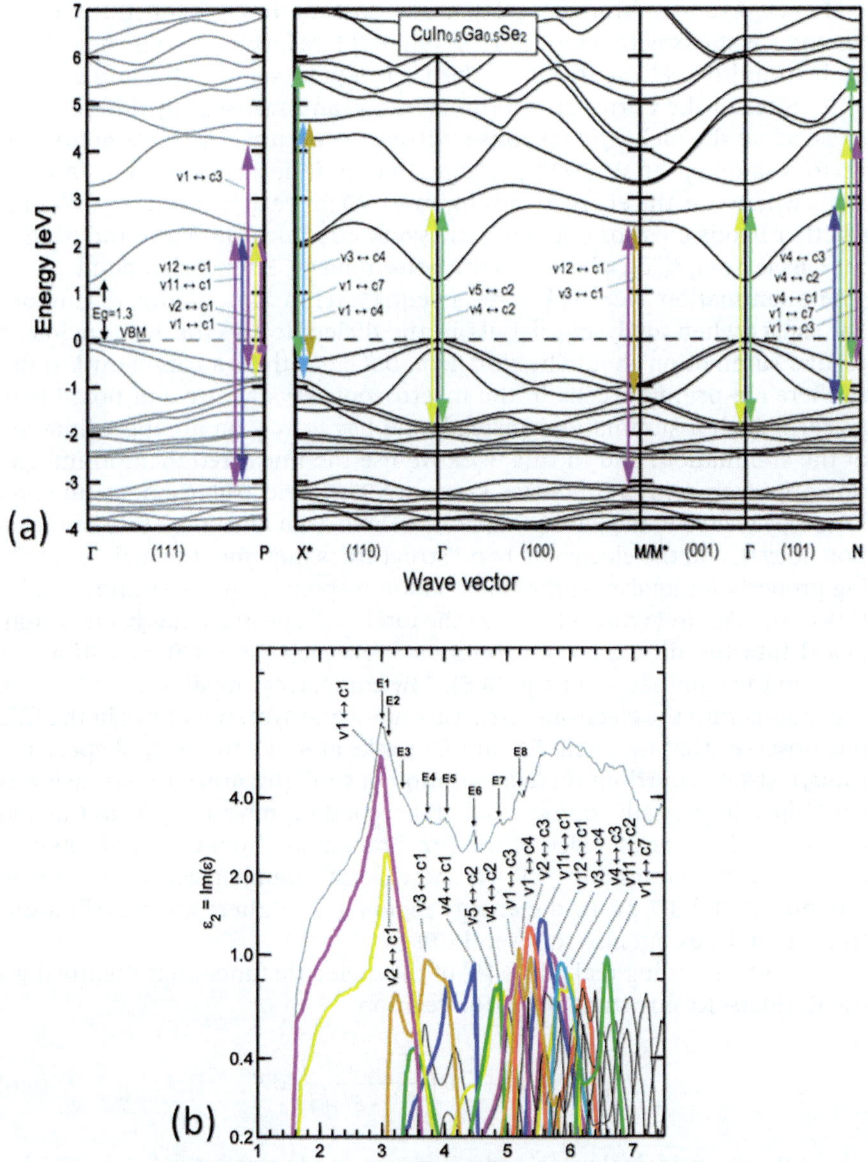

Figure 5.2 (a) The calculated electronic band structure of $CuIn_{0.5}Ga_{0.5}Se_2$ where the critical points are identified along the main symmetry directions. (b) Band-to-band analysis of the contribution to the total $\varepsilon_2(E)$ spectrum (thin black trace) and the individual critical points. The most important VB-to-CB transitions ($v_j \rightarrow v_{j'}$) are marked by thick coloured curves. Spin–orbit interaction is included and the band-to-band transitions involve a summation of the spin-up and spin-down-like contributions. Note: the vertical axis ε_2 is in the logarithmic scale. The results were obtained with the FPLAPW approach as implemented in the WIEN2k code, using the PBE+U potential with $U_d(\mathrm{Cu})$ = 6 eV. Reproduced from ref. 69 with permission from AIP Publishing.

The total, geometric averaged dielectric function is obtained as:

$$\varepsilon(E) = \frac{\left(\varepsilon_{xx}(E) + \varepsilon_{yy}(E) + \varepsilon_{zz}(E)\right)}{3} \tag{5.7}$$

The absorption coefficient is determined directly from the complex dielectric function through the relation

$$\alpha_{\alpha\beta}(E) = \frac{\sqrt{2}E}{\hbar c}\sqrt{\frac{\left|\varepsilon_{\alpha\beta}(E)\right| - \varepsilon_{1,\alpha\beta}(E)}{2}}, \tag{5.8}$$

where c is the speed of light. Moreover, the total geometric averaged absorption coefficient $\alpha(E)$ is obtained similarly as for the dielectric function.

Even though LDA and GGA often generate very reasonable band structures, they do underestimate the gap energies by, typically, 50%. A constant energy shift, Δ_g, of the CBs can sometimes simulate the correct gap. However, for direct gap materials with experimental gap energies of ~1.5 eV and smaller (which is often the case for solar cell materials), the underestimated gap energy can also imply an incorrect hybridisation of the CB and VB states at the gap edges.[64] This will result in incorrect band dispersions of the band edges, and thus incorrect effective electron and hole masses, also when the LDA (or GGA) generate a non-zero gap. In those cases, a constant energy shift of the CBs will not improve the incorrect hybridisation, and that can yield rather strange absorption spectra. The band curvature and the effective masses can be corrected substantially by weakening the hybridisation with any gap state-correcting method, even if the correct gap energy is not achieved.[64,70] Therefore, the LDA+U (or GGA+U) method with correction of only the d-like states can sometimes overcome this gap problem for Cu-based chalcogenides; this is illustrated for Cu_2ZnSnS_4 in Figure 5.3. The regular PBE potential yields E_g ~0.2 eV and that can cause the absorption coefficient to be too large in the low-energy region due to the strong VB–CB hybridisation, while PBE+U (despite the small gap) and HSE06 substantially improve the spectrum.

In addition to avoiding underestimating the gap energies in LDA and GGA for solar cell materials, one also has to consider accuracy with respect to other computational parameters. Many theoretical investigations on solar cell materials prefer to utilise hybrid functionals (often HSE06) in order to calculate the correct band gap energies, but, because hybrid functionals are both time and memory demanding, the number of **k**-points in the calculation of the dielectric function (see eqn (5.5)) is often relatively small. However, we have recently demonstrated that calculations with KS-eq in the DFT, employing the tetrahedron integration method, can require a very large **k**-mesh to accurately describe details in the absorption coefficient.[28] This is demonstrated for the calculation of $\varepsilon_2(E)$ in Cu_2SiS_3 in Figure 5.4. For a small **k**-mesh of 5 × 5 × 5 (implying $N_{\mathbf{k}}$ = 39 **k**-points in the IBZ) both HSE06 and PBE+U (corrected Δ_g to gap energy of HSE06) generate spectra that are similar in shape with a slowly increasing feature above the gap energy. When the

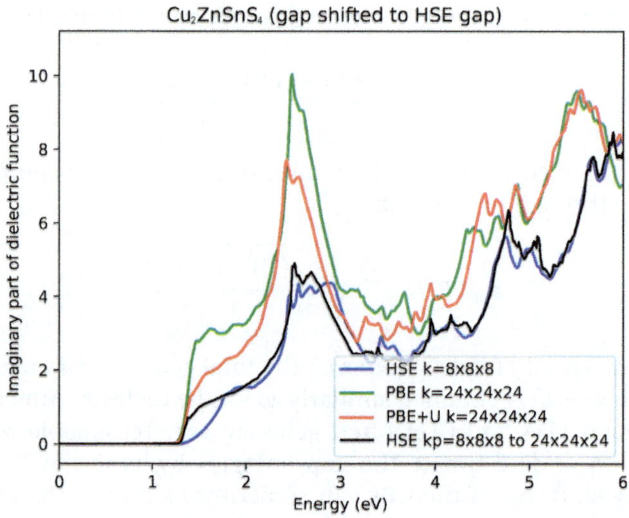

Figure 5.3 Calculated imaginary part $\varepsilon_2(E)$ of the dielectric function for kesterite Cu_2ZnSnS_4 using PBE, PBE+U (with $U_d(\text{Cu}) = 4.0$ eV) and HSE06 with the $\mathbf{k}{\cdot}\tilde{\mathbf{p}}$ method as described in ref. 72. The band gap energies have been shifted to agree with that of HSE06. A **k**-mesh with grid size 24 × 24 × 24 gives a good absorption onset feature also near the band gap energy.

k-mesh is increased to 20 × 20 × 20 ($N_\mathbf{k}$ = 2112 **k**-points) or even 30 × 30 × 30 ($N_\mathbf{k}$ = 6992 **k**-points) the PBE+U calculations reveal detail in the optical properties that are overlooked for the small **k**-mesh. That is, the imaginary part of the dielectric function (and consequently also of the absorption coefficient), is better described only when using a dense **k**-mesh, especially for energies just above the band gap energy.

Since it is not computationally realistic to perform HSE06 calculations with 3000 or 5000 **k**-points, we chose herein an approximated approach to both benefit from the rather accurate gap energies generated by the HSE06 approach and the possibility of using large **k**-meshes for PBE+U to reveal **k**-dependent details of the spectrum. The imaginary part of the dielectric function $\varepsilon_2(E)$ is calculated with both HSE06 and PBE+U for the same small **k**-mesh (typically 5 × 5 × 5 for a cubic primitive cell of eight atoms), and the spectrum calculated using PBE+U is shifted Δ_g to match the gap energy of HSE06. Then, $\varepsilon_2(E)$ is corrected with a large **k**-mesh (typically 30 × 30 × 30 for a few-atom primitive unit cell) using PBE+U:

$$\varepsilon_{2,\alpha\beta}(E) = \varepsilon_{2,\alpha\beta}(E;\text{HSE06 with small }\mathbf{k}\text{-mesh}) \cdot \frac{\varepsilon_{2,\alpha\beta}(E;\text{PBE}+\text{U}+\Delta_g \text{ with large }\mathbf{k}\text{-mesh})}{\varepsilon_{2,\alpha\beta}(E;\text{ PBE}+\text{U}+\Delta_g \text{ with small }\mathbf{k}\text{-mesh})}. \tag{5.9}$$

In the limit where the large **k**-mesh generates spectra that match those generated for the small **k**-mesh, the true HSE06 spectrum is generated.

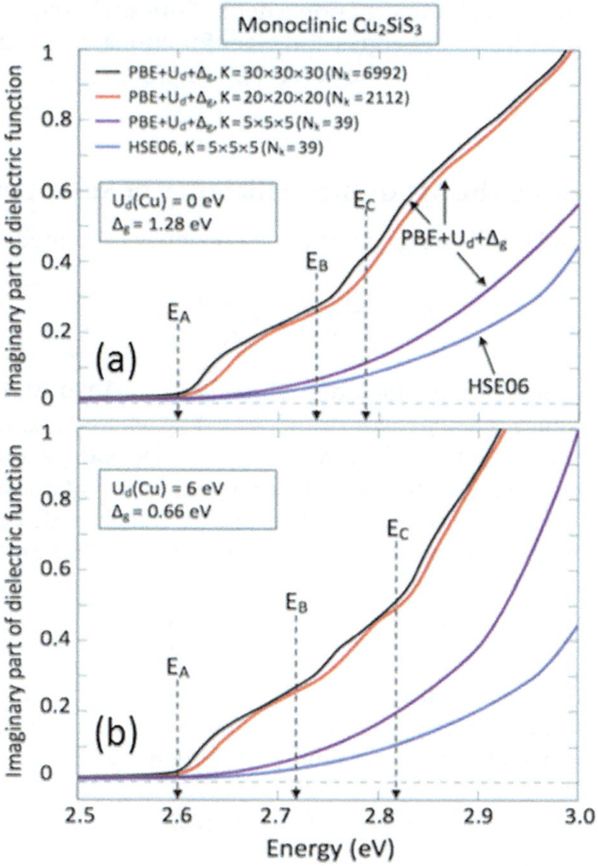

Figure 5.4 The imaginary part $\varepsilon_2(E)$ of the dielectric function for monoclinic Cu_2SiS_3 with the HSE06 and gap-corrected PBE+U computational approaches. All spectra have an onset at $E_g = E_A = 2.60$ eV, determined from the HSE06 calculation. First, we demonstrate that the PBE [*i.e.*, with no on-site Coulomb potential, thus U_d(Cu) = 0 eV; see (a)] generates qualitatively the same near-gap spectra as PBE+U [with U_d(Cu) = 6 eV; see (b)] near the band edge energy. Second, with a sparse 5 × 5 × 5 **k**-mesh (implying N_k = 39 **k**-points in the IBZ) both HSE06 (light blue lines) and PBE+U (purple lines) show qualitatively the same spectra with no indication of a double onset near E_g. Third, with a denser 20 × 20 × 20 **k**-mesh (2121 **k**-points; red lines) and 30 × 30 × 30 **k**-mesh (6992 **k**-points; black lines) the double onset is clear and the contributions from the three transitions E_A = 2.60 eV, E_B = 2.76 eV and E_C = 2.83 eV can be observed. Reproduced from ref. 28 with permission from Elsevier, Copyright 2017.

Also, in the cases where PBE+U and HSE06 generate similar spectra for any **k**-mesh size, the approach will of course generate the true HSE06 spectrum for a large **k**-mesh.

An alternative approach would be to use a much less computationally demanding **k·p** method[71] (or, even better, the modified $\mathbf{k}\cdot\tilde{\mathbf{p}}$ approach[72]) to post-calculate the optical properties on a dense **k**-mesh, based on the results (*i.e.*, the KS single-electron energies and eigenfunctions) from HSE06 with a sparse mesh. That approach is, however, not employed in here (apart from the graph in Figure 5.3).

5.2.3 Modelling the Maximum Efficiency of Solar Cells

The maximum solar cell energy conversion efficiency is defined as

$$\eta_{\max} = \frac{P_{\text{max-out}}}{P_{\text{in}}} = \frac{J_{\text{opt}} V_{\text{opt}}}{P_{\text{sun}}} \tag{5.10}$$

where P_{in} is the Sun's total irradiance which is $P_{\text{sun}} \approx 1000$ W m^{-2}, and the maximum efficiency is reached for $P_{\text{max-out}} = J_{\text{opt}} \cdot V_{\text{opt}}$, where J_{opt} and V_{opt} are the current per unit area (current density) and the device voltage, optimised for maximum power. We model the efficiency where the current density J and the device voltage V are balanced within the Shockley ideal diode equation,[73–75] involving current from absorption-induced excitations that are reduced by charge recombination:

$$J = J_{\text{sc}} - J_{\text{r}}^{0} \cdot q \cdot \left(\mathrm{e}^{qV/k_{\text{B}}T_{\text{c}}} - 1\right), \tag{5.11}$$

where q is here the elementary charge, k_{B} is the Boltzmann constant, and T_{c} is the device temperature, by default set to 300 K. The first term on the right hand side is the maximum short circuit current density

$$J_{\text{sc}} = q \cdot \int A(E) \cdot I_{\text{sun}}(E) \mathrm{d}E. \tag{5.12}$$

Here, it is considered the absorption of photons with energy E in a single-junction cell as described by SQ theory,[73] assuming air mass 1.5 solar spectral irradiance $I_{\text{sun}}(E)$ and an absorptivity

$$A(E) = 1 - \mathrm{e}^{-2 \cdot \alpha(E) \cdot d} \tag{5.13}$$

in the absorber film with thickness d. The factor 2 in the exponential arises from the assumption that the light can be either absorbed from the incoming sunlight direction or also reflected from the back surface, thus the effective layer thickness is $2d$. The second term in eqn (5.11) is the thermal recombination current J^{0}_{r} which can involve both radiative and non-radiative recombination:

$$J^{0}_{\text{r}} = J^{0}_{\text{rad}} + J^{0}_{\text{non-rad}} = J^{0}_{\text{rad}} + J^{0}_{\text{aug}}, \tag{5.14}$$

$$J_{\text{rad}}^{0} = \frac{1}{(2\pi \cdot c)^{2} \hbar^{3}} \cdot \int A(E) \cdot \frac{E^{2}}{\mathrm{e}^{E/k_{\text{B}}T_{\text{c}}} - 1} \mathrm{d}E, \tag{5.15}$$

$$J^{0}_{\text{aug}} = q \cdot C_{\text{a}} \cdot d \cdot n^{2}_{\text{i}} N_{\text{A}}. \tag{5.16}$$

The radiative recombination J^0_{rad} depends on the black-body spectrum of the device. The non-radiative recombination is here included in terms of the Auger recombination J^0_{aug} for a p-type absorber, and the effect depends mainly on the intrinsic free carrier concentrations n_i and the acceptor concentration.[74] Since we analyse the maximum efficiency of the materials, recombination *via* defects is not considered in this study. C_a is the total ambipolar Auger coefficient, which is rather independent of material (typically 10^{-32} – 10^{-30} cm^6 s^{-1} and is of the order of 10^{-31} for indirect transition in Si and Ge and GaAs).[76–79] We want to estimate the Auger effect using an approximate model that is material independent, and we therefore set $C_a = 10^{-31}$ cm^6 s^{-1} for all materials considered. The acceptor concentration is set to $N_A = 10^{16}$ cm^{-3}, which is a normal doping concentration. The square of the intrinsic carrier concentration can be expressed as $n_i^2 = n_a^2 \exp(E_g/k_BT_c)$, where $n_a = 2[m_a k_B T_c/(2\pi\hbar^2)]^{3/2}$ with the effective DOS mass $m_a = [(\gamma_e^{2/3} m_e^{DOS})(\gamma_h^{2/3} m_h^{DOS})]^{1/2}$, where γ_e is the number of CBM and m_e^{DOS} is the effective DOS mass of a single CBM (and corresponding for the holes in the VBM). Again, to construct a material-independent expression, we set $m_a = 1m_0$, which is a good value for indirect material such as Si (with six equivalent CBM) and somewhat overestimated for Ge (with three equivalent CBM). The value is, however, much larger than for direct materials, like GaAs and CIGS, but since the Auger effect is much weaker for these materials, using $m_a = 1m_0$ also for these direct gap materials does not affect the results in this work.

Within the SQ limit, an infinite thick solar cell absorber has the theoretical maximum efficiency of $\eta_{max} \approx 33\%$ for a direct band gap energy of $E_g = 1.2$–1.4 eV (Figure 5.5a), independent of absorber material. Since all light is absorbed in the infinitely thick layer (*i.e.*, $A(E) = 1$ in eqn (5.15)), the maximum efficiency η_{max} depends only on E_g. Thus, any material with proper gap energy

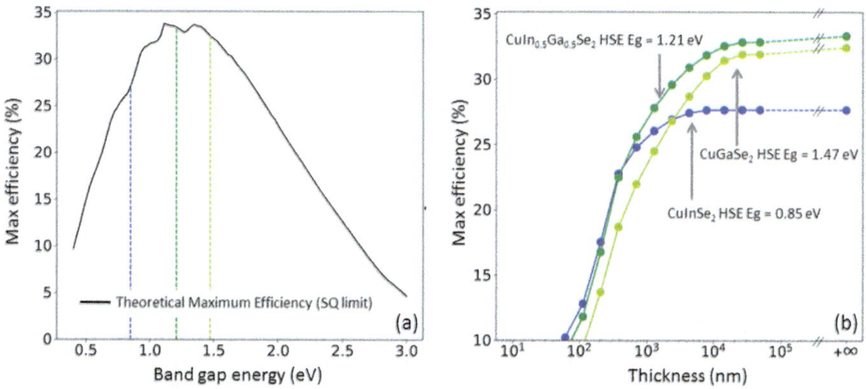

Figure 5.5 (a) The Shockley–Queisser (SQ) limit as a function of the band gap energy, E_g, for an infinitely thick solar cell absorber; thus $d \rightarrow \infty$. The maximum efficiency, η_{max}, is obtained for $E_g = 1.2$–1.4 eV. (b) The thickness-dependent maximum efficiency, η_{max}, for $CuIn_{1-x}Ga_xSe_2$ (≡ CIGS) for x = 0.0, 0.5 and 1.0 with the gap energies E_g = 0.85, 1.21 and 1.47 eV, respectively. Notice the logarithmic scale for the thickness (*i.e.*, the x-axis). Notice that the efficiencies drop substantially around 500–800 nm.

is, in this model, a suitable solar cell material. This is no longer true for solar cell structures with finite thickness d, where $A(E) < 1$. This is illustrated by the thickness-dependent η_{max} for $CuIn_1Se_2$, $CuIn_{0.5}Ga_{0.5}Se_2$, and $CuGaSe_2$ in Figure 5.5b. For sufficiently large d, the efficiencies are about η_{max} = 27%, 33%, and 32%, respectively, only as a consequence of the different gap energies; here, the alloy $CuIn_{0.5}Ga_{0.5}Se_2$, with E_g = 1.21 eV, has the largest η_{max}. However, for thinner absorber thicknesses, the efficiencies of the three compounds do not exactly follow each other because the actual energy dependence of the spectra of absorption coefficients $\alpha(E)$ differ for the three compounds. Furthermore, the drop in maximum efficiency at around 500–1000 nm is rather characteristic for traditional thin-film absorbers, and efficiency is often rather poor for $d < 500$ nm. Thus, in order to design materials with high efficiency for thickness $d < 100$ nm, special requests for materials with large low-energy absorption coefficients are needed.

5.3 Electronic and Optical Properties of the Cu-based Chalcogenides

$Cu_2ZnSn(S,Se)_4$ (*i.e.*, CZTSSe) has attracted increasing attention as it is an indium-free alternative to $Cu(In,Ga)Se_2$ (CIGS) for which the costly indium is an obstacle in large-scale commercial production. The first CZTS-based solar cell was made in 1988 by Ito and coworkers,[8] followed by more efficient devices by Friedlmeier *et al.*[9] and Katagiri *et al.*[80] Despite promising developments in efficiency, even higher efficiencies are required for CZTS to be a viable alternative to CIGS. Sulfur-based CZTS has a band gap energy of $E_g \approx 1.5$ eV, while the gap energy of selenium-based CZTSe has a gap energy of ~1.1 eV, and with Se–S alloying/grading the cell performance and the efficiency are optimised with respect to optical absorption. While ongoing extensive research on CZTSSe reveals interesting results, other Cu-based chalcogenides as alternatives to CIGS and CZTSSe are regarded as emerging materials. The basic crystalline structures of a variety of emerging Cu-based chalcogenides are described in Figure 5.1 and Table 5.1.

5.3.1 CZTS Cation Alloys

There is a search for alternative alloys to CZTSSe that can both stabilise the crystalline structure, with less native defects, and at the same time provide proper band gap energy and optoelectronic properties. Alloying in order to vary the gap energy can also be used to design device structures with a graded band profile of the absorber. In addition, alloying in order to increase the gap energy is a possibility for optimising the material for a top-cell absorber in tandem solar cells, where the band gap energy of the top-cell should be around 2 eV. In these cases, isovalent alloying is a natural choice for alteration that often preserves the crystalline structure and does not induce localised (in-gap) defect states. A potential example is to substitute the Sn cation in CZTSe or CZTS with the isovalent group IV elements

Table 5.1 Crystal structure parameters of the considered copper-based chalcogenides: space group names, lattice parameters (a, b, c, and non-trivial β angle) and cation–anion bond lengths δ(cation X-anion), as obtained from calculations with the HSE06 exchange-correlation potential. The Cu_2XSnS_4 (X = Mn, Fe and Ni) compounds have antiferromagnetic configurations that slightly distort the crystalline structure.

Compound	Space group	a (Å)	b (Å)	c (Å) β (°)	Bond lengths δ(X–anion) (Å)
$CuGaSe_2$	D_{2d}^{12}; 122	5.610	5.610	11.000	Cu: 2.42; Ga: 2.41
$CuInSe_2$	D_{2d}^{12}; 122	5.833	5.833	11.649	Cu: 2.42; In: 2.59
$Cu_2ZnSnSe_4$	S_4^2; 82	5.740	5.740	11.380	Cu: 2.45; Zn: 2.47; Sn: 2.55
Cu_2ZnSnS_4	S_4^2; 82	5.457	5.457	10.827	Cu: 2.33; Zn: 2.35; Sn: 2.41
Cu_2ZnGeS_4	S_4^2; 82	5.355	5.355	10.512	Cu: 2.33; Zn: 2.34; Ge: 2.23
Cu_2ZnSiS_4	S_4^2; 82	5.307	5.307	10.625	Cu: 2.35; Zn: 2.36; Si: 2.15
Cu_2CdSnS_4	D_{2d}^{11}; 121	5.622	5.622	10.819	Cu: 2.34; Cd: 2.54; Sn: 2.41
Cu_2BeSnS_4	S_4^2; 82	5.263	5.263	10.373	Cu: 2.28; Be: 2.11; Sn: 2.43
Cu_2MgSnS_4	D_{2d}^{11}; 121	5.489	5.489	10.731	Cu: 2.28; Mg: 2.45; Sn: 2.44
Cu_2CaSnS_4	D_{2d}^{11}; 121	5.809	5.809	10.253	Cu: 2.32; Ca: 2.69; Sn: 2.42
Cu_2SrSnS_4	C_3^2; 144	6.304	6.304	15.631	Cu: 2.30–2.37; Sr: 3.13–3.22; Sn: 2.40
Cu_2BaSnS_4	C_3^2; 144	6.401	6.401	15.868	Cu: 2.34–2.37; Ba: 3.23–3.25; Sn: 2.40
$Cu_2SrSnSe_4$	C_{2v}^{16}; 40	11.048	10.912	6.713	Cu: 2.43–2.53; Sr: 3.23–3.33; Sn: 2.54
$Cu_2BaSnSe_4$	C_{2v}^{16}; 40	11.180	11.371	6.779	Cu: 2.45–2.52; Ba: 2.37–2.44; Sn: 2.54
Cu_2MnSnS_4	D_{2d}^{11}; 121	5.473	5.473	10.727	Cu: 2.29; Mn: 2.44; Sn: 2.43
Cu_2FeSnS_4	D_{2d}^{11}; 121	5.432	5.432	10.603	Cu: 2.28; Fe: 2.35; Sn: 2.43
Cu_2NiSnS_4	S_4^2; 82	5.464	5.464	10.398	Cu: 2.30; Ni: 2.28; Sn: 2.45
Cu_2O	O_h^4; 224	4.237	4.237	4.237	Cu: 1.83
Cu_2SnS_3	C_s^4; 9	6.664	11.574	6.682 $\beta = 108.85$	Cu: 2.31, 2.40 Sn: 2.30, 2.48
Cu_2GeS_3	C_s^4; 9	6.445	11.317	6.439 $\beta = 108.42$	Cu: 2.30, 2.37 Sn: 2.18, 2.32
Cu_2SiS_3	C_s^4; 9	6.338	11.224	6.287 $\beta = 108.11$	Cu: 2.30, 2.38 Si: 2.09, 2.22
Cu_4SnS_4	D_{2h}^{16}; 62	14.486	7.874	5.557	Cu: 2.24, 2.29, 2.39, 2.43, 2.50 Sn: 2.43, 2.47, 3.08

(*continued*)

Table 5.1 *(continued)*

Compound	Space group	a (Å)	b (Å)	c (Å) β (°)	Bond lengths δ(X–anion) (Å)
Cu_4GeS_4	D_{2h}^{16}; 62	14.204	7.615	5.608	Cu: 2.21, 2.25, 2.38, 2.48, 2.52 Ge: 2.22, 2.28
Cu_4SiS_4	D_{2h}^{16}; 62	14.030	7.475	5.650	Cu: 2.20, 2.23, 2.26, 2.29, 2.38, 2.48, 2.53 Si: 2.09, 2.13, 2.18
Cu_4GeSe_4	D_{2h}^{16}; 62	14.705	8.235	5.695	Cu: 2.33, 2.40, 2.45, 2.50, 2.61, 2.53, 2.66 Ge: 2.40, 2.45
Cu_4SiSe_4	D_{2h}^{16}; 62	14.591	8.061	5.784	Cu: 2.32, 2.40, 2.43, 2.36, 2.51, 2.57, 2.61, 2.73 Si: 2.26, 2.29, 2.35
Cu_3SbS_3	D_2^4; 19	8.221	10.139	6.662	Sb: 2.46; Cu: 2.26–2.33
Cu_3SbSe_3	D_2^4; 19	8.383	10.740	6.866	Cu: 2.37, 2.39, 2.41, 2.40, 2.45, 2.58, 2.61 Sb: 2.58, 2.61
Cu_3BiS_3	D_2^4; 19	7.790	10.416	6.762	Cu: 2.26, 2.27, 2.28, 2.31, 2.34, 2.57, 2.59 Bi: 2.57, 2.59
$CuSbS_2$	D_{2h}^{16}; 62	6.031	3.993	14.060	Cu: 2.32, 2.41; Sb: 2.41, 2.56, 3.06
$CuSbSe_2$	D_{2h}^{16}; 62	6.346	4.204	14.614	Cu: 2.44, 2.52; Sb: 2.56, 2.69, 3.16
$CuSbTe_2$	D_{2h}^{16}; 62	6.803	4.522	15.327	Cu: 2.59, 2.66; Sb: 2.89, 2.79, 3.33
$CuBiS_2$	D_{2h}^{16}; 62	6.199	4.087	14.115	Cu: 2.32, 2.43; Bi: 2.65, 3.09
$CuBiSe_2$	D_{2h}^{16}; 62	6.491	4.303	14.700	Cu: 2.43, 2.53; Bi: 2.67, 2.79, 3.20
$CuBiTe_2$	D_{2h}^{16}; 62	6.954	4.621	15.232	Cu: 2.59, 2.68; Bi: 2.88, 2.30, 3.37
$CuBi_2O_4$	D_{4h}^{8}; 130	8.635	8.635	5.845	Cu: 2.01; Bi: 2.14, 2.30, 2.80

Ge or Si, thus forming the $Cu_2Zn(Sn,Si/Ge)(S/Se)_4$ alloys. Incorporation of Ge or Si in CZTSSe has been proposed as one method to increase the solar cell efficiency as well as the crystalline stability of high-quality CZTSSe.[14–20] It has been shown that the presence of Ge in CZTSe (*i.e.*, Se as the anion) can tune the gap energy towards the optimal value of ~1.3 eV[15,17,19] with an improved stability of the absorber layer in solar cells.[16] Alloying CZTSe with Si should require less alloying content (estimated to x ~0.3, including correction of HSE06 gap energy) to reach $E_g \approx 1.2$ eV compared with alloying with Ge, which requires a much larger alloying concentration ($x \sim 1$) to reach a similar gap energy. There is a lack of data regarding the expected solar cell efficiency for Si alloying of CZTS and CZTSe, which may indicate a lower crystalline quality due to too strong a relaxation effect, since Si is a much smaller atom than Sn (and Ge). Calculations by Shu *et al.*[17] reveal that formation enthalpies of the Ge-based alloys are smaller than those of Si-based alloys, which is also confirmed by our calculated mixing energy for $x = 0.5$, obtained from the total energies

$$\Delta E_{tot}(x) = E_{tot}(x) - [(1 - x)\, E_{tot}(0) + xE_{tot}(1)]. \tag{5.17}$$

This is presented in Figure 5.6a, upper panel. One can notice that the mixing energy for Sn–Ge alloying is roughly half of that for Sn–Si alloying, and only about 2 meV per atom. Therefore, more investigations have been presented for Sn–Ge alloying than for Sn–Si alloying.

In Figure 5.6a, lower panel, we present the variation of the fundamental band gap energy E_g with respect to the alloy composition for $Cu_2Zn(Sn,Ge/si)(S/Se)_4$. HSE06 underestimates the gap energies by roughly 0.2 eV. Overall, the results in Figure 5.6a agree with earlier first-principles calculations performed by Shu *et al.* All compounds have direct band gaps, and rather similar electronic band structures. The trend is that the gap energy increases almost linearly when the Si/Ge content increases:[21] the largest change in E_g is seen for the Si-based alloy, as Si normally generates larger gap energies compared with both Ge and Sn. However, there is a small non-linear contribution to the

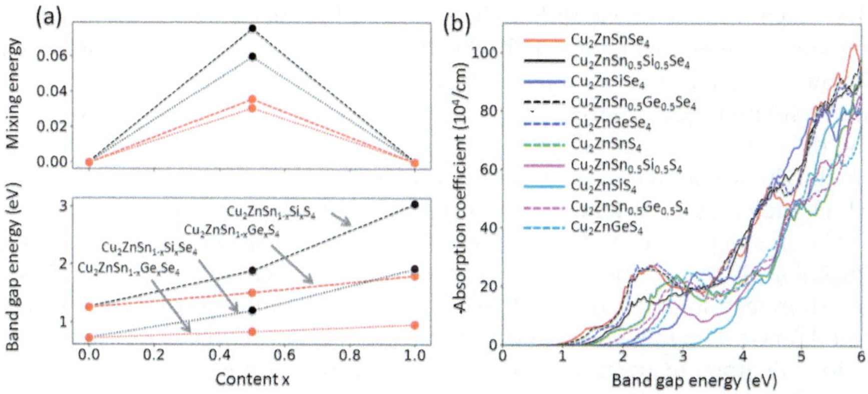

Figure 5.6 (a) Upper panel shows the mixing energy (eV/cell where cell has 16 atoms) of $Cu_2ZnSn_{1-x}Ge_xS_4$, $Cu_2ZnSn_{1-x}Si_xS_4$, $Cu_2ZnSn_{1-x}Ge_xSe_4$ and $Cu_2ZnSn_{1-x}Si_xSe_4$ as functions of alloy compositions x = 0, 0.5 and 1 computed by using the HSE06 hybrid functional. The lower panel presents the band gap energies E_g, where HSE06 underestimates the gap energies by about 0.1–0.4 eV for all compounds compared with experimental data. (b) Corresponding absorption coefficients $\alpha(E)$.

variation of the gap energies, which can be described with a bowing coefficient, b, in the quadratic function

$$E_g(x) = (1 - x)E_g(0) + xE_g(1) - b\cdot(1 - x)x \tag{5.18}$$

The fitted bowing parameters are about 0.18 eV for the $Cu_2Zn(Sn,Ge)(S/Se)_4$ alloys, and 0.65 or 0.77 eV for the $Cu_2Zn(Sn,Si)(S/Se)_4$ alloys, when also considering more alloy compositions.[21] This implies that the contribution from the quadratic term is at most 0.05 and 0.19 eV for the Sn–Ge and Sn–Si alloys, respectively. The bowing parameters are about twice that of previously reported values;[17] however, since the bowing parameters are small the effect on the composition-dependent band gap energy is moderate. Small bowing parameters indicate good miscibility, especially for Ge-based alloys, which again indicates that it is easier to obtain crystalline Sn–Ge alloying compared with the corresponding Sn–Si alloying.

Intriguingly, the gap energies for unalloyed $Cu_2ZnSi(S/Se)_4$ compounds are significantly larger than for the corresponding Si-rich alloys ($x \approx 1$) of $Cu_2Zn(Sn,Si)(S/Se)_4$. Thus, the quadratic formula to describe the composition dependence of $E_g(x)$ is not valid for ($x \approx 1$); this is specific for the Si-containing alloys, and not seen for the Ge-containing alloys. The origin of this effect is that the small content of Sn, in the otherwise unalloyed $Cu_2ZnSi(S/Se)_4$ compounds, will not form perfect alloy properties of CBM and VBM by hybridising with Si. Instead, Sn will form more localised Sn-like defect states near the CBM.[21] A higher concentration of Sn (larger than ~5% or $x < 0.95$) is required before Sn and Si form a more unified alloy.

The calculated absorption coefficients, $\alpha(E)$, for the alloys demonstrate that all compounds have rather similar optical properties apart from different gap energies; see Figure 5.6b. The high-frequency dielectric constant, ε_∞, follows the trend that a larger energy gap implies a smaller dielectric constant. Si-based quaternaries, especially, have a large E_g and small ε_∞. However, the differences in the dielectric constants between the alloys are relatively small. The difference in the static and high-frequency constant is $\varepsilon_0 - \varepsilon_\infty \approx 3$, which indicates a moderate and comparable ionicity for all compounds.

The results from these studies suggest that cation Sn–Ge alloying $Cu_2Zn(Sn,Ge)Se_4$ can be utilised to optimise the gap energy of $E_g \approx 1.3$ eV for thin-film technologies, but the theoretical analysis indicates that the effect is small for Ge alloying, and Si alloying may induce more defects. Furthermore, anion alloying of $Cu_2ZnSi(S,Se)_4$ is preferable for optimising the gap energy to ~2 eV for a top-cell material in tandem structures.

5.3.2 CZTS-like Compounds

For many of the Cu-based chalcogenides, the copper vacancy V_{Cu}^- is a shallow native acceptor with a very low formation energy, and in Cu-poor films vacancies can easily be formed in high concentration together with the compensated single donor. In CZTSSe, the most prominent single donor (together with copper interstitial Cu_i^+) is the cation antisite Zn_{Cu}^+. Thus, the compensated Cu vacancy is a two-atom complex (V_{Cu}^- + Zn_{Cu}^+) with a low formation energy. In addition, antisite defect pairs (ASPs) involving Zn_{Cu}^+ and the corresponding shallow antisite acceptors Cu_{Zn}^- can be formed without a change in crystalline stoichiometry.[81,82] This ASP from cation substitution (Cu_{Zn}^- + Zn_{Cu}^+) also has a low formation energy in CZTSSe, and it can appear in high concentrations (of the order of a percentage). It is expected that such Cu–Zn disorder induces band gap energy fluctuations in the material,[83] and it is thus important to investigate how to reduce the concentration of ASPs in CZTSSe. Yuan *et al.*[84] have suggested that Zn should be substituted by the large-size atom Cd, alternatively Cu should be substituted by Ag, to suppress the formation of antisite defects. They found that while Cd_{Cu}^+ in Cu_2CdSnS_4 had as low a formation energy as Zn_{Cu}^+ in CZTS, the corresponding antisite Zn_{Ag}^+ in Ag_2ZnSnS_4 had a significantly much higher formation energy. Therefore, investigations of CXTSSe (*i.e.*, where Zn is replaced by the element X) are highly relevant, both in order to understand quaternary Cu-based chalcogenides in general, and to develop a material that it more resistant to formation of ASPs and, thereby, a material that will be less affected by local variations in the band gap energy. Some of these CXTSSe compounds can be alternative emerging candidates as thin-film PV materials. Studies are also expected to reveal a fundamental understanding of the synthesis processes and the crystal stability of the materials, as well as of the defect physics of these materials.

The electronic properties of Cu_2XSnS_4 (CXTS), where X is a 3d transition metal or a group II alkaline earth metal (herein, X = Mn, Fe, Ni, Be, Mg, Ca, Sr, or Ba) are rather similar to those of CZTS (Figure 5.7). The CXTS compounds are analysed typically for the tetragonal kesterite (KS; space group S_4^2) or stannite (ST; space group D_{2d}^{11}) crystalline structures, but when those compounds are not thermodynamically stable, other crystalline structures (*e.g.*, trigonal phase) have been found to be energetically favourable. For Cu_2MnSnS_4, Cu_2FeSnS_4, and Cu_2NiSnS_4 with magnetic elements, we have also considered ferromagnetic and different antiferromagnetic configurations, and the most stable magnetic configuration is an antiferromagnetic phase described in ref. 85. The fundamental band gap energy E_g of Cu_2MnSnS_4 is experimentally estimated to be between 1.18 and 1.63 eV, depending on the heat treatment, and a solar cell based on Cu_2MnSnS_4 has an efficiency of 0.49%.[86] The band gap energies of synthesised Cu_2FeSnS_4 nanocrystals, Cu_2CoSnS_4, and Cu_2NiSnS_4 nanoparticles are ~1.33 or ~1.5 eV,[87,88] ~1.40,[89] and ~1.40 eV,[90,91] respectively. Moreover, kesterite Cu_2MgSnS_4 nanoparticles without a secondary phase have been synthesised using a hot-injection method.[92] From GGA calculations, Wang *et al.*[93] have found that kesterite Cu_2CdSnS_4 and Cu_2HgSnS_4 are thermodynamically stable, but kesterite Cu_2MgSnS_4 and especially kesterite Cu_2CaSnS_4 are not stable. Zhong *et al.*[94] found that stannite Cu_2MgSnS_4 (and $Cu_2MgSnSe_4$) are thermodynamically stable, but, similarly to Wang *et al.*, that kesterite and stannite phases of Cu_2CaSnS_4 (and $Cu_2CaSnSe_4$) are unstable with respect to competing compounds. Recent developments of devices based on emerging Cu_2BaSnS_4 (or Cu_2SrSnS_4) compounds and anion

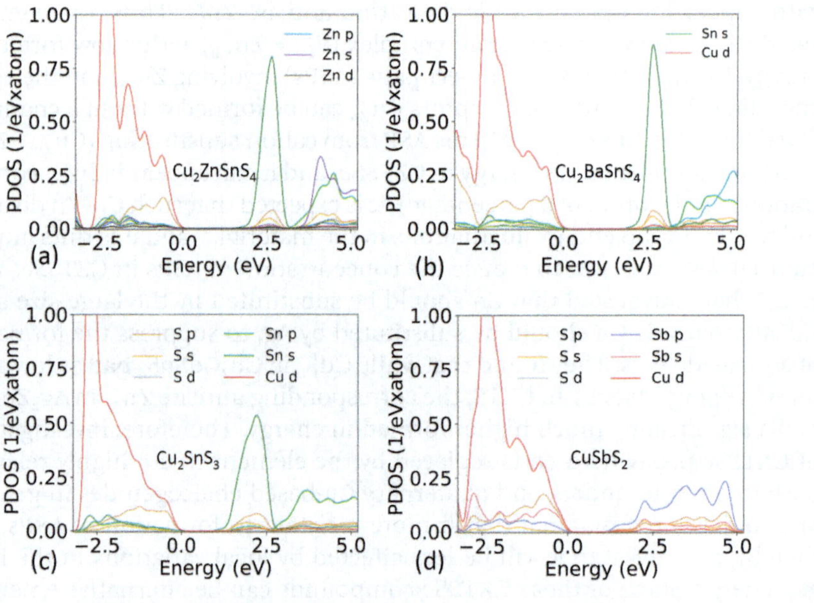

Figure 5.7 The atom and angular momentum-resolved DOS (*i.e.*, the PDOS) obtained with HSE06 and presented with a 0.1 eV post-calculated Lorentzian smearing and where the energy scales refer to the VBM. The PDOS has been weighted by $1/(2l+1)$ for better visibility of s- and p-like states. (a) Cu_2ZnSnS_4, (b) Cu_2BaSnS_4, (or $Cu_2BaSnSe_4$), (c) Cu_2SnS_3 and (d) $CuSbS_2$.

S/Se alloying have yielded, in early attempts, a solar cell efficiency of 1.6%.[95] Experimental and theoretical characterisation reveals that these compounds with large group II elements crystallise in trigonal or orthorhombic crystalline structures.[95,96]

One can estimate the thermodynamic stability of the compounds by calculating the total energy between different crystalline configurations, in conjunction with analyses of the total energies with respect to chemical decomposition, *i.e.*, $Cu_2XSnS_4 \rightarrow [2(CuS) + XS + SnS]$, $[Cu_2S + XS + SnS_2]$, $[Cu_2SnS_3 + XS]$, or a similar mixture of compounds. We have found[85] that the total energy differences between KS and ST phases of Cu_2XSnS_4, for X = Be, Mg, Ca, Sr, and Zn are 12, 2, 30, and 3 meV per atom, respectively. Cu_2BeSnS_4 seems to be a very stable crystal with respect to decomposition. In Cu_2MgSnS_4 (and also for Cu_2ZnSnS_4) the presence of secondary ST phases in the materials is possible. For Cu_2SrSnS_4 and Cu_2BaSnS_4 neither of the two tetragonal structures are thermodynamically stable, instead, a trigonal structure (space group C_3^2) is more stable. Intriguingly, while the trigonal phases of Cu_2SrSnS_4 and Cu_2BaSnS_4 are stable with respect to decomposition, trigonal Cu_2CaSnS_4 is unstable. Thus, the trend for S-based compounds is that lighter group II alkaline earth metal elements for X form tetragonal structures, somewhat heavier elements for X yield unstable crystalline structures, and heavy elements for X prefer a trigonal structure. The corresponding Se-based compounds with heavy group II elements, *i.e.*, $Cu_2SrSnSe_4$ and $Cu_2BaSnSe_4$, are more stable in an orthorhombic structure (space group $C^{16}{}_{2v}$). However, it is worth noticing that for materials and structures that balance between being stable or not, the choice of exchange-correlation potential in the calculation can give slightly different results; our results are for HSE06 calculations. While Cu_2ZnSnS_4 crystallises in a non-magnetic KS phase, compounds with X being a 3d transition metal crystallise in ST or KS phase, with a magnetic configuration. For Cu_2MnSnS_4, the differences in total energies between the most stable ST antiferromagnetic phase and other crystalline and magnetic configurations are between 0.3 and 4 meV per atom. The larger value is from the KS ferromagnetic phase, which is then rather unfavourable. For Cu_2FeSnS_4, the total energy differences between stable ST antiferromagnetic and other phases are between 1 and 2 meV per atom; here, the larger value is from the KS antiferromagnetic phase. This may indicate that the paramagnetic phase can exist depending on growth conditions and temperature for Cu_2MnSnS_4 and Cu_2FeSnS_4. Cu_2NiSnS_4 has a more distinct structure, where the total energy differences between the most stable KS antiferromagnetic phase and the other phases are between 6 and 100 meV per atom; this large energy difference indicates that the compound has a strong magnetic phase that governs its crystalline structure. Cu_2MnSnS_4 and Cu_2FeSnS_4 are clearly stable chalcogenides. Cu_2NiSnS_4 is theoretically also a stable compound from our analysis, although the decomposition energy is smaller, and this is supported by the fact that at least nanoparticles have been synthesised.[97,98]

The HSE06 band gap energies of the most stable crystalline compounds are E_g = 1.76, 1.48, 1.84, 1.80, 1.79, 1.40, 1.71, 1.26, 1.17, and 1.48 eV for CXTS with X = Be, Mg, Ca, Sr, Ba, Mn, Fe, Ni, and Zn, respectively, as obtained from the HSE06 calculations. It can be seen that $Cu_2BaSnSe_4$ and similar compounds have similar electronic structures and DOS (Figure 5.7), and the gap energies fall in a rather narrow energy region of about 1.6 ± 0.3 eV. We expect that these HSE06 gap energies should be ~0.2 eV smaller than experimental values, and the calculated energies are then in good agreement with available measured data for CXTS with X = Mg (E_g = 1.6 eV), Sr (2.1 eV), Ba (1.9 eV), Mn (1.2–1.6 eV), Fe (1.3 or 1.5 eV), Ni (1.4 eV), and Zn (1.4 eV).[86–88,90,91,95,99–101] Interestingly, the trigonal compounds with heavier group II elements (*i.e.*, Cu_2SrSnS_4 and Cu_2BaSnS_4) have electronic structures with a more indirect gap character, which can be advantageous for absorption but disadvantageous for device efficiency. As expected, the Cu d-like states hybridise strongly with S/Se p-like states in the VBs. The CB DOS of compounds with Sn as the group IV cation element exhibits a strong and rather localised Sn s-like band at the CBM (hybridised mainly with S/Se sp-like anion state). This is also observed when alloying Sn with the group IV element Ge, while corresponding Si-based compounds for Si s-like states are higher up in energy region and hybridise more with other CB states.[21,36] The CB DOS of all CXTS (except Cu_2NiSnS_4) is comparable with that of

CZTS, at least near the CBM, because there is mainly hybridisation between the Sn s-like and the S p-like states.[85] It is also noticeable that Ca and Mn d-like states hybridise with Sn p-like states in the energy range above 3 eV. The electronic structure of Cu_2NiSnS_4 is somewhat different compared with the other compounds, since its Ni d-like states strongly hybridise with Sn s-like states near the CBM. This results in an indirect band gap, which is in contrast to the other CXTS compounds considered. However, the lowest CB is flat and the indirect band gap energy of Cu_2NiSnS_4 is only about 10 meV smaller than the direct Γ-point gap energy.

One issue with the traditional CZTS material is that disorder from Cu–Zn ASPs can introduce band gap fluctuation if the defect concentration is high, that is, if the formation energy of the defects is sufficiently small with respect to the temperatures during device operation or during the growth of the material. The formation energy of defect D is determined from

$$\Delta H(D) = E_{\text{tot}}(\text{CXTS}:D) - E_{\text{tot}}(\text{CXTS}) + \Sigma_\alpha n_\alpha \cdot \mu_\alpha, \tag{5.19}$$

where $E_{\text{tot}}(\text{CXTS}:D)$ and $E_{\text{tot}}(\text{CXTS})$ are the total energies for CXTS with and without the defect D, respectively, $n_\alpha = -1$ (+1) when one atom α is added (removed), and μ_α is the chemical potential. The chemical potential is split into two terms: the total energy per atom of the solid $\mu_{\alpha,\text{sol}}$ for element α, and the remaining term $\Delta\mu_\alpha$ that can vary within an energy range depending on the external conditions.[102] For simplicity, we choose the α-rich conditions with $\Delta\mu_\alpha = 0$, since the focus is on ASPs for which $n_\alpha = 0$ and thereby the last term in eqn (5.19) is zero. A positive (negative) value of $\Delta H(D)$ implies an energy cost (gain) to form the defect. The formation energy of a compensated Cu vacancy is known to be small.[102] Also, the formation energy of the ASPs ($Cu_{Zn}^- + Zn_{Cu}^+$) is small, and our calculated value is 0.30 eV; see Table 5.4. Notice that the compound can contain high concentrations of ASP defects, making the structure disordered, although the stoichiometry remains the same as the perfect crystalline material.

For all compounds, the V_{Cu} and Cu_X defects induce rather shallow acceptor states at the VBM, and thus the Fermi level E_F is below the VBM. V_{Cu} is a very delocalised acceptor that has a minor effect on the VB states and acts almost as a hole carrier dopant. The X_{Cu} defect induces shallow donor states with a Fermi level in the CB, and the donor shifts the CBM energetically downwards, thus narrowing the band gap. Both Cu_X and X_{Cu} narrow the gap. The calculated change in the band gap energy of CZTS in Table 5.4 is similar to that reported before,[83,103] where the band gap energies are reduced by ~0.1 eV for Zn_{Cu}, and increased by ~0.1 eV for V_{Cu}. From the defect formation energies, we find that all considered CXTS have rather similar formation energy of V_{Cu} as for CZTS in the Cu-rich limit, indicating that it is easy to form non-stoichiometric materials. Moreover, the $\Delta H(D)$ of the other single point defects are relatively small (see also ref. 81), and an oppositely charged point can be compensated to form neutral complexes, such as (V_{Cu} + X_{Cu}) and (Cu_X + X_{Cu}). We find that the defect formation energies of (Cu_X + X_{Cu})

are small, and for Cu_2MgSnS_4, Cu_2MnSnS_4, Cu_2FeSnS_4, and Cu_2NiSnS_4, the defect formation energy is about 0.3 eV, which is similar to that for CZTS. We also confirm the earlier published results[82] that the ASP in Cu_2CdSnS_4 can be formed as easily as in CZTS despite the larger Cd atom. However, Cu_2BeSnS_4 has a relatively high defect formation energy of 0.58 eV. One reason for the larger $\Delta H(D)$ is that the atom size of Be is much smaller than the sizes of both Cu and Sn, and the relaxation effect is therefore stronger. This increase in the energy may seem to be small, but the relative change is a factor of two and the formation energy is only ~10 times k_BT for relevant temperatures. Since the defect concentration depends exponentially on the defect formation energy, *i.e.*, $[N(D)] \propto \exp(-\Delta H(D)/k_BT)$, small variations in $\Delta H(D)$ have a large effect on $[N(D)]$. We estimate that the defect concentration $[N(Cu_X + X_{Cu})]$ is as much as ~10^4 times lower in Cu_2BeSnS_4 than in CZTS and most of the other CXTS. Cu_2CaSnS_4 in its tetragonal phase has a very low formation energy for the $(Cu_X + X_{Cu})$ defect (0.13 eV), which can indicate an instability of this phase. For the corresponding trigonal phase, the corresponding defect formation energy is much larger (1.40 eV), and also the $(V_{Cu} + X_{Cu})$ defect has about 0.6 eV larger formation energy in the trigonal phase. This suggests that it should be possible to grow trigonal Cu_2CaSnS_4 with much fewer native defects compared with the other tetragonal Cu_2XSnS_4 compounds. Moreover, the compounds with the much larger Sr or Ba elements will discourage the formation of $(Cu_X^- + X_{Cu}^+)$ antisite disorder, at least for their trigonal crystalline structures.[96]

Since the $(Cu_X + X_{Cu})$ defect narrows the gap and clustering enhances this effect, non-uniform distributions or concentrations of Cu-X disorder may imply spatial fluctuation of the band gap energy in the material. However, the band gap energy can be stabilised by incorporating other neutral complexes that increase the gap energy, such as$(V_{Cu} + X_{Cu})$. Here, clustering of neutral $(Cu_X + X_{Cu})$ with neutral $(V_{Cu} + X_{Cu})$ can occur due to relaxation effects and preferred octet formation, as modelled by Huang *et al.*[103] Thus, a delicate synthesis process may form a material with a spatially uniform band gap, despite a huge concentration of native defects and complexes.

We can conclude that most CXTS compounds have a low formation energy for the compensated Cu vacancy $(V_{Cu} + X_{Cu})$ and the ASP $(Cu_X + X_{Cu})$. From this initial analysis there seems to be very little gain regarding the native defects in developing alternative (and environmentally friendly) tetragonal CXTS compounds. Further analysis of energy barriers to forming defects, as well as better understanding of the synthesis mechanism could, however, support approaches to design a material with lower concentrations of ASPs or a balance of defects that compensates the effects on the electronic structure. For tetragonal Cu_2BeSnS_4, however, the formation energy is sufficiently high, and we expect of the order of 10^4 times lower concentration of the defects at room and operating temperatures than for the other compounds. The disadvantage of ASPs involving atoms with different radii is the stronger relaxation effect that can result in a larger band gap narrowing for a given defect concentration. Moreover, Be is a toxic element and difficult to handle during the synthesis process, which makes it less favourable. CXTS compounds with Sr and Ba, *i.e.*, trigonal structures of Cu_2BaSnS_4 and Cu_2BaSnS_4, are an interesting class of large gap compounds for further developments. S/Se anion alloying can lower the band gap energy to the range $E_g \approx 1.5$–2.1 eV; however, disorder may occur due to the orthorhombic structure for Se-rich compounds.

5.3.3 Group II-free Compounds

Another way of avoiding formation of ASPs (Cu_{Zn} + Zn_{Cu}) and compensated Cu vacancies (V_{Cu} + Zn_{Cu}) is to design materials that exclude elements with two sp-like valence electrons, like the group II elements Zn, Mg, *etc.*, or the 3d transition metals Mn, Fe, *etc.* For example, ternary Cu-based group IV or group V chalcogenides and their alloys can be considered. In such compounds, there is no natural native single donor that bonds easily with the single acceptor.

From that perspective, monoclinic Cu_2SnS_3 (CTS), with Sn as the group IV element, is a potential emerging material. The compound has already been shown to give reasonable solar cell conversion efficiencies of 4.63% and 4.29%.[27,104] The main potential advantage of CTS over the quaternary absorbers CIGS and CZTS is a relatively broad single-phase region and simpler fabrication due to fewer chemical constituents.[26] CTS can form with different crystal structures, such as tetragonal, cubic, monoclinic, and triclinic phases, but all the commonly observed crystal structures are based on the same zinc-blende pattern with tetrahedral coordination.[105] Similar to the traditional Cu-based chalcogenides, the Cu d-like states dominate the VB DOS and hybridise strongly with S/Se p-like states, and, similarly to CXTS, the energetically lower CB region is dominated by Sn s-like states that hybridise with S/Se sp-like states (Figure 5.7).

Intriguingly, monoclinic CTS exhibits a so-called "double absorption onset", with a first photon absorption onset at the band gap energy E_g = 0.90–0.93 eV and a second distinct onset at about E_g + 0.1 eV.[24,27,106] It has been demonstrated experimentally that the double onset is an intrinsic feature of CTS and it does not arise from the presence of other CTS phases or other compounds.[107] However, it was puzzling that the earlier theoretical study of the electronic structure and the optical properties of intrinsic CTS did not report this double onset.[105,108–110] The possibility that the phenomenon was a consequence of the high concentration of defects in the CTS thin films was discussed. By combining spectroscopic ellipsometry with first-principles DFT calculations of the optical properties, we have instead demonstrated[28] that the double onset originates from direct transitions at the Γ-point from three energetically close-lying VBs to a single CB in the defect-free CTS. The double onset can be predicted by DFT calculations only when using a very dense **k**-mesh to reveal the fine details in the band structure. Hence, structural imperfection, such as a secondary phase, is not needed to explain the absorption spectrum. With the large **k**-mesh, the shape of both the real and the imaginary parts of the dielectric function are improved considerably in the low-energy region (*i.e.*, below 1.5 eV). The sensitivity of the calculations is such that the linear tetrahedron integration method is not sufficiently good to capture small variations in the parabolic shape of the electronic band structures. This is demonstrated for Cu_2SiS_3 in Figure 5.4.

In principle, this inaccuracy is an issue when analysing any absorber material with several VBs, but in most materials the measurements do not reveal multiple distinct onsets. The difference for CTS is that the material has a somewhat special band structure near the band edges, especially at the VBM; see Figure 5.8. One observes that the material has a direct gap at the Γ-point. The energy state of the lowest CB is rather normal with a clear energy dispersion and the CBM has the irreducible representation Γ_1 of the C_2 point group. Also, the VBM energy state has Γ_1 symmetry. This describes the allowed polarisation of the allowed electric dipole-induced transitions for that **k**-state, considering that the *y*-direction is the crystalline symmetry rotation axis for the unique axis *b*. Since the topmost VB is very flat along the (010) direction (and also along the (100) direction; not displayed), the absorption of photons can be large near the gap energy, *i.e.*, $E_g \leq E < E_g + 0.5$ eV. Moreover, similarly to CZTS the lowest CB has a strong contribution of Sn s-like character, while the uppermost VB has primarily Cu d_{z2}-like and S p_z-like character. Along the (010) direction ((001) direction), the CB contains also S p_y-like (S p_z-like) character. This will generate an anisotropy of the dielectric response with a strong and in-plane *z*-polarised dielectric response for $E_g \leq E < E_g + 0.15$ eV. The second uppermost VB has Γ_2 symmetry at the Γ-point with primarily Cu d_{xy}-like and S p_y-like character, while the third VB has Γ_1 symmetry with primarily Cu d_{x2-y2}-like and S p_x-like character. Of the three topmost bands, it is the third VB that has the most Sn s-like character away from the Γ-point, however the contribution is small (not displayed). Noticeably, along the (001) direction, the topmost VB (with Γ_1 symmetry at VBM) and second VB (with Γ_2 symmetry) cross about 18% away from the Γ-point, while the third VB (with Γ_1 symmetry) "interacts" and bends downwards. However, the character of the VBs changes symmetry, an effect which is typical for these types of band–band interactions. Moreover, the character of the VB states strongly affects the optical transitions. In the top panel in Figure 5.8, we present the optical matrix elements

$$M^{\alpha\alpha}_{v_j \to c_1} = \left| \left\langle \psi_{kc_1} \mid \hat{p}_\alpha \, \psi_{kv_j} \right\rangle \right|^2, \tag{5.20}$$

describing the probability for optical transitions from the VB v_j to the CB c_1. One observes that close to the Γ-point the transitions from the topmost VB (blue lines) have a mainly $M^{zz}_{v_1 \to c_1}$ contribution. The other terms in the optical matrix are less than 0.03 and are, therefore, not presented in Figure 5.8. Transitions from the second VB (red lines) have a mainly $M^{yy}_{v_2 \to c_1}$ contribution, while transitions from the third VB (black lines) have a mainly

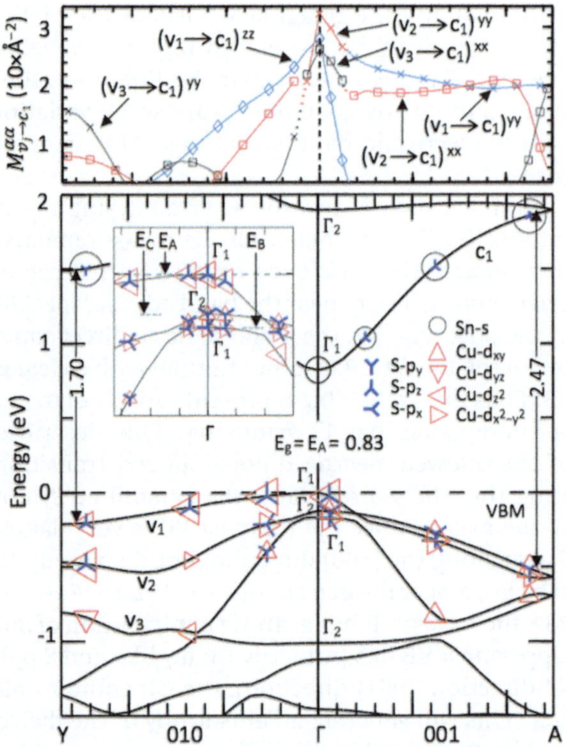

Figure 5.8 (a) Electronic band structure of monoclinic Cu_2SnS_3 (CTS) along two main symmetry directions, *i.e.*, (010) and (001), computed with the HSE06 potential. At the Γ-point VBM, there are three closely lying VBs (see inset figure) that contribute to the observed double absorption onset near the band gap energy. The marks indicate the main character of the energy states. The upper panel shows the optical interband matrix elements for electric dipole-induced transitions. Each line colour refers to a specific VB for the transition $v_j \to c_1$ and each marker type refers to a specific polarisation of the matrix element $M^{\alpha\alpha}$. Blue, red and black lines represent transitions from the uppermost (v_1), the second uppermost (v_2) and the third uppermost (v_1) VBs, respectively, to the lowest CB (c_1). Square, cross and diamond markers represent *x*-, *y*- and *z*-polarised transitions, respectively. Reproduced from ref. 28 with permission from Elsevier, Copyright 2017.

$M^{xx}_{v_3 \to c_1}$ contribution. It is also clear how the character of the band states affects the contribution of $M^{\alpha\alpha}_{v_j \to c_1}$ when a band crosses or interacts with another band.

The character of the bands, and the corresponding optical matrix elements, is the origin of the anisotropy of the dielectric function $\varepsilon_2(E)$ spectrum, see Figure 5.9. The computed dielectric response spectra are in good agreement with the experimental spectra from monoclinic CTS in the whole energy region 0 to 6 eV.[28] In the very low-energy region ($E_g \leq E < E_g + 0.15$ eV) the polarisation in the *z*-direction completely dominates, while for higher photon

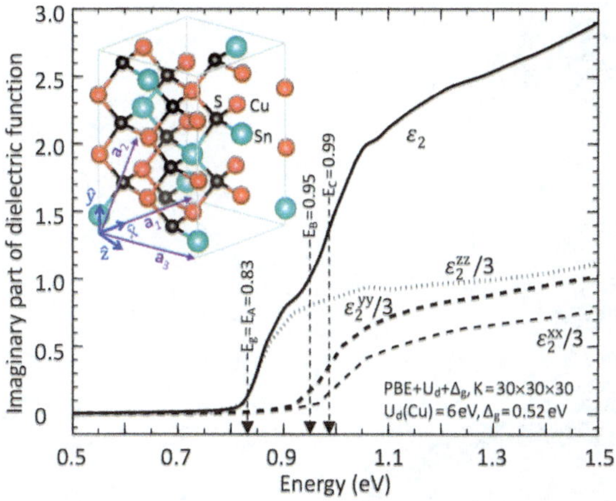

Figure 5.9 Polarisation dependency of the dielectric function of CTS from the PBE+U approach with corrected gap energy. The calculated dielectric response $\varepsilon_2(E)$ is very anisotropic in the near-band gap spectral region. The primitive cell of the crystal is defined by the lattice vectors $\mathbf{a}_1$, $\mathbf{a}_2$ and $\mathbf{a}_3$ and the orientation of the polarisation is described in the Cartesian coordinate space. Reproduced from ref. 28 with permission from Elsevier, Copyright 2017.

energies the polarisation in the *y*-direction (and partly in the *x*-direction) also contributes to the dielectric response. The gap energy between VBM (*i.e.*, v_1) and CBM (c_1) is directly associated with the first absorption onset at around 0.9–1.0 eV. The second and the third uppermost valence bands (v_2 and v_3) have maxima also at the Γ-point, with the energies 0.12 and 0.16 eV below the VBM. These two bands are thus responsible for the second onset about 0.15 eV above the gap energy in the dielectric function $\varepsilon_2(E)$ spectrum. The three Γ-point transition energies are thus $E_A = E_g$ (equal to 0.83 eV in the HSE06 calculation), $E_B = E_g + 0.12$ eV (or 0.95 eV), and $E_C = E_g + 0.16$ eV (or 0.99 eV). The energy split of the VBs near the VBM in conjunction with the anisotropic optical matrix elements explains why monoclinic CTS exhibits such a clear double onset phenomenon. In principle there are three absorption onsets, but the E_B and E_C are (almost) indistinguishable. Another observation is that the topmost VB is very flat along the (010) direction and the direct energy gap at the Y-point is only 1.70 eV = E_g + 0.93 eV. The corresponding energy at the A-point is 2.47 eV. Optical transitions at these **k**-states generate the peaks in the dielectric function for energies 1.8 to 2.5 eV. However, **k**-states in other directions in the IBZ also contribute to the spectrum in this energy range. For instance, the energy between the VBM and the second lowest CB is only 1.90 eV at the Γ-point.

Due to the sensitivity of the optical response for energies closely above the gap energy, we utilise the approach in eqn (5.9) to generate the dielectric functions with a dense **k**-mesh. We present the results of the total average $\varepsilon(E) = \varepsilon_1(E) + i\varepsilon_2(E)$ for several Cu-based chalcogenides in Figure 5.10. The

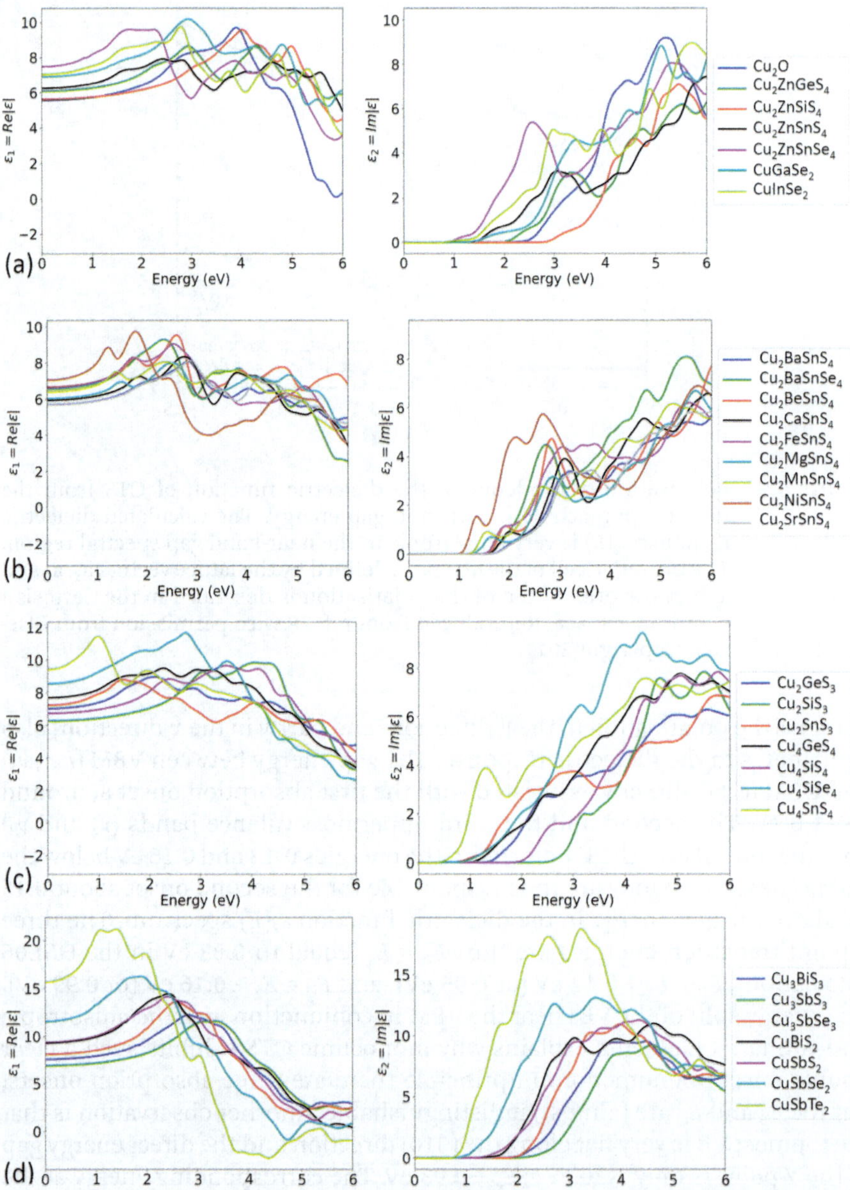

Figure 5.10 The real part ($\varepsilon_1(E)$; left panels) and the left part ($\varepsilon_2(E)$; right panels) of the complex dielectric functions $\varepsilon(E) = \varepsilon_1(E) + i\varepsilon_2(E)$, obtained with a **k**-mesh-corrected HSE06 approach and with a 0.1 eV post-prepared Gaussian broadening. The ionic contribution for $E \approx 0$ eV from electron–optical phonon interaction is not presented. The compounds are grouped in (a) traditional, (b) CZTS-like, (c) Cu-(Si/Ge/Sn), and (d) Cu-(Sb/Bi) chalcogenides.

corresponding dielectric constants are presented in Table 5.2, which can be compared with dielectric constants for traditional thin-film materials CIGS, CdTe and GaAs and also with other selected materials in Table 5.3. The related refractive index is a material property that needs to be considered when designing devices with these compounds.

Overall, the different compounds have fairly similar spectra. Binary Cu_2O is a wide-gap material and can therefore be expected to have a small, high-frequency dielectric constant, but the data ($\varepsilon_\infty \approx 5.6$) are not significantly smaller than for many other compounds ($\varepsilon_\infty \approx 6$–8 for CXTS and CXTSe) with a much smaller gap. This establishes that Cu_2O is an interesting wide-gap solar cell material for a top-cell structure. Moreover, the spectra of CIGS, CXTS, and CXTSe (Figure 5.10a and b) are all comparable and the main difference is the onset of the response, *i.e.*, the optical gap energy. The peak in $\varepsilon_2(E)$ in the energy region 2–4 eV roughly follows the gap energy: $Cu_2ZnSnSe_4$ has a large $\varepsilon_\infty = 7.7$ and has a small gap $E_g = 0.90$ eV, while Cu_2SiSnS_4 has a small $\varepsilon_\infty = 5.6$ and has a large gap $E_g = 2.89$ eV. The dielectric response of Cu_2NiSnS_4 deviates slightly from the other spectra as it has very strong Ni d-like states at the CBM. This enhances the response with a large dielectric constant of $\varepsilon_\infty \approx 7.0$. However, Cu_2NiSnS_4 has a small and indirect gap which is disadvantageous. The dielectric function of CTS (in Figure 5.10c) is $\varepsilon_\infty \approx 7.4$ for $E_g = 0.83$ eV, which is similar to that for $Cu_2ZnSnSe_4$; also, their spectra of $\varepsilon_2(E)$ in the energy region 2–3 eV are similar. The electron–optical phonon interaction affects the dielectric function in the very low-energy region, typically below 0.2 eV. For the CIGS, CXTS, and CXTSe compounds, as well as CTS, the static dielectric constant increases by about 2–3 to a value of about $\varepsilon_0 \approx 9$–12. The calculated dielectric constants normally agree rather well with the experimental data, at least considering the accuracy of theoretically determined gap energies and the measurement approach for various qualities of the materials.

Because the band gap energy of CTS is rather small (~0.9 eV) alternative ternary chalcogenides Cu_2XS_3 (where X = Sn, Ge, and Si) and their cation alloys have been suggested. Another option is anion S–O alloying, but since oxygen has a rather different atomic radius and electronegativity compared with sulfur, this is expected to be less thermodynamically advantageous. Instead, since Ge and Si are group IV elements, like Sn, it is expected that the band gap energies can be tailored by alloying Sn in CTS with Ge or Si. By alloying Sn with Ge, Umehara *et al.* fabricated a solar cell using graded $Cu_2Sn_{1-x}Ge_xS_3$, with a gap energy of ~1.0 eV, and a cell efficiency of 6.7% was reported.[35] Araki *et al.* synthesised a solar cell using Cu_2GeS_3 (CGS) as an absorber, which had an efficiency of 1.70%, and the gap of CGS is between 1.5 and 1.6 eV.[111] The band gap energy of Cu_2SiS_3 is as high as 2.56 eV, but on alloying Sn with Si the band gap energy of $Cu_2Sn_{0.5}Si_{0.5}S_3$ decreased to 1.40 eV according to measurements by Toyonaga *et al.*[112] To our knowledge, there is no PV cell fabricated based on compounds alloying Sn with Si in CTS.

Table 5.2 The direct, $E_{g,dir}$, indirect, $E_{g,ind}$, and Γ-point, $E_{g,\Gamma}$, band gap energies, obtained from the HSE06 calculations. The static ε_0 and high-frequency ε_∞ dielectric constants are presented in the three Cartesian directions *x*, *y* and *z*.

	Band gap energies (eV)			Static and high-frequency dielectric constants					
Compound	$E_{g,dir}$	$E_{g,ind}$	$E_{g,\Gamma}$	$\varepsilon_{0,xx}$	$\varepsilon_{0,yy}$	$\varepsilon_{0,zz}$	$\varepsilon_{\infty,xx}$	$\varepsilon_{\infty,yy}$	$\varepsilon_{\infty,zz}$
$CuGaSe_2$	1.47		1.47	9.6	9.6	10.2	7.0	7.0	7.0
$CuInSe_2$	0.85		0.85	11.8	11.8	13.0	7.4	7.4	7.4
$Cu_2ZnSnSe_4$	0.90		0.90	10.5	10.5	11.5	7.7	7.7	7.7
Cu_2ZnSnS_4	1.48		1.48	8.8	8.8	9.4	6.3	6.3	6.2
Cu_2ZnGeS_4	2.08		2.08	8.4	8.4	8.9	6.1	6.1	6.1
Cu_2ZnSiS_4	2.89		2.89	7.9	7.9	8.3	5.6	5.6	5.7
Cu_2CdSnS_4	1.22		1.22	9.4	9.4	20.7	6.1	6.1	6.6
Cu_2BeSnS_4	1.76		1.76	8.8	8.8	9.5	6.7	6.7	6.7
Cu_2MgSnS_4	1.48		1.48	8.6	8.6	8.3	6.0	6.0	6.6
Cu_2CaSnS_4	1.84		1.84	8.7	8.7	8.5	5.9	5.9	6.1
Cu_2SrSnS_4	1.81	1.80	1.81	11.0	11.0	10.1	5.7	5.7	5.6
Cu_2BaSnS_4	1.80	1.79	1.80	9.6	9.6	9.1	5.7	5.7	5.6
$Cu_2SrSnSe_4$	1.59	1.55	1.59	12.3	12.4	11.8	6.8	6.3	7.4
$Cu_2BaSnSe_4$	1.65		1.65	11.4	11.3	11.0	6.6	6.3	6.8
Cu_2MnSnS_4	1.40		1.40	8.9	8.9	8.6	6.3	6.3	6.8
Cu_2FeSnS_4	1.71		1.71	8.8	8.8	8.4	6.6	6.6	6.9
Cu_2NiSnS_4	1.26	1.17	1.26	9.6	10.0	8.5	7.0	7.4	6.6
Cu_2O	2.06		2.06	6.8	6.8	6.8	5.6	5.6	5.6
Cu_2SnS_3	0.83		0.83	10.5	10.8	12.2	7.4	7.5	7.4
Cu_2GeS_3	1.42		1.42	9.1	9.3	10.7	7.0	7.1	7.2
Cu_2SiS_3	2.58		2.58	8.2	8.2	9.0	6.1	6.2	6.2
Cu_4SnS_4	0.52	0.42	0.52	12.8	33.5	12.1	9.5	10.9	8.6
Cu_4GeS_4	1.14	1.10	1.14	11.6	11.0	9.8	8.1	8.0	7.2
Cu_4SiS_4	1.95	1.82	1.95	11.4	40.0	8.9	7.0	6.7	6.4
Cu_4SiSe_4	0.87	0.65	0.87	12.9	16.4	11.6	8.8	9.1	8.6
Cu_3SbS_3	2.13	2.06	2.13	8.0	11.5	9.5	6.2	7.4	7.1
Cu_3SbSe_3	2.01	1.85	2.01	9.1	14.6	11.4	7.3	8.7	8.3
Cu_3BiS_3	1.88	1.75	1.88	11.6	13.4	10.8	7.3	7.9	7.5
$CuSbS_2$	1.37	1.25	1.40	11.6	41.6	14.1	8.5	10.8	8.9
$CuSbSe_2$	1.05	0.97	1.05	12.4	57.0	25.6	9.8	14.2	19.8
$CuSbTe_2$	0.92	0.63	0.92	13.4	103.1	19.0	11.9	20.7	14.5
$CuBiS_2$	1.61	1.35	1.61	12.4	36.0	13.9	8.8	10.1	8.8
$CuBiSe_2$	1.18	1.12	1.18	13.4	46.5	16.0	10.2	13.1	10.6
$CuBiTe_2$	1.04	0.84	1.04	14.2	65.4	19.2	12.3	18.2	13.9
$CuBi_2O_4$	1.24	1.18	1.24	18.8	18.8	22.0	5.4	5.4	4.9

Very similar to group IV alloying of kesterite CZTS, we find that the compounds $Cu_2Sn_{1-x}Ge_xS_3 = Cu_2(Sn,Ge)S_3$ and $Cu_2Sn_{1-x}Si_xS_3 = Cu_2(Sn,Si)S_3$ with $x = 0$, 0.5, and 1, have overall a rather similar band dispersion and PDOS, especially near the VBM;[36] see Figure 5.7. The calculations also demonstrate that all the alloys have similar absorption coefficients, apart from a shift in the onset energy. With Ge alloying, the band gap energy can be increased

Table 5.3 Same as in Table 5.2 for selected complementary compounds.

Compound	Band gap energies (eV)			Static and high-frequency dielectric constants					
	$E_{g,dir}$	$E_{g,ind}$	$E_{g,\Gamma}$	$\varepsilon_{0,xx}$	$\varepsilon_{0,yy}$	$\varepsilon_{0,zz}$	$\varepsilon_{\infty,xx}$	$\varepsilon_{\infty,yy}$	$\varepsilon_{\infty,zz}$
$CuIn_{0.5}Ga_{0.5}Se_2$	1.21		1.21	11.4	11.4	12.6	7.1	7.1	7.1
GaAs	1.24		1.24	12.7	12.7	12.7	10.3	10.3	10.3
CdTe	1.45		1.45	10.4	10.4	10.4	6.6	6.6	6.6
Cu_3N	1.71	0.99	4.84	7.6	7.6	7.6	7.2	7.2	7.2
Cu_3P	1.58		5.37	6.2	6.2	6.2	6.1	6.1	6.1
$BaSi_2$	1.40	1.29	1.70	18.4	19.4	17.9	10.9	11.1	11.2
$BaGe_2$	1.15	1.00	1.50	20.1	21.6	20.3	12.3	13.0	12.8
$SrSi_2$	1.22	1.06	1.50	18.6	20.2	19.3	11.9	12.1	12.6
$SrGe_2$	1.06	0.78	1.28	21.4	24.1	22.8	13.7	14.6	14.7
GeSe	1.96	1.67	2.59	21.7	15.4	16.8	11.0	8.5	9.8
SnS	1.85	1.50	2.94	23.9	38.3	24.3	10.3	11.3	8.6
Sb_2Se_3	0.87	0.72	2.23	99.9	95.4	12.7	31.2	11.5	9.1
Bi_2S_3	1.01	0.79	2.30	89.8	83.7	21.9	19.8	11.2	13.5
Bi_2Se_3	1.06	0.68	1.63	92.1	95.2	26.8	27.5	13.6	18.6
$ZnSnN_2$	1.24		1.24	10.3	10.3	10.6	5.7	5.7	5.6
$ZnSnP_2$	1.71		1.71	11.2	11.2	11.3	8.8	8.8	8.7
Zn_3P_2	1.19		1.19	28.2	28.2	17.7	10.0	10.0	9.8

linearly by ~0.6 eV, and with Si alloying the gap energy can be increased by as much as ~1.8 eV. Overall, the trend of the calculated gap energies is comparable to the one from experimental measurements by Toyonaga *et al.*, Umehara *et al.*, and Araki *et al.*[35,111,112] Similarly to CZTS, at very high concentrations of Si the alloy exhibits an abrupt increase of the gap energy. For the $Cu_2(Sn,Ge)S_3$ alloy, near the CBM the Sn or/and Ge s-like states mainly hybridise with S p-like states. For compounds of $Cu_2(Sn,Si)S_3$ ($x = 0$, 0.5, and 1), Sn and Si s-like states of $Cu_2Sn_{0.5}Si_{0.5}S_3$ are not obviously mixed compared with Sn and Ge s-like states in $Cu_2Sn_{0.5}Ge_{0.5}S_3$, but a hybridisation with S p-like states is found. This weaker hybridisation between the Sn and Si orbitals can explain the non-linear increase of the gap energy for high Si concentrations, and it may also indicate some difficulties with forming thermodynamically stable Sn–Si alloys. Furthermore, we find that, for all these compounds, the effective electron masses are small and apparently isotropic. The mass values are all between $0.15m_0$ and $0.25m_0$ in the (010) direction, and between 0.13 m_0 and $0.22m_0$ in the (001) direction;[36] thus, the masses are comparable to the electron mass of CZTS ($m_e \approx 0.18m_0$).[10] The trend is that the alloys with larger band gaps have somewhat larger effective electron masses, which is consistent with CIGS and CZTSSe.[8,10] The effective hole masses of the topmost VBs are strongly anisotropic in the two directions. For instance, the hole mass in CTS is $1.85m_0$ in the (010) direction, but $0.11m_0$ in the (001) direction.[36] The larger hole mass often implies that the compounds have a better electron mobility than hole

Table 5.4 Defect formation energy $\Delta H(D)$ for $\Delta\mu_\alpha = 0$ and the induced shift $\Delta E_g(D)$ of the gap energy for native defects in tetragonal or trigonal Cu_2XSnS_4 (X = Be, Mg, Ca, Mn, Fe, Ni, Cd and Zn); all values are in units of eV. The defect concentration is about 1.5%. Total magnetic moment M_t, units in μ_B, is given for magnetic compounds including the defects, where the local magnetic moment is ~4.4 μ_B for Mn, ~3.5 μ_B for Fe and ~1.5 μ_B for Ni. Reprinted from ref. 85 with the permission of AIP Publishing.

			Defect type D, where X is the group II element					
Compound	Phase	Energy (eV)	V_{Cu}	V_X	X_{Cu}	Cu_X	(V_{Cu} + X_{Cu})	(Cu_X + X_{Cu})
Cu_2ZnSnS_4	$S^2{}_4$; 82	ΔH	1.42	3.54	0.45	1.64	−0.09	0.30
		ΔE_g	0.06	−0.03	−0.13	−0.12	0.08	−0.13
Cu_2CdSnS_4	$D^{11}{}_{2d}$; 121	ΔH	1.43	4.83	−0.75	2.89	−1.26	0.34
		ΔE_g	0.08	0.01	−0.10	−0.12	0.09	−0.09
Cu_2BeSnS_4	$S^2{}_4$; 82	ΔH	1.50	4.14	0.19	2.33	−0.28	0.58
		ΔE_g	0.00	−0.12	−0.22	−0.26	0.07	−0.30
Cu_2MgSnS_4	$D^{11}{}_{2d}$; 121; AFM	ΔH	1.61	4.85	−0.52	2.83	−1.07	0.34
		ΔE_g	0.07	−0.07	−0.10	−0.22	0.10	−0.18
Cu_2CaSnS_4	$D^{11}{}_{2d}$; 121	ΔH	1.14	5.92	−1.86	3.86	−2.68	0.13
		ΔE_g	−0.07	−0.14	−0.22	−0.34	0.02	−0.29
	$C^2{}_3$; 144	ΔH	1.27	5.69	−1.47	4.66	−2.05	1.40
		ΔE_g	0.04	−0.08	−0.74	−0.41	0.14	−0.29
Cu_2MnSnS_4	$D^{11}{}_{2d}$; 121; AFM	ΔH	1.60	4.37	−0.21	2.35	−0.54	0.33
		ΔE_g	0.05	−0.35	−0.21	−0.37	0.06	−0.19
		M_t	0.1	3.0	4.6	4.0	5.0	0.0
Cu_2FeSnS_4	$D^{11}{}_{2d}$; 121; AFM	ΔH	1.64	5.94	−1.67	3.81	−2.01	0.35
		ΔE_g	−0.02	−0.83	−1.08	−0.26	0.01	−0.17
		M_t	0.3	2.0	2.8	2.8	3.7	0
Cu_2NiSnS_4	$S^2{}_4$; 82; AFM	ΔH	1.57	3.09	−0.22	1.08	0.60	0.36
		ΔE_g	−0.07	−0.18	0.00	−0.22	−0.15	−0.53
		M_t	0.1	2.0	1.0	2.8	2.0	0.0

mobility. The effective electron masses were determined directly from the curvature of the energy bands through

$$m(\mathbf{k})_{\alpha\beta} = \pm\frac{\hbar^2}{\partial^2 E_j(\mathbf{k})/\partial k_\alpha \partial k_\beta} \tag{5.21}$$

where the plus/minus sign is associated with the electrons and holes, respectively.

The Cu–group IV chalcogenides can also be formed in the Cu_4(Sn/Ge/Si)(S/Se)$_4$ stoichiometry, which is then Cu- and S/Se-rich (or Sn/Ge/Si poor). Overall, the DOS of these phases agrees with their corresponding Cu_2(Sn/Ge/Si)S_3 compounds (not presented in the figures). For example, efforts have been paid to the monoclinic and orthorhombic phase Cu_4SnS_4, for thermoelectric devices and photovoltaic applications.[32,113–116] High-quality thin films can be prepared by, for example, by annealing Cu_4SnS_4 deposited from a chemical bath onto chemically cleaned soda-lime glass substrates.[33] However, the gap energy calculated previously was rather small and direct: ~0.49 eV for the orthorhombic phase.[32] This is close to the indirect gap energy $E_g \approx 0.42$ eV that we calculated for orthorhombic Cu_4SnS_4, with a direct Γ-point gap energy of 0.52 eV. Thus, the Sn-poor Cu_4SnS_4 composition has a considerably smaller gap energy than the Cu_2SnS_3 composition. Early measurements yielded band gap energies as high as 1.2–1.8 eV.[117–119] However, diffuse reflectance data by Choudhury *et al.*[32] and spectral transmittance curves by Chalapathi *et al.*[33] indicate that the direct optical band gap of thin films was 0.92 or 1.0 eV for the orthorhombic structure. Thus, further investigations are needed to establish the optoelectronic properties of this composition. Nonetheless, we find that the gap increases to ~1.10 eV for Cu_4GeS_4 and to 1.82 eV for Cu_4SiS_4, and the trend is that these materials have indirect gaps and the band gap energies are smaller than the $Cu_2(Sn/Ge/Si)S_3$ composition. The Se-based compound of Cu_2SiSe_3 has a ~0.2 eV lower gap energy of 0.65 eV, and it is still an indirect band gap material.

The optical properties in terms of the dielectric functions for the compounds $Cu_2(Sn/Ge/Si)(S/Se)_3$ and $Cu_4(Sn/Ge/Si)(S/Se)_4$ are presented in Figure 5.10c. One can see that the materials span a large energy region for E_g onset, and a large variation in high-frequency dielectric constants ($\varepsilon_\infty \approx 6–10$).

$CuSbS_2$ (CSS) and $CuSbSe_2$ (CSSe) as well as $CuBiS_2$ (CBS) and $CuBiSe_2$ (CBSe), with the group V elements Sb or Bi, are interesting alternatives to the CTS-based compounds for potential Cu-based chalcogenides free of group II elements. The compounds can be crystallised in an orthorhombic layered structure with the composition $CuBiS_2$, in addition to an orthorhombic composition Cu_3BiS_3 which is somewhat Bi-poor. CBS and CSS have particularly strong absorption coefficients near the gap energy,[38–40,43,45,50] which can make them of special importance for ultrathin inorganic solar cells.[45] The experimental values of the band gap energies for CSS, CSSe, and CBS are ~1.5 eV,[42] ~1.09 eV,[120] and ~1.65 eV,[37] respectively. No experimental gap is available for CBSe, but the gap energy was previously calculated to be ~1.1 eV.[50,121] Calculations estimate[45] that all the considered compounds have indirect gaps, and the fundamental gap energies for CSS, CSSe, CBS, and CBSe are E_g = 1.25, 0.97, 1.35, and 1.12 eV, respectively. The direct gap energies are about 0.05–0.25 eV larger. Thus, utilising Sb or Bi cations leads to fairly similar gap energies (slightly smaller for Bi-based compounds), while utilising Se anions yields a 0.3–0.4 eV smaller gap compared with that for S. This is in contrast to Ge or Sn cation alloying in CTS which more strongly affects the gap energies, but it is similar to (although slightly smaller than) the effect due to S/Se anion alloying in other Cu-based chalcogenides. The solar cell energy conversion efficiency is η = 3% for devices based on $CuSbS_2$ or $CuSbSe_2$.[122,123] In addition, the Bi-poor phase Cu_3BiS_3 can be prepared by chemical bath deposition, sputtering, co-evaporation, or sulfurisation of Bi–Cu metal precursors, and the reported measured band gap energy ranges from 1.0 to 1.6 eV.[46–49] Theoretically, the band gap energy is indirect, with E_g = 1.75 eV, but the direct gap energy is only ~0.1 eV larger.[50] In contrast to the $Cu(Sb/Bi)S_2$ compounds, replacing Bi with Sb in the $Cu_3(Bi/Sb)S_3$ phases increases the gap energy, here by as much as 0.3 eV. Thus, Cu_3SbS_3 compounds can be utilised as wide-gap materials with $E_g \approx 2$ eV. Cu_3SbSe_3 has a 0.2 eV smaller gap energy than its S-based compound.

Also for CSS, CSSe, CBS, and CBSe, chalcogenide alloying can be utilised to optimise and fine-tune the gap energy. Since CSSe has a somewhat large gap, we also explored alloys with Te. The calculated fundamental gap energies are 0.63 eV for $CuSbTe_2$ (CSTe) and 0.84 eV for $CuBiTe_2$ (CBTe), and the direct gap energies are about 0.2–0.3 eV larger than their indirect gaps, which is then disadvantageous compared with Cu(Sb/Bi)Se_2. We have found that the band gap energies (similar to other chalcogenide anion alloying) follow a roughly linear relationship with the alloy composition.[45] In CuSb(Se,Te)$_2$ the direct gap at the Γ-point $E_{g,\Gamma}$ is about 0.4 eV larger than the indirect gap $E_{g,ind}$, while the corresponding energy difference is about 0.7 eV in CuBi(S,Se)$_2$. In both the CuSb(Se,Te)$_2$ alloys and the CuBi(S,Se)$_2$ alloys, the direct $E_{g,dir}$ gap energy is only ~0.2 eV larger than the fundamental indirect band gap. This is a rather small energy difference, and that is important for enabling efficient photovoltaics. However, an indirect gap material is known to suffer from a strong Auger effect. The direct gap governs the sunlight absorption and also, mainly, the radiative recombination, while the indirect gap governs the Auger non-radiative recombination. Therefore, optimising the direct gap energy by forming bands with small (or negative) energy difference $\Delta_g = E_{g,ind} - E_{g,dir}$ is the route for tailor making highly efficient solar cell materials.

For these Cu(Sb/Bi)(S/Se/Te)$_2$ compounds, the electronic band structures exhibit strikingly flat energy dispersions for both the CBs and VBs.[45] In contrast to compounds containing group IV elements Sn and Ge with s-like CB character, the Cu–Sb/Bi chalcogenides do not generate a strong group V s-like DOS peak at the lower energy region of the CB DOS; see Figure 5.7. Instead, the CBs are dominated by relatively localised Sb/Bi p-like states and that generates a strong DOS close to the CB edges. This effect yields rather flat lowest CBs. Flat energy dispersion can imply large effective masses, which can be disadvantageous for the lifetime of the carriers. However, the effective electron masses at the CBM are relatively small for these compounds, for instance: $m_x \approx 0.23m_0$, $m_y \approx 0.16m_0$, and $m_z \approx 0.40m_0$ for $CSTe_2$; and $m_x \approx 0.19m_0$, $m_y \approx 0.83m_0$, and $m_z \approx 0.50m_0$ for $CBSe_2$. These values are comparable to the corresponding calculated effective masses for CIGS (~$0.1m_0$) and Si ($m_x = m_y \approx 0.2m_0$; $m_z \approx 1.0m_0$), while considerably larger than that of GaAs (~$0.08m_0$).[64,70,124] It is worth mentioning that the mass values are valued only in the vicinity of the CBM where the band is parabolic (at most 50 meV above CBM in these materials). However, even with a small effective electron mass of ~$0.1m_0$, the quasi-Fermi level is only ~20 meV above the CBM, even for a high free electron concentration of 10^{17} cm^{-3}.[125] Furthermore, similarly to traditional CIGS and CZTS, the VBs in these Cu–(Sb/Bi) chalcopyrites have the characteristic Cu-d–anion-p hybridisation that helps to form localised band structures near the VB edges.

Since the topmost VBs are flat, the DOS near the VB edges is high compared to the DOS of CIGS and CZTSSe. The flat CBs and VBs in CuSb(Se,Te)$_2$

and $CuBi(S,Se)_2$ compounds is an advantage since it implies a strong onset of the optical response. This is demonstrated in Figure 5.10d where the imaginary part $\varepsilon_2(E)$ of the dielectric function is considerably larger than that for the other compounds, especially for low energies near the gap energy E_g. Since $\varepsilon_2(E)$ is large for a wide energy range up to 5–6 eV, the real part $\varepsilon_1(E)$ also becomes large. That is reflected in the high-frequency dielectric constant. While that constant $_{has}$ ranges $\varepsilon_\infty \approx 6$–8 in the CXTZ compounds and $\varepsilon_\infty \approx 6$–10 in the Cu–group IV chalcogenides, the values for the Cu–group V chalcogenides are in the range $\varepsilon_\infty \approx 7$–12 (and ~16 for the small gap $CuBiTe_2$). If one compares compounds with similar band gap energies ($E_g \approx 1.3$ eV) it becomes more evident: $\varepsilon_\infty = 7.1$ for CIGS, 6.3 for CZTS, 7.1 for $Cu_2(Sn,Ge)S_3$, whereas $\varepsilon_\infty = 9.4$ for CBS, which is close to the value for GaAs. It is also noticeable that $Cu(Sb/Bi)(S/Se/Te)_2$ have much larger and more anisotropic static dielectric functions, and some compounds become high-κ materials. This indicates a strong ionic character for the materials, but also that the materials might be less thermodynamically stable.

5.3.4 Absorption Coefficient and Theoretical Maximum Device Efficiency

The absorption coefficient $\alpha(E)$ is determined directly from the dielectric function through eqn (5.8). One can therefore expect that Cu–group IV chalcogenides will have slightly stronger absorption than CXTS compounds in the low-energy region, and that Cu–group V chalcogenides will have even stronger absorption. Indeed, the absorption coefficients for the Cu(Sb/Bi)$(S/Se/Te)_2$ compounds are significantly larger than the corresponding calculated values for the traditional absorbers CIGS and CZTS; see Figure 5.11. This larger $\alpha(\omega)$ is especially notable in the energy region E_g to $E_g + 1.5$ eV. For instance, at the photon energy $\hbar\omega = E_g + 1$ eV, the CSSe and CBS have ~7 times larger absorption coefficients than those of CIGS and CZTS. That is a significant improvement in the optical properties. The $CuSbTe_2$ compound has a very pronounced absorption also close to the gap energy; however the material has a small indirect gap ($E_{g,ind} = 0.63$ eV) with a relatively large direct gap $\Delta_g = E_{g,ind} - E_{g,dir} \approx 0.3$ eV, which is a disadvantage for its application as a solar cell material. All materials reach $\alpha(E) = 10^5$ cm^{-1} for an energy less than ~2 eV above the band gap energy. The commercialised CIGS do not have an especially large absorption coefficient, and CZTS has a slightly larger absorption coefficient in the low-energy region. This demonstrates that optical properties are only one of several material properties required for high-efficiency PV devices; crystalline quality/stability, defect physics, carrier mobility, as well as band alignment and interface physics also play an important role.

From the absorption coefficients and the gap energies, the theoretical maximum device efficiency can be derived for the absorber material within the SQ model. Since $\alpha(E)$ has units of cm^{-1}, the thickness of the material determines how much of the light is absorbed, and we therefore determine the efficiency for film thickness d; see Section 5.2.3. The maximum efficiency

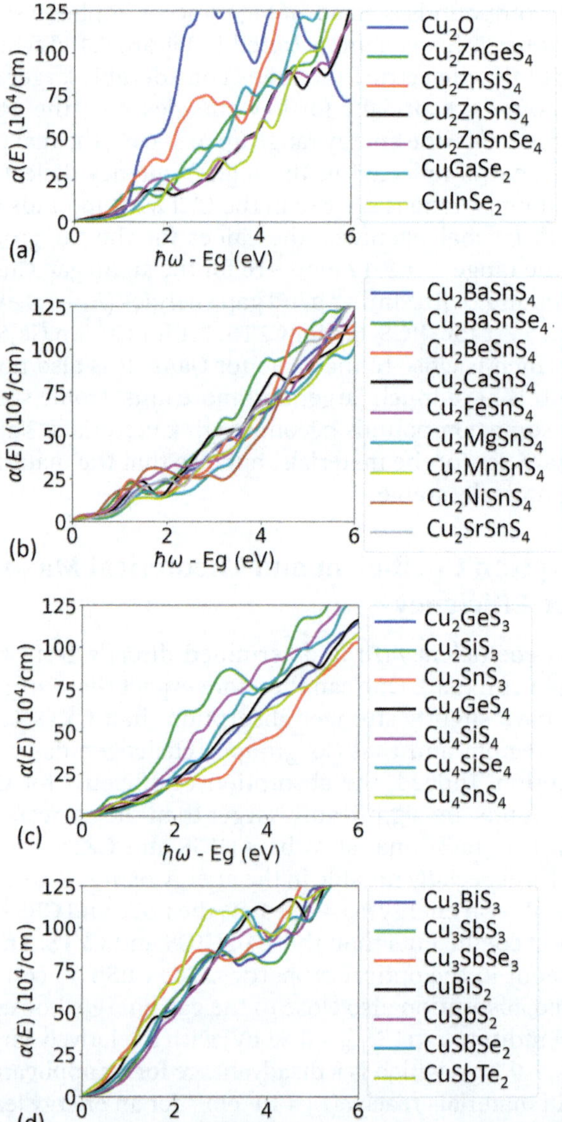

Figure 5.11 The absorption coefficient $\alpha(E)$, obtained directly from $\varepsilon(E)$ in Figure 5.10 and the energy scale refers to the fundamental gap energy (*i.e.*, $\hbar\omega - E_g$) to better compare the spectra of compounds with different gap energies. The compounds are grouped in (a) traditional, (b) CZTS-like, (c) Cu-(Si/Ge/Sn), and (d) Cu-(Sb/Bi) chalcogenides.

η_{max} for different film thickness $d < 1000$ nm is displayed in Figure 5.12. We also compare the corresponding results for traditional or other selected thin-film materials (Figure 5.12e). For thickness larger than ~1000 nm, most absorbers have sufficiently good efficiency. If the thickness is increased even further (*i.e.*, $d \rightarrow \infty$) then η_{max} will become independent of the absorption

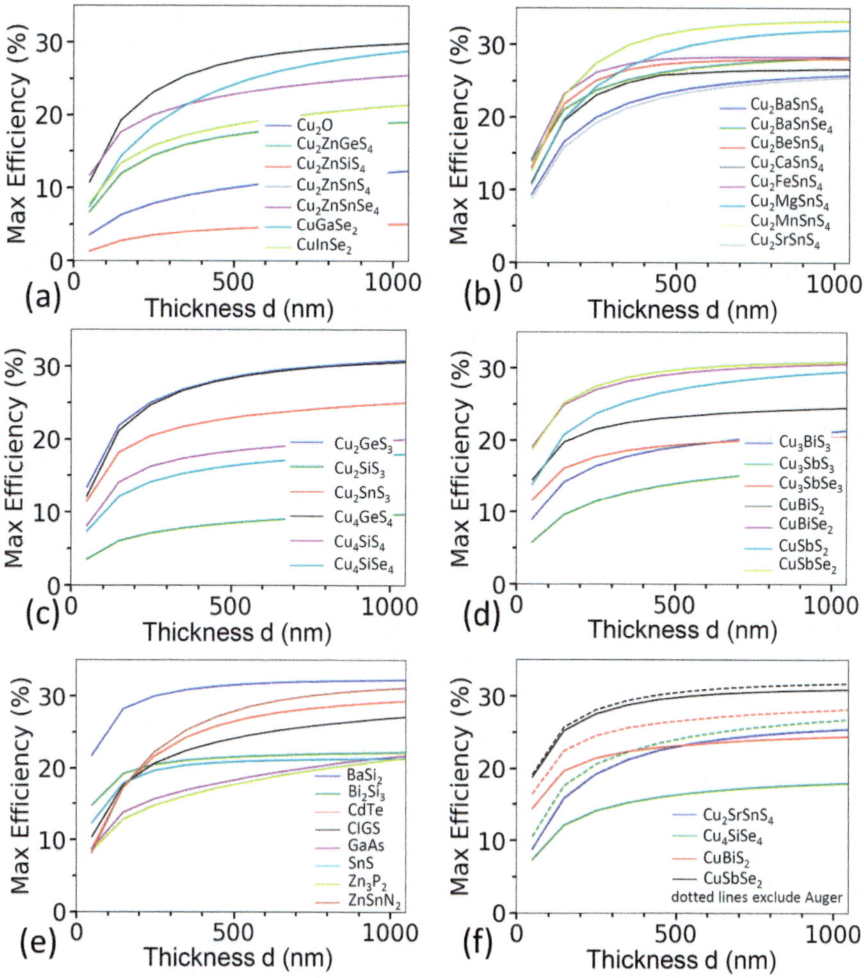

Figure 5.12 The maximum solar energy efficiency η_{max} as a function of film thickness d. The compounds are grouped in (a) traditional, (b) CZTS-like, (c) Cu-(Si/Ge/Sn), and (d) Cu-(Sb/Bi) chalcogenides, and (e) various other known absorbers. In (f), the Auger effect for indirect compounds is exemplified.

coefficient, and it will only depend on the gap energy, with $\eta_{max} \approx 32$–34% for CIGS, CZTSSe, CdTe, and GaAs. For the CXTS compounds, CZTS and, especially, Cu_2MnSnS_4 exhibit very good optoelectronic properties. Cu_2O has a very good absorption coefficient, but because the band gap energy is large ($E_g = 2.06$ eV) the maximum efficiency is poor ($\eta_{max} \approx 15\%$ for $d = 2000$ nm). Cu_2ZnGeS_4 (also with $E_g = 2.08$ eV) is a better wide-gap material for top-cell thin-film technologies, with $\eta_{max} \approx 20\%$ for $d = 2000$ nm. Also, the gap energy of $Cu_2BaSn(S/Se)_4$ is too large to be interesting for regular solar cells, but $Cu_2(Ba/Sr)SnS_4$ (with $E_g \approx 1.8$ eV with HSE06 and ~2.0 eV experimentally) is potentially interesting as a top-cell material, as $\eta_{max} \approx 26\%$ for $d = 2000$ nm.

One observes that the efficiency of GaAs devices decreases when the thickness decreases, and CdTe shows a typical drop in efficiency for d < 500 nm. This efficiency drop is true also for CIGS and CZTS, the CXTS compounds and the Cu–group IV chalcogenides. For thickness less than 100 nm, the efficiency is less than 20% for these compounds. In order to attain high efficiencies with thin films, it is important to have a strong absorption coefficient in the low-energy region. That is the case for the Cu–group V chalcogenides in Figure 5.11d. With flat uppermost VBs, but especially with flat bottommost CBs with a multi-valley structure, these materials have the potential to be highly efficient even when applied as ultrathin films. The issue with this family of compounds and their alloys is that they are indirect gap materials and the Auger effect decreases their efficiencies by about 1 to 5%. This decrease due to the non-radiative recombination effect is fairly constant with respect to thickness d, and, therefore, the Auger effect is less dominant for smaller thicknesses where, instead, the efficiency drop is due to the lower absorptivity. That means, for thin films with d < 500 nm, the important material property for retaining a high efficiency is the absorption coefficient in the low photon energy region (E_g to about E_g + 1 eV). CSSe, CBS, and CBSe maintain their efficiencies when the film thickness is narrowed down to around 100 nm. The gap energy of CBS is too large ($E_{g,dir}$ = 1.6 eV), and its efficiency is significantly lower than the other alloy compounds. CSSe and CBSe, however, have more suitable gap energies and the efficiencies are close to 25% down to 100–200 nm with these materials and remain remarkably high even for d = 50–100 nm, with $\eta_{max} \approx 26\%$ for d = 200 nm and $\eta_{max} = 19\%$ for d = 50 nm.

One reason for the high efficiencies for these materials is that the energy difference $\Delta_g = E_g - E_{g,dir}$ is small, only 0.06 eV for CBSe. The efficiency would be even larger if the energy difference $\Delta_g = E_g - E_{g,dir}$ was decreased; this is discussed in ref. 45. The Auger effect is small if Δ_g is smaller than ~0.1 eV, since it is suppressed by the radiative recombination that determines the voltage drop. If Δ_g goes to zero, the Auger effect decreases to zero and the maximum efficiency increases, reaching the efficiency corresponding to a direct gap material. This means that the computational accuracy affects details in the band curvatures and it may therefore be difficult to predict the material properties, but it also means that the performance of the materials can depend on crystalline quality; for example, defects can affect Δ_g. In Figure 5.12f the efficiency is presented both with the Auger effect included (solid lines) and excluded (dotted lines, with same colours as for the full calculation). The effect depends strongly on Δ_g and the maximum efficiency increases typically by 3–5% when there is no Auger effect. However, for CSSe with Δ_g = 0.08 eV the Auger effect is less than 1% for thin films (and the effect is negligible in Cu_2SrSnS_4 with Δ_g = 0.01 eV). Alternatively, the Auger recombination can be reduced by minimising the carrier concentration since $J^0{}_{aug} \propto n_i{}^2 N_A$. That can be achieved by using materials with a large fundamental gap, but too large a gap is disadvantageous for optical absorption. For p-type materials, the efficiency increases by approximately 1–2% if the doping concentration decreases to $N_A = 10^{15}$ cm^{-3}. However, for intrinsic materials, the Auger model becomes $J^0_{aug} \propto n_i^2 p$, which depends even more strongly on Δ_g and on the voltage drop, and this can result in an even smaller efficiency.[126]

Alternatively, the carrier concentration can be decreased with smaller effective DOS masses.[45] The effective DOS mass employed in this work is $m_a = 1m_0$, which is estimated to be somewhat too large, especially for CSTe. If one instead uses a smaller mass of $m_a = 0.4m_0$ (and with $\Delta_g \approx 0.2$ eV) the maximum efficiency of $CSTe_2$ increases by ~1.5%. However, trying to optimise the electronic band structure to obtain even smaller DOS masses also means more dispersion of the band structure and thereby a smaller absorption coefficient. Therefore, to tailor make materials for even thinner solar cells, the concept of multi-valley engineering (to achieve large joint DOS and large $\alpha(\omega)$) balanced with band edge engineering for small effective masses (with low carrier concentration $n_i^2 \propto [(\gamma_e{}^{2/3} m_e^{DOS})\cdot(\gamma_h{}^{2/3} m_h^{DOS})]^{3/2}$ and long lifetime) can be applied. Here, better models that include non-parabolic energy dispersion together with a better description for the absorption and recombination processes are required to more accurately describe the behaviour of the materials. As with the other chalcogenides, anion alloying can be used to tune the gap energy to achieve the maximum efficiency, but the alloying must have almost no impact on the energy difference Δ_g. There are also other emerging solar cell materials that have a high efficiency for $d < 100$ nm; for example, $BaSi_2$, which has a suitable gap energy and a sufficiently small Δ_g. Its electronic properties and defect physics have been explored theoretically by Kumar *et al.*[127]

The overall results demonstrate that Cu–group V chalcogenides, *e.g.*, CuSb(Se,Te)$_2$ and CuBi(S,Se)$_2$ alloys with optimised E_g, are an alternative emerging class of materials that are suitable for ultrathin inorganic photovoltaics.[45] The advantage of the thin-film absorber layer is the shorter path required to transport the minority carriers. We can thus conclude that there are several emerging Cu-based, group II-free chalcogenides that have superior optical properties over the more traditional solar cell absorbers. The compounds with group IV Sn and especially the group V Sb and Bi are suitable for ultrathin inorganic solar cells, and it is possible to (theoretically) achieve a single-junction solar cell with a maximum efficiency of $\eta_{max} > 25\%$, even for film thickness $d < 100$ nm, if the Auger recombination is avoided or diminished. The challenge is, of course, to be able to develop and tune the synthesis to fabricate materials with sufficiently good quality, without devastating native defects. Nonetheless, the results presented in this chapter can be used as a guide when designing future materials for ultrathin solar cells.

5.4 Conclusions

This chapter demonstrates how computational methods can be used to predict and evaluate the structural, electronic, and optical properties of solar energy capture materials. We have focused on several emerging copper-based chalcogenides, employing DFT-based atomistic first-principles computational methods, and compared the material properties of these chalcogenides with traditional as well as other proposed compounds for thin-film inorganic PV applications. We have demonstrated that it is possible to develop an inorganic PV device with an ultrathin photon-absorbing film (*i.e.*, with thickness $d < 100$ nm). The material should exhibit a proper band gap energy E_g, as well as having a very high absorption coefficient $\alpha(E)$ in the lower energy region of the spectrum; that is, for photons with energies near the gap energy $E_g \leq E < (E_g + 2\ \text{eV})$. However, attention regarding the material quality and modifications in the band dispersion is required in order to minimise non-radiative recombination. To tailor make materials for very thin, inorganic PV technologies one should consider: (1) an optimised band gap energy for the specific film thickness. The fundamental optoelectronic performances can easily be fine-tuned by alloying with isovalent elements. In addition, we suggest that (2) materials are designed with multi-valley band edges that absorb sunlight at several valleys in the IBZ. This positions the VBM and CBM away from the Γ-point, out in the IBZ. To further enhance strong absorption in the lower energy region of the spectrum, the materials should have (3) flat energy dispersions to achieve a high absorption coefficient for the low photon energies, $E = \hbar\omega = E_g$ to E_g + 2 eV. Thinner films require high absorption in an even more narrow energy region, approximately $\hbar\omega = E_g$ to E_g + 1 eV. However, (4) the band structure should have a direct fundamental gap to avoid Auger recombination; thus, both the CBM and VBM should be located at the same **k**-point but away from the Γ-point. The material should, preferably, have (5) small DOS electron and hole masses to reduce the intrinsic carrier concentration, which in turn reduces the Auger effect. In addition, (6) with small effective electronic masses the materials will exhibit high minority carrier mobility. The preferred requirements of a flat band dispersion for high sunlight absorption and a small mass for low band filling and high mobility are often competing characteristics, and the materials have to be fine-tuned to optimise the balance between them. In addition to this, the material should, of course, as for regular thin-film PVs, (7) involve only Earth-abundant, non-toxic, and inexpensive elements, and the devices should be inexpensive to fabricate and handle. The material must also be (8) thermodynamically stable with no devastating in-gap defect states, be able to be doped, and have suitable band alignment at the interfaces.

Acknowledgement

We thank the Research Council of Norway (projects 243642, 221469, and 251131) as well as the Swedish Foundation for Strategic Research for financial support. We acknowledge access to high performance computer resources at the University of Oslo (UiO) in Norway operated by the University Center for Information Technology (USIT) provided through the National Infrastructure for High Performance Computing (NOTUR), as well as at the KTH Royal Institute of Technology in Stockholm, operated by the Center for High

Performance Computing (PDC), and at Linköping University also in Sweden, operated by the National Supercomputer Centre (NSC), through the Swedish National Infrastructure for Computing (SNIC). We also acknowledge the Partnership for Advanced Computing in Europe (PRACE) for awarding access to resource MareNostrum based in Spain at the Barcelona Supercomputing Center – Centro Nacional de Supercomputación (BSC–CNS), and access to the resource Archer based in the UK at Edinburgh, with support from the Distributed European Computing Initiative (DECI) in PRACE aisbl. We acknowledge scientific discussions with Dr Andrea Crovetto at Technical University of Denmark (DTU) Denmark, Dr Sukgeun G. Choi now at University of California at Santa Barbara, USA, and teams led by Prof. Marika Edoff and Prof. Charlotte Platzer Björkman at Uppsala University in Sweden.

References

1. N. S. Pujari, *et al.*, *International Technology Roadmap for Photovoltaic: 2017 Results*, 9th edn, ITRPV, 2018.
2. *Best Research-Cell Efficiencies, Rev. 10-30-2017*, National Renewable Energy Laboratory, Denver, 2017.
3. *Clean Electricity from Photovoltaics*, ed. M. D. Archer and R. Hill, Imperial College Press, London, 2001.
4. S. Siebentritt, M. Igalson, C. Persson and S. Lany, *Prog. Photovoltaics*, 2010, **18**, 390.
5. *Handbook of Photovoltaic Science and Engineering*, ed. A. Luque and S. Hegedus, John Wiley & Sons Ltd, Chichester, 2011.
6. *Solar Cells*, ed. A. McEvoy, T. Markvart and L. Castañer, Elsevier, Oxford, 2nd edn, 2013.
7. T. Feurer, P. Reinhard, E. Avancini, B. Bissig, J. Lockinger, P. Fuchs, R. Carron, T. P. Weiss, J. Perrenoud, S. Stutterheim, S. Buecheler and A. N. Tiwari, *Prog. Photovoltaics*, 2017, **25**, 645.
8. K. Ito and T. Nakazawa, *Jpn. J. Appl. Phys.*, 1988, **27**, 2094.
9. T. M. Friedlmeier, N. Wieser, T. Walter, H. Dittrich and H. W. Schock, *14th European Conference of Photovoltaic*, Bedford, 1997, p. 1242.
10. C. Persson, *J. Appl. Phys.*, 2010, **107**, 053710.
11. *Thin Film Solar Cells from Earth Abundant Materials*, ed. S. Kodigala, Elsevier, Oxford, 2013.
12. *Copper Zinc Tin Sulphide-based Thin Film Solar Cells*, ed. K. Ito, John Wiley & Sons, Ltd, Chichester, 2015.
13. S. Li, S. Zamulko, C. Persson, N. Ross, J. Larsen and C. Platzer-Bjorkman, *Appl. Phys. Lett.*, 2017, **110**, 021905.
14. G. Q. Yao, H. S. Shen, E. D. Honig, R. Kershaw, K. Dwight and A. Wold, *Solid State Ionics*, 1987, **24**, 249.
15. H.-R. Liu, S. Chen, Y.-T. Zhai, H. J. Xiang, X. G. Gong and S.-H. Wei, *J. Appl. Phys.*, 2012, **112**, 093717.
16. S. Bag, O. Gunawan, T. Gokmen, Y. Zhu and D. B. Mitzi, *Chem. Mater.*, 2012, **24**, 4588.

17. Q. Shu, J.-H. Yang, S. Chen, B. Huang, H. Xiang, X.-G. Gong and S.-H. Wei, *Phys. Rev. B*, 2013, **87**, 115208.
18. M. Hamdi, A. Lafond, C. Guillot-Deudon, F. Hlel, M. Gargouri and S. Jobic, *J. Solid State Chem.*, 2014, **220**, 232.
19. D. B. Khadka and J. Kim, *J. Phys. Chem. C*, 2015, **119**, 1706.
20. X. Zhang, D. Rao, R. Lu, K. Deng and D. Chen, *AIP Adv.*, 2015, **5**, 057111.
21. S. Zamulko, R. Chen and C. Persson, *Phys. Status Solidi B*, 2017, **254**, 1700084.
22. L. O. Grondahl, *Rev. Mod. Phys.*, 1933, **5**, 141.
23. B. P. Rai, *Sol. Cells*, 1988, **25**, 265.
24. D. M. Berg, R. Djemour, L. Gütay, G. Zoppi, S. Siebentritt and P. J. Dale, *Thin Solid Films*, 2012, **520**, 6291.
25. N. Aihara, A. Kanai, K. Kimura, M. Yamada, K. Toyonaga, H. Araki, A. Takeuchi and H. Katagiri, *Jpn. J. Appl. Phys.*, 2014, **53**, 5S1.
26. L. L. Baranowski, P. Zawadzki, S. Christensen, D. Nordlund, S. Lany, A. C. Tamboli, L. Gedvilas, D. S. Ginley, W. Tumas, E. S. Toberer and A. Zakutayev, *Chem. Mater.*, 2014, **26**, 4951.
27. M. Nakashima, J. Fujimoto, T. Yamaguchi and M. Izaki, *Appl. Phys. Express*, 2015, **8**, 042303.
28. A. Crovetto, R. Chen, R. B. Ettlinger, A. C. Cazzaniga, J. Schou, C. Persson and O. Hansen, *Sol. Energy Mater. Sol. Cells*, 2016, **154**, 121.
29. A. C. Lokhande, R. B. V. Chalapathy, M. He, E. Jo, M. Gang, S. A. Pawar, C. D. Lokhande and J. H. Kim, *Sol. Energy Mater. Sol. Cells*, 2016, **153**, 84.
30. D. Tiwari, T. Koehler, R. Klenk and D. J. Fermin, *Sustainable Energy Fuels*, 2017, **1**, 899.
31. Q. Chen, X. Dou, Z. Li, Y. Ni, J. Chen, F. Zhou, Y. Yamaguchi and S. Zhuang, *Optik*, 2014, **125**, 3217.
32. A. Choudhury, S. Mohapatra, H. Yaghoobnejad Asl, S. H. Lee, Y. S. Hor, J. E. Medvedeva, D. L. McClane, G. E. Hilmas, M. A. McGuire, A. F. May, H. Wang, S. Dash, A. Welton, P. Boolchand, K. P. Devlin, J. Aitken, R. Herbst-Irmer and V. Petříček, *J. Solid State Chem.*, 2017, **253**, 192.
33. U. Chalapathi, B. Poornaprakash and S.-H. Park, *Sol. Energy*, 2017, **155**, 336.
34. X. Chen, M. Onoda, H. Wada, A. Sato, H. Nozaki and R. Herbst-Irmer, *J. Solid State Chem.*, 1999, **145**, 204.
35. M. Umehara, S. Tajima, Y. Aoki, Y. Takeda and T. Motohiro, *Appl. Phys. Express*, 2016, **9**, 072301.
36. R. Chen and C. Persson, *Phys. Status Solidi B*, 2017, **254**, 1700111.
37. S. H. Pawar, A. J. Pawar and P. N. Bhosale, *Bull. Mater. Sci.*, 1986, **8**, 423.
38. Y. Rodriguez-lazcano, M. T. S. Nair and P. K. Nair, *J. Cryst. Growth*, 2001, **223**, 399.
39. P. S. Sonawane, P. A. Wani, L. A. Patil and T. Seth, *Mater. Chem. Phys.*, 2004, **84**, 221.
40. J. Zhou, G. Q. Bian, Q. Y. Zhu, Y. Zhang, C. Y. Li and J. Dai, *J. Solid State Chem.*, 2009, **182**, 259.
41. A. Rabhi, M. Kanzari and B. Rezig, *Thin Solid Films*, 2009, **517**, 2477.

42. D. Colombara, L. M. Peter, K. D. Rogers, J. D. Painter and S. Roncallo, *Thin Solid Films*, 2011, **519**, 7438.
43. M. Kumar and C. Persson, *J. Renewable Sustainable Energy*, 2013, **5**, 031616.
44. M. Kumar and C. Persson, *Energy Procedia*, 2014, **44**, 176.
45. R. Chen and C. Persson, *EPJ Photovoltaics*, 2017, **8**, 85504.
46. V. Estrella, M. T. S. Nair and P. K. Nair, *Semicond. Sci. Technol.*, 2003, **18**, 190.
47. N. J. Gerein and J. A. Haber, *Chem. Mater.*, 2006, **18**, 6297.
48. F. Mesa, A. Dussan and G. Gordillo, *Phys. Status Solidi C*, 2010, **7**, 917.
49. D. Colombara, L. M. Peter, K. Hutchings, K. D. Rogers, S. Schäfer, J. T. R. Dufton and M. S. Islam, *Thin Solid Films*, 2012, **520**, 5165.
50. M. Kumar and C. Persson, *Appl. Phys. Lett.*, 2013, **102**, 062109.
51. P. Hohenberg and W. Kohn, *Phys. Rev.*, 1964, **136**, B864.
52. W. Kohn and L. J. Sham, *Phys. Rev.*, 1965, **140**, A1133.
53. G. Kresse and J. Hafner, *Phys. Rev. B*, 1993, **47**, 558.
54. G. Kresse and D. Joubert, *Phys. Rev. B*, 1999, **59**, 1758.
55. P. Blaha, K. Schwarz, G. K. H. Madsen, D. Kvasnicka and J. Luitz, *WIEN2k, An Augmented Plane Wave + Local Orbitals Program for Calculating Crystal Properties*, Karlheinz Schwarz, Techn. Universität Wien, Austria, 2001, ISBN 3-9501031-1-2.
56. J. P. Perdew, K. Burke and M. Ernzerhof, *Phys. Rev. Lett.*, 1996, **77**, 3865.
57. J. P. Perdew, A. Ruzsinszky, G. I. Csonka, O. A. Vydrov, G. E. Scuseria, L. A. Constantin, X. Zhou and K. Burke, *Phys. Rev. Lett.*, 2008, **100**, 136406.
58. V. I. Anisimov, J. Zaanen and O. K. Andersen, *Phys. Rev. B*, 1991, **44**, 943.
59. J. Heyd, G. E. Scuseria and M. Ernzerhof, *J. Chem. Phys.*, 2003, **118**, 8207.
60. A. V. Krukau, O. A. Vydrov, A. F. Izmaylov and G. E. Scuseria, *J. Chem. Phys.*, 2006, **125**, 224106.
61. C. Persson and A. Zunger, *Appl. Phys. Lett.*, 2005, **87**, 211904.
62. C. Persson, C. L. Dong, L. Vayssieres, A. Augustsson, T. Schmitt, M. Mattesini, R. Ahuja, J. Nordgren, C. L. Chang, A. Ferreira da Silva and J.-H. Guo, *Microelectron. J.*, 2006, **37**, 686.
63. G. Baldissera and C. Persson, *J. Appl. Phys.*, 2016, **119**, 045704.
64. C. Persson and S. Mirbt, *Braz. J. Phys.*, 2006, **36**, 286.
65. W. Setyawan, R. M. Gaume, S. Lam, R. S. Feigelson and S. Curtarolo, *ACS Comb. Sci.*, 2011, **13**, 382.
66. M. Gajdoš, K. Hummer, G. Kresse, J. Furthmüller and F. Bechstedt, *Phys. Rev. B*, 2006, **73**, 045112.
67. C. Ambrosch-Draxl and J. O. Sofo, *Comput. Phys. Commun.*, 2006, **175**, 1.
68. P. E. Blöchl, O. Jepsen and O. K. Andersen, *Phys. Rev. B*, 1994, **49**, 16223.
69. S. G. Choi, R. Chen, C. Persson, T. J. Kim, S. Y. Hwang, Y. D. Kim and L. M. Mansfield, *Appl. Phys. Lett.*, 2012, **101**, 261903.
70. C. Persson, R. Ahuja and B. Johansson, *Phys. Rev. B*, 2001, **64**, 033201.
71. C. Persson and C. Ambrosch-Draxl, *Comput. Phys. Commun.*, 2007, **177**, 280.

72. (a) K. Berland and C. Persson, *Comput. Mater. Sci.*, 2017, **134**, 17; (b) K. Berland and C. Persson, *J. Appl. Phys.*, 2018, **123**, 205703.
73. W. Shockley and H. J. Queisser, *J. Appl. Phys.*, 1961, **32**, 510.
74. A. Cuevas, *Energy Procedia*, 2014, **55**, 53.
75. K. L. Chopra and S. R. Das, *Thin Film Solar Cells*, Springer, New York, 1983.
76. S. Dominici, H. Wen, F. Bertazzi, M. Goano and E. Bellotti, *Appl. Phys. Lett.*, 2016, **108**, 211103.
77. D. Steiauf, E. Kioupakis and C. G. Van de Walle, *ACS Photonics*, 2006, **108**, 211103.
78. J. Dziewior and W. Schmid, *Appl. Phys. Lett.*, 1977, **31**, 346.
79. R. A. Sinton and R. M. Swanson, *IEEE Trans. Electron Devices*, 1987, **ED–34**, 1380.
80. H. Katagiri, K. Saitoh, T. Washio, H. Shinohara, T. Kurumadani and S. Miyajima, *Proceedings of Technology Digest 11th International Photovoltaic Science Engineering Conference*, 1999, p. 647.
81. S. Chen, J.-H. Yang, X. G. Gong, A. Walsh and S.-H. Wei, *Phys. Rev. B*, 2010, **81**, 245204.
82. B. G. Mendis, M. D. Shannon, M. C. J. Goodman, J. D. Major, A. A. Taylor, D. P. Halliday and K. Durose, *J. Phys.: Conf. Ser.*, 2013, **471**, 012014.
83. J. J. S. Scragg, J. K. Larsen, M. Kumar, C. Persson, J. Sendler, S. Siebentritt and C. Platzer-Björkman, *Phys. Status Solidi B*, 2016, **253**, 247.
84. Z.-K. Yuan, S. Chen, H. Xiang, X.-G. Gong, A. Walsh, J.-S. Park, I. Repins and S.-H. Wei, *Adv. Funct. Mater.*, 2015, **25**, 6733.
85. R. Chen and C. Persson, *J. Appl. Phys.*, 2017, **121**, 203104.
86. L. Chen, H. Deng, J. Tao, H. Cao, L. Huang, L. Sun, P. Yang and J. Chu, *RSC Adv.*, 2015, **5**, 84295.
87. L. Li, X. Liu, J. Huang, M. Cao, S. Chen, Y. Shen and L. Wang, *Mater. Chem. Phys.*, 2012, **133**, 688.
88. X. Zhang, N. Bao, K. Ramasamy, Y. H. A. Wang, Y. Wang, B. Lin and A. Gupta, *Chem. Commun.*, 2012, **48**, 4956.
89. B. Murali, M. Madhuri and S. B. Krupanidhi, *Cryst. Growth Des.*, 2014, **14**, 3685.
90. T.-X. Wang, Y.-G. Li, H.-R. Liu, H. Li and S.-X. Chen, *Mater. Lett.*, 2014, **124**, 148.
91. A. Kamble, K. Mokurala, A. Gupta, S. Mallick and P. Bhargava, *Mater. Lett.*, 2014, **137**, 440.
92. M. Wei, Q. Du, R. Wang, G. Jiang, W. Liu and C. Zhu, *Chem. Lett.*, 2014, **43**, 1149.
93. C. Wang, S. Chen, J. H. Yang, L. Lang, H. J. Xiang, X. G. Gong, A. Walsh and S.-H. Wei, *Chem. Mater.*, 2014, **26**, 3411.
94. G. Zhong, K. Tse, Y. Zhang, X. Li, L. Huang, C. Yang, J. Zhu, Z. Zeng, Z. Zhang and X. Xiao, *Thin Solid Films*, 2016, **603**, 224.
95. D. Shin, B. Saparov, T. Zhu, W. P. Huhn, V. Blum and D. B. Mitzi, *Chem. Mater.*, 2016, **28**, 4771.
96. F. Hong, W. Lin, W. Meng and Y. Yan, *Phys. Chem. Chem. Phys.*, 2016, **18**, 4828.

97. T.-X. Wang, Y.-G. Li, H.-R. Liu, H. Li and S.-X. Chen, *Mater. Lett.*, 2014, **124**, 148.
98. A. Kamble, K. Mokurala, A. Gupta, S. Mallick and P. Bhargava, *Mater. Lett.*, 2014, **137**, 440.
99. M. Wei, Q. Du, R. Wang, G. Jiang, W. Liu and C. Zhu, *Chem. Lett.*, 2014, **43**, 1149.
100. P. M. P. Salomé, J. Malaquias, P. A. Fernandes, M. S. Ferreira, A. F. da Cunha, J. P. Leitão, J. C. González and F. M. Matinaga, *Sol. Energy Mater. Sol. Cells*, 2012, **101**, 147.
101. J. Llanos, C. Mujica, V. Sánchez and O. Peña, *J. Solid State Chem.*, 2003, **173**, 78.
102. C. Persson, S. Lany, Y.-J. Zhao and A. Zunger, *Phys. Rev. B*, 2005, **72**, 035211.
103. D. Huang and C. Persson, *Thin Solid Films*, 2013, **535**, 265.
104. A. Kanai, K. Toyonaga, K. Chino, H. Katagiri and H. Araki, *Jpn. J. Appl. Phys.*, 2015, **54**, 08KC06.
105. Y.-T. Zhai, S. Chen, J.-H. Yang, H.-J. Xiang, X.-G. Gong, A. Walsh, J. Kang and S.-H. Wei, *Phys. Rev. B*, 2011, **84**, 075213.
106. N. Aihara, A. Kanai, K. Kimura, M. Yamada, K. Toyonaga, H. Araki, A. Takeuchi and H. Katagiri, *Jpn. J. Appl. Phys.*, 2014, **53**, 05FW13.
107. J. de Wild, E. V. C. Robert, B. El Adib and P. J. Dale, *MRS Proc.*, 2015, **1771**, 151.
108. L. Xi, Y. B. Zhang, X. Y. Shi, J. Yang, X. Shi, L. D. Chen, W. Q. Zhang, J. H. Yang and D. J. Singh, *Phys. Rev. B*, 2012, **86**, 155201.
109. P. Zawadzki, L. B. Lauryn, H. W. Peng, E. S. Toberer, D. S. Ginley, W. Tumas, Z. Andriy and S. Lany, *Appl. Phys. Lett.*, 2013, **103**, 253902.
110. A. Shigemi, T. Maeda and T. Wada, *Phys. Status Solidi B*, 2015, **252**, 1230.
111. H. Araki, K. Chino, K. Kimura, N. Aihara, K. Jimbo and H. Katagiri, *Jpn. J. Appl. Phys.*, 2014, **53**, 05FW10.
112. K. Toyonaga and H. Araki, *Phys. Status Solidi C*, 2015, **12**, 753.
113. M. Hasaka, T. Aki, T. Morimura and S.-I. Kondo, *Energy Convers. Manage.*, 1997, **38**, 855.
114. M. I. Aliev, D. G. Arasly and T. G. Dzhabrailov, *Sov. Phys. Solid State*, 1983, **25**, 2009.
115. Y. Goto, Y. Kamihara and M. Matoba, *J. Electron. Mater.*, 2014, **43**, 2202.
116. A. Suzumura, N. Nagasako, Y. Kinoshita, M. Watanabe, T. Kita and R. Asahi, *Mater. Trans.*, 2015, **56**, 858.
117. K. Anuar, S. M. Ho, W. T. Tan, M. S. Atan, D. Kuang, H. Jelas and N. Saravanan, *Mater. Sci.*, 2008, **14**, 101.
118. A. Kassim, Z. Kuang, A. Sharif, T. Wee, H. S. Min and S. Nagalingam, *Arabian J. Sci. Eng.*, 2010, **35**, 83.
119. D. Avellaneda, M. T. S. Nair and P. K. Nair, *J. Electrochem. Soc.*, 2010, **157**, D346.
120. D. Tang, J. Yang, F. Liu, Y. Lai, J. Li and Y. Liu, *Electrochim. Acta*, 2012, **76**, 480.
121. D. J. Temple, A. B. Kehoe, J. P. Allen, G. W. Watson and D. O. Scanlon, *J. Phys. Chem. C*, 2012, **116**, 7334.

122. W. Septina, S. Ikeda, Y. Iga, T. Harada and M. Matsumura, *Thin Solid Films*, 2014, **550**, 700.
123. A. W. Welch, L. L. Baranowski, P. Zawadzki, S. Lany, C. A. Wolden and A. Zakutayev, *Appl. Phys. Express*, 2015, **8**, 082301.
124. C. Persson, *Appl. Phys. Lett.*, 2008, **93**, 072106.
125. R. Chen and C. Persson, *J. Appl. Phys.*, 2012, **112**, 103708.
126. A. Cuevas, *Energy Procedia*, 2014, **55**, 53.
127. M. Kumar, N. Umezawa and M. Imai, *Appl. Phys. Express*, 2014, **7**, 071203.

Subject Index